Multidisciplinary Design Optimization of Complex Structures Under Uncertainty

In the realm of engineering structures design, the inevitability of uncertainties poses a significant challenge. Uncertainty-Based Multidisciplinary Design and Optimization (UBMDO) stands out for its dual ability to precisely quantify the impact of uncertain variables and harness the potential of multidisciplinary design and optimization, thereby attracting considerable attention. From basic theory to advanced applications, this book helps readers achieve more efficient and reliable design optimization in complex systems through rich case studies and practical technical guidance.

The book systematically expounds the fundamental theories and methods of UBMDO, encompassing crucial techniques such as uncertainty modeling, sensitivity analysis, approximate modeling, and uncertainty-based optimization. It also introduces various uncertainty analysis methods, such as stochastic, non-probabilistic, and hybrid approaches, aiding readers in comprehending and managing uncertainty within systems. Through diverse practical engineering cases in fields like machinery, aerospace, and energy, it illustrates the specific application and implementation process of the UBMDO method. Rich graphics, algorithms, and simulation results augment the practicality and applicability of the theoretical knowledge. Furthermore, it explores in depth the future development trends and challenges of UBMDO, sparking innovative thinking and research interests among readers in this field.

Multidisciplinary Design Optimization of Complex Structures Under Uncertainty caters to a diverse audience: Engineers specializing in multidisciplinary design optimization are given the tools to master uncertainty management, and researchers in related fields will gain important theoretical insights and practical guidance in uncertainty analysis. Additionally, scholars and educators can utilize the book as a comprehensive resource for advanced courses, enabling students to grasp the latest UBMDO applications. Decision makers and managers handling complex systems can extract methods from the book, facilitating improved risk assessment, and strategic development through uncertainty-based optimization.

Structural Damage, Fatigue and Fracture

This series promotes scientific issues associated with damage, fatigue and fracture phenomena applied to all kind of materials, structural components and structures, as well as their engineering practical applications.

Multidisciplinary Design Optimization of Complex Structures Under Uncertainty *Debiao Meng and Shun-Peng Zhu*

Multidisciplinary Design Optimization of Complex Structures Under Uncertainty

Debiao Meng and Shun-Peng Zhu

CRC Press is an imprint of the
Taylor & Francis Group, an **informa** business

First edition published 2025
by CRC Press
2385 NW Executive Center Drive, Suite 320, Boca Raton FL 33431

and by CRC Press
4 Park Square, Milton Park, Abingdon, Oxon, OX14 4RN

CRC Press is an imprint of Taylor & Francis Group, LLC

ISBN: 978-1-032-73561-0 (hbk)
ISBN: 978-1-032-73563-4 (pbk)
ISBN: 978-1-003-46479-2 (ebk)

DOI: 10.1201/9781003464792

Typeset in Sabon
by Newgen Publishing UK

Contents

Foreword

As modern engineering technology continues to advance, we are constantly faced with challenges in structural design optimization, uncertainty being one of them. Uncertainty exists throughout the entire life cycle of a product. It comes from material variability, imprecision in the manufacturing process, unpredictability of the external environment, and so on. Furthermore, today's engineering equipment design often involves multiple disciplines, multiple factors, and multiple constraints. One of the most challenging problems in modern engineering design is how to achieve multidisciplinary and multi-objective optimization design in a complex environment full of uncertainty. In this context, Professor Debiao Meng and Professor Shun-Peng Zhu wrote the monograph *Multidisciplinary Design Optimization of Complex Structures Under Uncertainty Conditions* based on the needs of modern complex engineering structure design.

This monograph is rigorous in structure and rich in content. It builds a comprehensive and in-depth framework from theory to practice and conducts a holistic exploration of multidisciplinary design optimization based on uncertainty. It focuses on multidisciplinary design optimization under uncertainty, integrating the basic principles of optimization, modeling methods, applications in aviation equipment, applications in energy equipment, and applications in propulsion equipment. At the same time, it summarizes the research of predecessors and the author's own efforts in this field. Some of the methods described in the monograph can effectively improve the efficiency and accuracy of engineering design, which brings new vitality to the development of large-scale advanced manufacturing industries.

The long-term development of engineering equipment in all countries of the world today needs to be guided by this kind of monograph summarizing the methodology, which will help reduce economic losses and social risks caused by design defects. Especially in the context of unpredictable global environmental changes and increasing resource constraints, ensuring the safety and reliability of engineering structures and achieving efficient resource utilization and environmental sustainability has become an urgent task for us. This monograph not only provides us with advanced tools and strategies to address these challenges but also promotes innovation and technological progress in the global manufacturing industry.

Therefore, this monograph is an excellent work that combines theory, practicality, and readability, providing engineers and scholars with a rare reference book and study guide. Its publication will strongly promote the progress and high-level development of the global manufacturing industry in the field of multidisciplinary design optimization of complex equipment structures.

Dr. Kyung K. (KK) Choi
Mechanical Engineering Department
The University of Iowa, Iowa City, IA
June 5, 2024

Preface

Multidisciplinary Design Optimization (MDO) plays an increasingly important role in solving complex system design problems with the rapid development of science and technology and the increasing complexity of engineering systems. These systems often involve multiple disciplines and require us to consider various factors and constraints when designing them. However, the systems are full of uncertainties, which may come from many aspects, such as data, model, and environment, significantly complicating multidisciplinary design optimization. Uncertainty-Based Multidisciplinary Design Optimization (UBMDO) can solve these problems.

By explicitly considering uncertainty, the UBMDO method can fully predict and evaluate potential risks at the design stage, resulting in a more robust and reliable design solution. This not only improves the quality and performance of the product but also reduces the cost increase and reputation loss caused by design defects.

This book, *Multidisciplinary Design Optimization of Complex Structures Under Uncertainty*, provides a comprehensive overview and in-depth discussion of the latest research advances in the field of UBMDO. It not only provides experts with cutting-edge theories and methods but also provides beginners with a solid foundation for entering the field. Through this book, readers can gain a deep understanding of the basic principles, latest technologies, and application examples of UBMDO in different engineering fields.

The book consists of ten chapters, constructing a comprehensive framework from theory to practice and exploring the multidisciplinary design optimization based on uncertainty. Chapter 1 is an introduction that clarifies the background and progress of this research field, laying a foundation for the subsequent chapters. Chapters 2–4 focus on UBMDO's basic theory and methods. Chapter 2 elaborates on the fundamentals of multidisciplinary design optimization and methods for uncertainty modeling. Chapter 3 introduces approximation methods, which make optimization design more feasible in modern engineering by reducing computational load while maintaining necessary accuracy. Chapter 4 comprehensively introduces uncertainty analysis methods, including stochastic, non-probabilistic, and hybrid uncertainty analysis methods. Chapter 5 turns to specific methods for deterministic MDO, emphasizing the potential of MDO in reducing design costs and shortening design time. Chapters 6–9 demonstrate the practicality and effectiveness of the UBMDO method in different application scenarios through specific case studies. Chapter 6 outlines the application of UBMDO methods in various contexts. Chapters 7–9 provide detailed explanations of the specific application and implementation steps of the UBMDO method through examples of the design of a flip-driven mechanism, a turbine rotor support structure, and a water-cooled blower blade. Chapter 10 summarizes and looks forward to the book, pointing out the future development directions and challenges in UBMDO.

At the end of this preface, we would like to thank the CRC Press for its support, enabling such high-quality academic work to be presented to the readers. We are also grateful to the National Natural Science Foundation of China, Guangdong Basic and Applied Basic Research Foundation, and the Students Go Abroad for Scientific Research and Internship Funding Program of the University of Electronic Science and Technology of China. Their support has been instrumental in making this book a reality. We would like to acknowledge the contributions of the following individuals: Shiyuan Yang, Hui Ma, Peng Nie, and Yipeng Guo from the School of Mechanical and Electrical Engineering, University of Electronic Science and Technology of China; Yuting Zhang, Yuexiang Zhang, Hongjun Guo, and Qianyu Huang from Glasgow College, University of Electronic Science and Technology of China; Yongqiang Guo and Lidong Pan from Beijing Research Institute of Mechanical & Electrical Technology Ltd.; and Xinkai Guo from the Institute of Electronic and Information Engineering of UESTC in Guangdong. We are grateful for their willingness to share their knowledge and experiences.

We would like to offer our most sincere and profound acknowledgment to Professor Hong-Zhong Huang, founder of the reliability team at the University of Electronic Science and Technology of China. Without Professor Huang's steadfast devotion to the advancement of our field and his relentless pursuit of knowledge, this team would not have achieved what it has achieved, and this book would never have been existed. His tireless efforts and unyielding spirit have breathed life into our work, infusing it with a sense of purpose and direction that has propelled us toward unprecedented heights.

If the readers encounter any issues or have any questions regarding the content of this book, we welcome feedback and inquiries. Please feel free to contact us at any time (dbmeng@uestc.edu.cn, zspeng2007@uestc.edu.cn).

Debiao Meng
Shun-Peng Zhu
December 2023, Chengdu, China

Acknowledgments

We acknowledge the support and funding provided by the National Natural Science Foundation of China (Grant No. 12232004), Guangdong Basic and Applied Basic Research Foundation (Grant No. 2022A1515240010 and 2022A1515240011), Sichuan Science and Technology Program (Grants No. 2022YFQ0087 and 2022JDJQ0024), China Scholarship Council (No. 202406070025), and Students Go Abroad for Scientific Research and Internship Funding Program at the University of Electronic Science and Technology of China. We also extend our gratitude to all the individuals who have contributed to this book in various ways:

Shiyuan Yang
School of Mechanical and Electrical Engineering,
University of Electronic Science and Technology of China, China

Hui Ma
School of Mechanical and Electrical Engineering,
University of Electronic Science and Technology of China, China

Peng Nie
School of Mechanical and Electrical Engineering,
University of Electronic Science and Technology of China, China

Yipeng Guo
School of Mechanical and Electrical Engineering,
University of Electronic Science and Technology of China, China

Yuting Zhang
Glasgow College,
University of Electronic Science and Technology of China, China

Yuexiang Zhang
Glasgow College,
University of Electronic Science and Technology of China, China

Hongjun Guo
Glasgow College,
University of Electronic Science and Technology of China, China

Qianyu Huang
Glasgow College,
University of Electronic Science and Technology of China, China

Yongqiang Guo
China Academy of Machinery Beijing Research Institute of Mechanical & Electrical Technology Co., Ltd.

Lidong Pan
China Academy of Machinery Beijing Research Institute of Mechanical & Electrical Technology Co., Ltd.

About the authors

Debiao Meng is an associate professor in the School of Mechanical and Electrical Engineering at the University of Electronic Science and Technology of China, China. He is also a visiting professor at the University of Bologna and a visiting research fellow at the Zigong Innovation Centre of Zhejiang University. He obtained his BSc (2007) in Mechanical Engineering from Northwest A&F University and Ph.D. from UESTC with a major in Mechanical Engineering (2014). From 2013 to 2015, he was a joint Ph.D. student in the Department of Civil and Environmental Engineering at Vanderbilt University, United States; and from 2017 to 2019, he was a visiting scholar at Rutgers University, United States. His main research field is uncertainty-based multidisciplinary design and optimization of complex engineering systems. He has published more than 40 journal and proceedings papers on system reliability analysis, multidisciplinary design optimization, surrogate modeling, and uncertainty quantification and propagation.

Shun-Peng Zhu is a professor in the School of Mechanical and Electrical Engineering at the University of Electronic Science and Technology of China, China. He received his Ph.D. in Mechanical Engineering from UESTC in 2011, worked as a PIF Fellow of Politecnico di Milano, Italy, from 2016 to 2018 and was a research associate at the University of Maryland, United States, in 2010. His research, published in scholarly journals and edited volumes, over 150 peer-reviewed book chapters, journals, and proceedings papers, explores the aspects of fatigue design, structural reliability, probabilistic failure modeling, multi-physics damage modeling, and life prediction.

He received second prize of Natural Science Awards of the Sichuan Province and Chinese Society of Aeronautics and Astronautics in 2022, four Most Cited Paper Awards of *FFEMS Journal* since 2018, Award of Merit of ESIS-TC12 in 2019, Top 100 Most Influential International Academic Paper Award of China in 2018, Polimi International Fellowship in 2016, Hiwin Doctoral Dissertation Award in 2012, Best Paper Awards of five international conferences. He has been featured as Elsevier's Most Cited

Chinese Researcher in the field of safety, risk, reliability, and quality since 2018, World's top 2% scientists since 2020, and serves as Co-Editor-in-Chief of the *International Journal of Structural Integrity*, guest editor, editorial board member of several international journals, Springer, and Taylor & Francis book series.

Chapter 1

Introduction

1.1 THE BACKGROUND OF UBMDO

The significant industrial revolution in human history started in the 1860s, with Britain as the starting point. Since then, human beings have entered the industrial society from the agricultural society. During this period, machinery underwent unprecedented development. Their core part is often no longer only one institution but a system composed of multiple institutions. The performance of mechanical products depends more on the integrity and correlation of the system. Modern complex mechanical equipment design is a multidisciplinary integration of structure, control, drive, information interaction, and so on. The research method in the design stage is the integration of multi-field and multidisciplinary. Therefore, the factors that affect the design have become more and more. This makes the staff in different disciplines to participate in the optimization design simultaneously. They need to comprehensively consider the coupling effect between various disciplines and use multiple professional knowledge to optimize the mechanical system.

Multidisciplinary Design Optimization (MDO) is a special methodology for dealing with complex and coupled engineering system design problems. The MDO method was first born in the aerospace field. In 1982, J. Sobieszczanski-Sobieski at NASA first used the idea of linear decomposition to solve large-scale structural optimization problems. This practice is the prototype of the MDO method [1]. In 1991, AIAA published a white paper on MDO research, which marked the official birth of MDO as an independent research field [2]. After more than 30 years of development, research on MDO has spawned many research directions. The MDO research system contains many different aspects, and their relationship is shown in Figure 1.1.

Since Sobieszczanski-Sobieski proposed the concurrent subspace optimization method in 1988 [3], different MDO methods have been proposed. Single-stage optimization techniques and multi-stage optimization techniques are the two categories into which MDO approaches are currently divided.

DOI: 10.1201/9781003464792-1

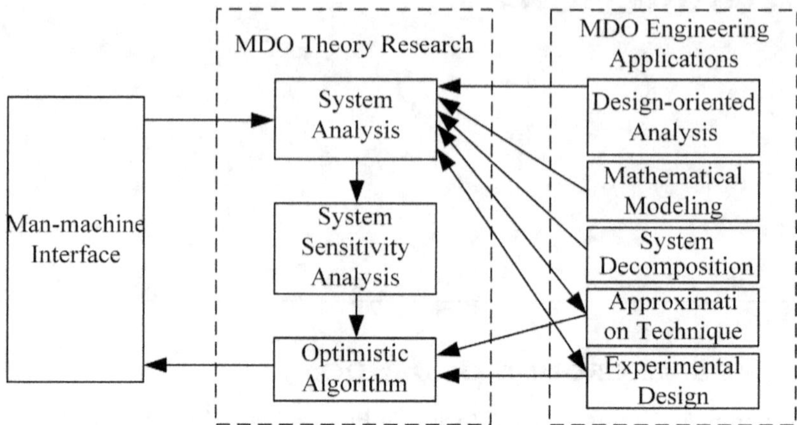

Figure 1.1 MDO technology architecture.

A single-stage optimization method usually contains an optimizer and directly applies a non-hierarchical optimization strategy. Multiple Discipline Feasible (MDF) [4], Individual Discipline Feasible (IDF), All at Once (AAO) [5], and Multidisciplinary Design Optimization based on Independent Subspaces (MDOIS) belong to such methods. Hannapel and Vlahopoulos [6] proposed a new MDO method based on a set-based design strategy. In the optimization process, this method adjusts the design space by changing the constraints and uses a set of multiple sets of design variable values to optimize the design. This method can replace the traditional single-point optimization to search for future design preferences.

The introduction and widespread adoption of the MDO method have spurred numerous scholars to engage in research within this field. Li et al. [7] proposed an MDO method for cooling turbine blades based on reliability, considering the coupling between aerodynamics, heat transfer, and strength. In the meantime, the Kriging model is introduced to reduce the calculation amount of multidisciplinary reliability analysis. Meng et al. [8] combined the subset simulation-based reliability analysis method with MDO to improve the computational efficiency of the reliability-based MDO problem. Wang et al. [9] proposed a reliability assessment model considering interval uncertainty and fuzzy uncertainty for various uncertainties in engineering. In this case, they developed a Sequential Optimization and Reliability Assessment (SORA) method for multidisciplinary systems to decouple the model. Jaron et al. [10] reduced the noise emission of an aeroengine fan with a bypass ratio of 19 and optimized the aerodynamic efficiency in MDO using the mixed noise prediction method. Chen et al. [11] proposed a coupling analysis-assisted gradient-enhanced kriging method, which uses a multidisciplinary feasible framework to obtain the value of the coupling variables to improve the quality and efficiency of solving global MDO problems.

In summary, the MDO method has achieved many results in the development process. It can comprehensively analyze different disciplines in various forms, such as parallel interconnection, serial interconnection, or hierarchical solution, under the premise of fully considering the system design objectives. At the same time, it can still meet the design requirements of different disciplines. This is of great significance for industrial design.

In addition to the synthesis of disciplines, there are also many uncertainties in complex mechanical systems. For example, the characteristics of materials, working environment, and working load are uncertain. In practical engineering, it is proved that if the optimization value is pursued one-sidedly and the influence of uncertain factors is ignored, the safety margin of mechanical equipment will be small, and the possibility of failure will be increased. Taking Japan's advanced land observation satellite in April 2011 as an example, its solar cell array tensile spring did not reserve enough deformation tolerance at the time of design, and the expansion and contraction of the solar cell array at low temperatures were underestimated. This leads to problems in the solar power supply system, and it cannot supply power to the satellite. These are the possible effects of uncertainty.

Given the above problems, Uncertainty-Based Design (UBD) and related Non-Deterministic Approaches have been rapidly developed. The research in this direction mainly solves two types of problems. One is to improve the robustness of mechanical systems, reduce the sensitivity of performance to uncertainty, and ensure the stability of mechanical systems. The second is to improve the reliability of the mechanical system and reduce the probability of failure and failure so that the system can meet the expected demand under some uncertain conditions. Based on these two objectives, Uncertainty-Based Design Optimization (UBDO) is mainly divided into two categories: Robust design optimization and reliability design optimization. These methods have been successfully applied in aerospace, civil construction, and other fields. In the process of application development, due to the complexity of the design system and the influence of various uncertain factors, the Uncertainty-Based MDO (UBMDO) method came into being.

Since Dantzig [12] considered the influence of uncertainty in linear programming problems in the 1950s, many design optimization methods under uncertainty have been proposed [13, 14]. Their research results are widely used in aerospace, civil engineering, and other fields [15–18]. The growing intricacy of engineering systems has made it more difficult to guarantee the overall performance of the system's safety and dependability due to the design optimization method's consideration of uncertainty in a single discipline [19–21]. The impact of variables causing uncertainty on the propagation between coupling disciplines makes UBMDO one of the important research directions of MDO. To improve the efficiency and accuracy of evaluating the influence of uncertainty on complex engineering systems and make the optimization design results closer to the actual requirements, this book will systematically introduce the development history and essential theories of UBMDO.

1.2 THE REVIEW OF UBMDO

The UBMDO process is based on system modeling and uncertainty analysis. Its main flow chart is shown in Figure 1.2.

The analysis steps include two main parts and multiple detailed steps.

Part 1. Uncertain system modeling: This is the preparatory work for UBMDO analysis, which mainly includes system and uncertainty modeling.

Step 1. System modeling. It mainly models the system of the research object and its constituent disciplines. Then, the design optimization problem is mathematically abstracted, and a mathematical model is established. The modeling mainly involves the optimization variables, design space, system parameters, optimization objectives, and constraints.

Step 2. Uncertainty modeling. Considering the uncertainty of the system is a major focus of UBMDO analysis. Here, the appropriate uncertainty mathematical method should be selected to describe and quantify the uncertainty factors involved in the design optimization of the system. In modern complex mechanical systems, there are many uncertain factors. Therefore, after completing the uncertainty modeling, generally, the sensitivity analysis method must be used for significance analysis. This filters out the factors

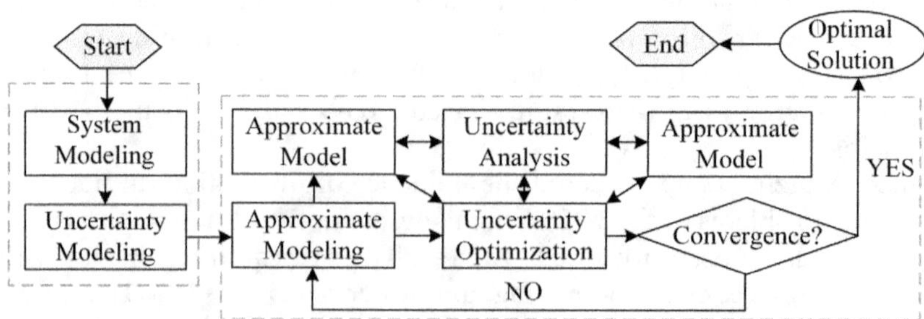

Figure 1.2 UBMDO general solution process diagram.

that negatively impact the system performance, thereby reducing the computational complexity in the UBDO process.

Part 2. UBMDO process solution: This is the organization and implementation process of UBMDO. Since UBMDO involves multidisciplinary coupling and coordinated optimization, a reasonable organization of coordinated optimization among disciplines can improve the solution efficiency of the UBMDO method. The main contents include approximation modeling, uncertainty analysis, and optimization under uncertainty.

Step 3. Approximate modeling. The high-precision disciplinary model analysis of modern mechanical systems produces a considerable calculation. If the original model is directly used for uncertainty analysis and optimization, it will bring terrible computational costs. Approximate modeling is to participate in uncertainty analysis and optimization by establishing a surrogate model of a high-precision model. This aims to grasp the problem's main contradiction to reduce the computational cost as much as possible while obtaining the results within the required accuracy. At the same time, for MDO problems, constructing a low-complexity surrogate model of coupling variables is also an important method to achieve discipline decoupling to support discipline autonomy and parallel optimization. Therefore, approximate modeling is a necessary preparation step for uncertainty analysis and optimization.

Step 4. Uncertainty analysis. The primary objective is to conduct a quantitative analysis of the system performance distribution within the given design scheme, considering the uncertainties inherent in the optimization process. Subsequently, the robustness of the design scheme and its reliability under various constraints will be evaluated.

Step 5. Uncertainty optimization. In the optimization search process, the stability and reliability of each search point are judged by nested calling uncertainty analysis. Therefore, the scheme is optimized according to the judgment requirements in the design space.

According to the above analysis, the key points of UBMDO include uncertain system modeling, sensitivity analysis, surrogate model, uncertainty analysis, uncertainty optimization, multidisciplinary analysis, and so on.

1.2.1 Theoretical research progress of UBMDO

Since the 1990s, many scholars have researched uncertainty analysis and multidisciplinary optimization. Numerous UBMDO techniques have been put forth since then. The theoretical progress of UBMDO mainly includes the following aspects.

1.2.1.1 Uncertain system modeling

As mentioned above, uncertain system modeling is a vital preparation process for UBMDO.

System modeling includes not only the modeling of physical models in practical engineering problems but also the mathematical abstract modeling of physical models. In the system modeling for the MDO process, the process is generally gradually refined from macro to micro. First, system/discipline level and component level model is established, and then the model of discipline coupling relationship is established. Then, all levels' optimization variables, objectives, and conditions are set reasonably. Finally, the MDO model of the system is established. MDO-oriented modeling methods mainly include process modeling, variable complexity modeling, and digital design-oriented modeling methods. From Zhang et al. [22] and other studies on the relevant methods give a specific introduction.

Uncertainty modeling involves representing uncertain factors in multidisciplinary optimization using mathematical language and various mathematical techniques. The specific modeling methods employed may vary depending on the sources of uncertainty that impact the factors. Uncertainty factors are mainly divided into epistemic uncertainty and aleatory uncertainty. Epistemic uncertainty can also be called subjective uncertainty. It is mainly due to the lack of subjective knowledge of designers or the lack of information knowledge sources, which leads to the inability to describe certain uncertain variables and parameters accurately. Therefore, epistemic uncertainty needs to be characterized by selecting corresponding mathematical methods according to specific characteristics. Aleatory uncertainty can also be called objective uncertainty. The majority originate from the system's intrinsic objective variability and its operational environment, which can be described directly based on the probability method. The widely used uncertainty modeling methods mainly include stochastic programming, fuzzy programming, possibility theory, interval analysis methods, and so on.

1.2.1.2 Sensitivity analysis

Sensitivity analysis is a technique for examining and assessing how changes in system parameters or external factors will affect the state or output of a system model. Sensitivity analysis is frequently used in optimization techniques to investigate the stability of the best solution in cases where the initial data is erroneous or fluctuates. You can use UBMDO's sensitivity analysis to determine how system design parameters or variables affect system performance. It can also include the quantification of the model itself and the degree of influence of various variables in the model input on the model output change. Through this method, the importance of each uncertainty in the system can be sorted to filter out the unimportant factors. This can reduce the complexity of the UBMDO analysis.

Wang et al. [23] systematically introduced various methods of sensitivity analysis. It includes methods for a single discipline, such as analytical method, version analytical method, manual derivation method, complex variable method, finite difference method, automatic differential method, and so on. It also includes multidisciplinary methods, such as the optimal sensitivity analysis method for hierarchical decomposition systems and the global sensitivity equation method for non-hierarchical decomposition systems. In particular, the random uncertainty methods include differential, response surface, Fourier amplitude sensitivity test, variance decomposition, and so on [24–27]. The method based on sampling simulation is the most straightforward and most feasible. It only needs to sample the value space of uncertain variables and analyze or simulate the sample points systematically to obtain the sample data. Then, we can flexibly carry out three-point distribution, correlation analysis, statistical analysis, regression analysis, and variance decomposition to obtain sensitivity information. Therefore, this method has substantial flexibility and practicability and has the most comprehensive application range.

However, dense sampling in the whole space will lead to unbearable computational costs. Based on this, the Elementary Effects (EE) method is developed [28, 29]. This method calculates the corresponding system performance by increasing the value of each uncertain factor. In this way, the statistical method can obtain the sensitivity information, and it is possible to achieve a balance between calculation accuracy and cost. There are relatively few studies on the sensitivity analysis method of epistemic uncertainty, mainly using evidence theory to describe uncertainty. On this basis, the method of sampling simulation is developed [30, 31].

1.2.1.3 Approximate modeling

Approximate modeling, a practical approach leveraging surrogate models, efficiently trims computational expenses, aiding in discipline decoupling and autonomous optimization. It serves as a linchpin in resolving the computational and organizational intricacies in UBMDO. In UBMDO, many practical methods are used, such as the response surface method, spline function interpolation approximation method, radial basis function interpolation approximation method, Kriging function interpolation approximation method, and so on. The approximate modeling method combining artificial intelligence and machine learning has also developed in recent years. Related research [32] concluded that the polynomial-based response surface and Kriging model are appropriate for handling approximation problems of modest to moderate size, and the neural network approach is more feasible for handling large-scale, highly nonlinear approximation problems. Among them, the Radial Basis Function Neural Network (RBFNN) can combine the characteristics of unbiased estimation at sample points. It also can provide an excellent nonlinear approximation of neural networks and has strong practicability for complex MDO models.

In the conventional RBFNN model, the transfer function employed by the hidden layer neurons is a radial basis function. The center point of this basis function is the same as the training sample point [33, 34]. After the shape parameter is given, a simple matrix operation can directly calculate the weight coefficient. Therefore, the parameter optimization of the RBFNN model in the first core problem is essentially a shape parameter optimization problem. For the study of shape parameter optimization, there are three main methods:

a. The shape parameters of all neuron basis functions are the same, so the independent shape parameters contain only one variable. It can be calculated by empirical formulas or optimized directly [35, 36].
b. The training sample points are clustered, and the neurons whose basis function center points are located in the same cluster have the same shape parameters [37]. The shape parameters of each cluster are estimated according to the spatial distribution characteristics of the cluster sample points and the distribution characteristics of the exact model response values.
c. Shape parameters can be estimated as independent variables using empirical formulas or directly optimized through methods like gradient descent, fuzzy rule adjustment, minimum description length, genetic algorithms, differential evolution, and particle swarm optimization. Among the above three methods, the first method is the simplest. However, its approximation accuracy is minimal due to the reduction of the model design freedom of RBFNN. The second method is more reasonable than the first method. However, this empirical method is often not universal. The third method maintains the modeling flexibility of RBFNN and has the best approximation ability. However, when there are many neurons in the network's hidden layer, it will lead to a large amount of calculation, and the quality of the results is not high.

In UBMDO, an integral application of the approximation model is to substitute the high-precision MDO model. Consequently, the precision of the approximate response values and the congruity of gradients between the approximate model and the precise model are of paramount importance. This significance is notably pronounced when dealing with profoundly nonlinear multimodal optimization issues. The second core problem of the above RBFNN approximation modeling is the sequential adding point strategy. It is based on a small number of initial training sample points and then explores how to sequentially sample the precise model domain based on the built model information and the currently available sample point information according. In this way, the accurate model information can be supplemented in a targeted manner, and the approximation accuracy of the approximate model can be gradually improved until it converges to the expected requirements. This helps to reasonably determine the number of training sample points to construct an approximate model that meets the accuracy requirements at a small cost. The sequential addition strategy used to improve the global approximation accuracy includes maximum entropy criterion, mean square error integral criterion, maximum and minimum conversion distance method, gradient or second derivative method, maximum sequential modeling cumulative change method, and so on. These methods have advantages and disadvantages, most requiring independent validation samples for calculation. Therefore, how to add points to the low-precision region of the current approximation model without independent verification samples is still a problem that needs to be studied and solved in the future.

1.2.1.4 Uncertainty analysis

The content of uncertainty analysis includes the influence of system input, external environment, and system uncertainty on system output. Uncertainty analysis can be divided into intrusive and non-intrusive types according to different implementation methods. The intrusive type refers to modifying the system control equation and adding uncertainties. This can directly incorporate the impact of uncertainty into the system model. Non-intrusive means that the system model is regarded as a black box, and uncertainty analysis is performed only based on input and output. This method does not need to modify the system model and can be calculated directly based on the original model and program code. It can avoid the coupling correlation between the method research and more than a dozen objects and has a broader application range than the invasive method.

Uncertainty analysis can also be divided into probability and non-probability methods according to the different types of uncertainty. Probability theory, a well-established branch of mathematics, has a long development history, resulting in a comprehensive foundational framework. This theory finds widespread application in practical engineering contexts, particularly in addressing uncertainty. However, developing non-probabilistic methods to handle cognitive uncertainty is relatively underexplored. Fewer methods comprehensively consider the mixed effects of random and epistemic uncertainties. There are mainly upper and lower limit analysis methods of reliability based on probability theory and fuzzy set, probability analysis based on one-time reliability method, and unified uncertainty analysis based on interval analysis of evidence theory. The basic idea of the above methods is to quantify the impact of random and cognitive uncertainty separately. Then, the two are nested together to analyze the impact of mixed uncertainty. The discipline model in UBMDO generally has the problem of high computational cost, which brings great difficulties to the analysis of mixed uncertainty.

1.2.1.5 Uncertainty optimization

The design under uncertainty has two primary purposes. One is to improve the robustness of the product; the other is to improve the reliability of the product. The purpose is to reduce the product's sensitivity to uncertainty and the possibility of failure. Stochastic linear programming has been proposed in the last century. It is mainly used to deal with optimization algorithms with parameter uncertainty. The field of optimization has seen significant advancements in linear programming, nonlinear programming, and discrete optimization. These developments have led to the emergence of hybrid optimization techniques, encompassing both discrete and continuous variables. Additionally, stochastic integer programming, stochastic nonlinear programming, and fuzzy stochastic programming have been introduced to address uncertain optimization problems. These methods are collectively referred to as stochastic programming.

In practical engineering applications, many designers have quantitative requirements for the robustness and reliability of the research object. Therefore, the optimization methods based on these two requirements have been studied extensively. Robust design optimization needs to consider the performance robustness of the design objectives. Therefore, it is a multi-objective optimization problem in engineering. Its current widely used methods include weighted summation, preference-based planning, compromise, genetic algorithm, and evolutionary algorithm.

In reliability-based optimization design, reliability analysis is required for each search point. This will lead to a sharp increase in computational complexity. Therefore, more approximate solutions are used in engineering applications to convert deterministic constraints concerning reliability into roughly equivalent ones. This transforms the uncertainty optimization problem into a deterministic optimization problem, reducing computational complexity. The related methods include the worst-case analysis method, the angle space analysis method, the change mode analysis method, the reliable design space method, and the approximate limit state equation method. However, in these deterministic optimizations after approximate transformation, the actual reliability of the design scheme is not analyzed. Therefore, the optimization scheme produced by the transformation method has limited accuracy and reliability. Another method transforms the traditional double-layer nested loop of optimization reliability analysis into a single-layer analysis. Optimization and reliability analysis can be decoupled by sequential execution, that is, single-layer sequential optimization. The single-layer fusion optimization method is another way to turn this fusion into an optimization problem. All of these techniques are called single-layer techniques. The main idea of the single-layer sequential optimization method is to directly decouple the reliability analysis from the outer optimization cycle and then perform the optimization, cycle, and uncertainty analysis sequentially. In this way, a single-layer cycle can be formed. In each single-layer cycle, the reliability constraints are transformed into deterministic constraints based on the information obtained from the uncertainty analysis of the previous cycle. This can transform uncertainty optimization into deterministic optimization. After the deterministic optimization is completed, the uncertainty analysis of the optimization scheme is carried out. The analysis results will be used to guide the next deterministic optimization.

Du and Chen [38] proposed the SORA algorithm. This method will first solve the inverse maximum possible point corresponding to the preset reliability at the current design point. Then, the limit state equation is translated until the inverse maximum possible point is moved to the boundary condition of the original deterministic constraint. The translated limit state equation is an equivalent deterministic constraint for the subsequent cycle of optimization. The constraint condition can use the first-order Taylor expansion at the maximum possible point as the deterministic constraint condition.

The single-layer fusion optimization method mainly aims at the problem that the inner loop is based on the maximum possible point for uncertainty analysis. It simplifies the inner loop optimization search for the maximum possible point into an equivalent maximum possible point simple calculation formula. Then, this can be used as a constraint condition to act on the outer layer optimization. Chen et al. [39] proposed a single loop single vector algorithm. It will approximately calculate the maximum possible point of each constraint condition of the current cycle according to the direction cosine of the maximum possible point of the limit state equation in the previous cycle and the preset safety factor. To improve the calculation accuracy of the maximum possible point in the fused single-layer optimization, Harish et al. [40] proposed to use the maximum possible point as the outer optimization variable. At the same time, the first-order Karush-Kuhn-Tucker necessary condition of the optimal solution of the original inner maximum possible point optimization problem is taken as the outer optimization constraint condition. This can ensure that the maximum possible point obtained by the outer layer optimization is equivalent to the maximum possible point obtained by the inner layer optimization. These single-layer methods have advantages over traditional optimization-reliability analysis double-layer nested loop optimization. However, since the current primary research is based on random uncertainty, these methods cannot be used for mixed uncertainty.

1.2.1.6 UBMDO process

The UBMDO process organizes uncertain MDO problems in a computer environment. It can be divided into two categories: The single-layer optimization process and the optimization process based on the decomposition and coordination of disciplines. They will be introduced separately below.

The realization of the single-layer optimization process is based on the single-layer sequential optimization method to solve the MDO and uncertainty analysis. In the single-layer optimization process, the information obtained from the uncertainty analysis in the previous cycle is first obtained. Then, based on this information. The reliability constraints are converted into corresponding deterministic constraints. In this way, the UBMDO problem can be transformed into a deterministic MDO problem. After the deterministic MDO is completed, the uncertainty analysis of the optimization scheme is carried out. The analysis results will be used to guide the next deterministic MDO. The deterministic MDO can be solved directly by the deterministic MDO process, which has been widely studied, for example, the AAO method [41], Simultaneous Analysis and Design (SAND) method [42], IDF method [43], MDF method [44], and Bi-Level Integrated Synthesis System (BLISS) method [44]. Using these methods can improve the efficiency of solving deterministic MDO.

In the study of the UBMDO process based on discipline decomposition and coordination, the deterministic MDO process is mainly used for reference. Applying the discipline decomposition and coordination strategy to the discipline organization of the UBMDO problem, the complex UBMDO problem is solved into several sub-problems with moderate difficulty. Simultaneously, each independent sub-problem in uncertainty optimization, after decomposition, can be efficiently addressed in a parallel fashion within a distributed computing environment. This further reduces the calculation time. McAllister et al. integrated the expected value/variance probability estimation method based on the first-order Taylor expansion into the CO framework and proposed the uncertain CO process [45].

In the above optimization process, the single-layer optimization process transforms UBMDO into two independent steps: Deterministic MDO and uncertainty analysis. Thus, the vast computational cost caused by the double-layer nested loop is reduced. At the same time, the solution of deterministic MDO can be directly solved using the existing deterministic MDO optimization process. Thus, the efficiency of optimization can be further improved. However, the decoupling separation of deterministic MDO and uncertainty analysis will lag the influence of uncertainty analysis results on MDO optimization. This will lead to lower convergence efficiency.

The deterministic constraints used in MDO have limited approximation accuracy for its equivalent uncertainty constraints. This equivalence is generally conservative. Therefore, the optimization effect and efficiency are difficult to guarantee. Decomposition and coordination-based optimization primarily borrow from the deterministic multi-level MDO optimization process based on discipline solution coordination. This process can decompose the overall UBMDO problem into various disciplines. This makes the control of each uncertain design optimization sub-problem within the range of computational cost. At the same time, the distributed parallel solution of each sub-problem can further reduce the calculation time, thereby improving the solution efficiency of UBMDO.

Much research has been done on UBMDO people around the world. Gu et al. [46] suggested a new robust collaborative optimization design method based on an implicit uncertainty analysis method to assess the uncertainty of system response in a two-level optimization framework. Agarwal et al. [47] proposed a new decomposition strategy for the problem of the high computational cost of MDO under single-stage uncertainty. This method replaces the original multidisciplinary analysis with SAND decisions, improving the optimization efficiency. Ahn and Kwon [48] combined the single-stage RBMDO with the BLISS strategy to improve the optimization efficiency by approximating the limit state equation and using the multidisciplinary optimization and reliability analysis method. Kokkolaras et al. [49] considered the uncertainty in multidisciplinary and multi-level systems. They used the advanced mean value to measure the probability distribution of subsystems at each level in the system. Then, a target hierarchical decomposition method was proposed based on reliability.

Chiralaksanakul and Mahadevan [50] used the quantile function to establish the approximation problem of the MDO problem considering reliability. The deterministic method is used to solve the approximate problem, separate the reliability analysis in UBMDO, and improve the computational efficiency. Under the framework of BLISS, Chen et al. [51] proposed an MDO method to deal with mixed random and cognitive uncertainty using probability theory and evidence theory. Fu et al. [52] suggested a sequential multidisciplinary reliability analysis method based on the comprehensive reliability index. They provided a comprehensive reliability index with a clear geometric meaning under multi-source uncertainty. Based on probability theory, the convex model, and Performance Measurement Assessment (PMA), this approach can offer a decoupling strategy. Han and Deng [53] summarized and classified the sources of uncertainty factors and established a robust MDO model considering input variable uncertainty and model error. Acar et al. [54] reviewed and summarized the treatment practice of uncertainty in structural and multidisciplinary system design optimization under uncertainty and summarized and analyzed various design optimization methods under uncertainty. Through some engineering examples, the application of specific technologies is demonstrated to readers.

1.2.2 Application and development of UBMDO

Since UBMDO has unique advantages in solving the MDO problem of complex system uncertainty, its application in designing complex mechanical equipment is also one of the research hotspots this year. More than 20 universities, such as Stanford University and Georgia Institute of Technology, have researched the next generation of high-speed civil aircraft projects. The innovation and expansion of the UBMDO method provide theoretical support for the overall design of aircraft in the future [55]. In NASA's 'Opportunities and Challenges of UBMDO Application in Aircraft Design' published in 2002, the needs and difficulties of this research direction were analyzed in depth. It provides ideas and directions for applying UBMDO theory in aircraft design.

To avoid the high computational cost and limitations brought by the numerical model, Li et al. [56] discussed the uncertainty analysis of assembly quality characteristics and the determination of key quality influencing factors under multi-source uncertainty. Based on the surrogate model and data sampling, Monte Carlo Simulation (MCS) is used to analyze the uncertainty and sensitivity of the assembly process, and the relevant parameters are determined. An assembly example verifies the efficiency of the method. Xie [57] carried out the MDO of turbine rotor at random intervals with uncertainty. This paper introduces a multidisciplinary optimization approach that integrates random and interval uncertainties. The study employs the SORA strategy within the RBMDO mathematical model, leveraging PMA to decouple deterministic optimization and reliability analysis. Liu et al. [58] proposed a multidisciplinary design method using a network engineering environment based on a multi-hierarchical integrated product design data model. This method uses technical methods such as network services, ontology, data flow, and evidence theory to enable designers to participate in UBMDO in a decentralized and dynamic manner. This is more in line with the current distributed industrial system. Yue [59] carried out a multidisciplinary division and optimization modeling of the bulldozer working device, including two disciplines: Mechanism and structure. They used the PSO algorithm to solve the optimization problem under the framework of MDF. In the optimization scheme, this algorithm increases the force arm of the propeller head rotation shaft when the bulldozer shovel cylinder works and obtains a more reasonable working stress distribution. These methods further improve the efficiency of the solution.

1.3 CHAPTER CONTENTS

This book is a subject explanation book, and the content is modular. This is to make readers read any part of the content as much as possible without being affected by the content of other chapters. Therefore, this book divides the issues discussed into several vital sections. This allows readers to find relevant information faster when solving specific problems. However, at the same time, the chapter content of this book is relevant and progressive. This section is an introductory guide for novice learners, offering a structured approach to comprehending the subject matter. It will provide a brief overview of the key topics covered in each chapter of the book.

1.3.1 Chapter 2: The basic knowledge of UBMDO

1.3.1.1 Aim

This chapter aims to introduce the basic knowledge of UBMDO. It contains the basic logic of UBMDO and the pre-order knowledge that needs to be mastered.

1.3.1.2 Outline

This chapter mainly introduces the basic knowledge of UBMDO. The pre-order knowledge includes MDO theory and uncertainty modeling theory, which is necessary to understand this book's key UBMDO theory. Then, it introduces the basic theory of UBMDO, which helps read the following chapters.

1.3.2 Chapter 3: Approximation method

1.3.2.1 Aim

This chapter aims to introduce the basic theory of approximation methods used in optimization design and the MDO method based on the approximation model.

1.3.2.2 Outline

This chapter mainly introduces the approximation method. This is a method developed by applied statistics in engineering. This development primarily addresses the issue of excessive computational load during the optimization process. The primary objective is to minimize computational overhead while maintaining a requisite level of precision. Therefore, the approximate method is indispensable in modern optimization design. This chapter focuses on the basic idea and theory of the approximate model. The specific treatment of different approximate models can be found in more professional and detailed literature.

1.3.3 Chapter 4: Uncertainty analysis methods

1.3.3.1 Aim

This chapter aims to introduce various uncertainty analysis methods used in optimization design.

1.3.3.2 Outline

The uncertainty analysis method introduced in this chapter is a significant focus of this book. Practical engineering has many uncertain factors, so the uncertainty analysis method is critical in UBMDO. The main content of this chapter is to introduce the uncertainty analysis method of the system. In addition to the basic theory of uncertainty analysis, this chapter introduces three branches of uncertainty analysis methods: the stochastic uncertainty analysis method, the non-probabilistic uncertainty analysis method, and the mixed uncertainty analysis method.

1.3.4 Chapter 5: Deterministic MDO method

1.3.4.1 Aim

This chapter aims to introduce the deterministic MDO method involved in the optimization process of large-scale projects.

1.3.4.2 Outline

This chapter mainly introduces the specific method logic of deterministic multidisciplinary optimization, which is one of the key contents of this book. The MDO method has the potential to achieve optimization and reduce design costs and design time. This chapter mainly introduces the single-stage and multi-stage optimization methods in MDO. The specific sub-methods of each method are introduced in more detail. This knowledge has a significant effect on the UBMDO understanding in Chapter 6.

1.3.5 Chapter 6: UBMDO method

1.3.5.1 Aim

This chapter aims to introduce the specific content of the UBMDO method, which is the most essential content of the book.

1.3.5.2 Outline

This chapter mainly introduces different kinds of UBMDO methods used in different situations. Based on the previous chapters, it is an executable sequence composed of system analysis, approximate modeling, uncertainty analysis, and uncertainty optimization. This chapter includes the UBMDO method based on probability theory, non-probability UBMDO, UBMDO based on mixed uncertainty, and their sub-methods.

1.3.6 Chapter 7: Overall design of UBMDO-based flip drive mechanism

1.3.6.1 Aim

This chapter mainly introduces the overall design idea and result analysis of a flip drive mechanism based on the UBMDO method.

1.3.6.2 Outline

This chapter systematically introduces the flip drive mechanism based on the UBMDO method. It mainly includes its subject division and optimization definition, uncertainty modeling, and result analysis. In this case, two UBMDO methods are used for related research, and the final optimization design results are analyzed and explained.

1.3.7 Chapter 8: Overall design of hydraulic turbine rotor bracket based on UBMDO

1.3.7.1 Aim

This chapter mainly introduces the overall design idea and result analysis of a hydraulic turbine rotor bracket structure based on the UBMDO method.

1.3.7.2 Outline

This chapter systematically introduces the structure of the hydraulic turbine rotor bracket based on the UBMDO method. This example realizes the MDO of the rotor through the coupling relationship between the components in the rotor system. It analyses and quantifies uncertainty's influence and finally establishes the UBMDO model. This is an excellent example of the detailed analysis of the multi-source uncertainty of the rotor mechanism. This helps readers better understand the application of the UBMDO method in specific projects.

1.3.8 Chapter 9: Overall design of ducted fan blades based on UBMDO

1.3.8.1 Aim

This chapter mainly introduces the overall design idea and result analysis of a ducted fan blade based on the UBMDO method.

1.3.8.2 Outline

This chapter systematically introduces the water-ducted fan blades designed based on the UBMDO method. In this example, the multi-source uncertainty is considered, and the resonance of the fan blade during operation is analyzed using the finite element method. Finally, according to the comparison of simulation results, the optimization results obtained by the MDO method considering multidisciplinary coupling have better results.

1.3.9 Chapter 10: Development prospect of UBMDO

1.3.9.1 Aim

This chapter mainly looks forward to the future development direction.

1.3.9.2 Outline

This chapter reviews the progress of theoretical research and the application development of UBMDO. The future development of UBMDO is also one of the key contents of this book. Constructing an accurate UBMDO model and improving the efficiency of modeling, and solving is still the main problem that will be solved in the future. Combining the UBMDO method and advanced technology can better promote the development of theory and application. This chapter will more precisely explain the direction of the UBMDO method in future development and application.

REFERENCES

[1] Sobieszczanski-Sobieski J., James B. B., Dovi A. R. (1985). Structural optimization by multilevel decomposition. AIAA Journal, 23(11): 1775–1782.

[2] Sobieszczanski-Sobieski J. (1995). MDO: An emerging new engineering discipline. Advances in Structural Optimization, 25: 483–496.

[3] Sobieszczanski-Sobieski J. (1988). Optimization by decomposition: A step from hierarchic to non-hierarchic systems. Second NASA/Air Force Symposium on Recent Advances in Multidisciplinary Analysis and Optimization, 3031: 51–78.

[4] Cramer E. J., Dennis Jr, J. E., Frank P. D., Lewis R. M., Shubin G. R. (1994). Problem formulation for multidisciplinary optimization. SIAM Journal on Optimization, 4(4): 754–776.

[5] Park G. J. (2007). Analytic Methods for Design Practice. London, England: Springer Science & Business Media.

[6] Hannapel S., Vlahopoulos, N. (2014). Implementation of set-based design in MDO. Structural and Multidisciplinary Optimization, 50: 101–112.

[7] Li L., Wan H., Gao W., Tong F., Li H. (2019). Reliability based MDO of cooling turbine blade considering uncertainty data statistics. Structural and Multidisciplinary Optimization, 59(2): 659–673.

[8] Meng D., Li Y. F., Huang H. Z., Wang Z., Liu Y. (2015). Reliability-based MDO using subset simulation analysis and its application in the hydraulic transmission mechanism design. Journal of Mechanical Design, 137(5): 051402.

[9] Wang L., Xiong C., Yang Y. (2018). A novel methodology of reliability-based MDO under hybrid interval and fuzzy uncertainties. Computer Methods in Applied Mechanics and Engineering, 337: 439–457.

[10] Jaron R., Moreau A., Guérin S., Enghardt L., Lengyel-Kampmann T., Otten T., Nicke E. (2022). MDO of a low-noise and efficient next-generation aero-engine fan. Journal of Turbomachinery, 144(1): 011004.

[11] Chen X., Wang P., Dong H., Zhao X., Xue D. (2021). Coupled-analysis assisted gradient-enhanced kriging method for global MDO. Engineering Optimization, 53(6): 1081–1100.

[12] Dantzig G. B. (1955). Linear programming under uncertainty. Management Science, 1(3–4): 197–206.

[13] Sahinidis N. V. (2004). Optimization under uncertainty: State-of-the-art and opportunities. Computers & Chemical Engineering, 28(6–7): 971–983.

[14] Schuëller G. I., Jensen H. A. (2008). Computational methods in optimization considering uncertainties – An overview. Computer Methods in Applied Mechanics and Engineering, 198(1): 2–13.

[15] Dhillon B. S., Belland J. S. (1986). Bibliography of literature on reliability in civil engineering. Microelectronics Reliability, 26(1): 99–121.

[16] Tong Y. C. (2001). Literature Review on Aircraft Structural Risk and Reliability Analysis. Melbourne, Australia: DSTO Aeronautical and Maritime Research Laboratory.

[17] Frangopol D. M., Maute K. (2003). Life-cycle reliability-based optimization of civil and aerospace structures. Computers & Structures, 81(7): 397–410.

[18] Padula S. L., Gumbert C. R., Li W. (2006). Aerospace applications of optimization under uncertainty. Optimization and Engineering, 7: 317–328.

[19] Uebelhart S. A. (2006). Non-deterministic design and analysis of parameterized optical structures during conceptual design. PhD thesis of Massachusetts Institute of Technology.

[20] DeLaurentis D. A. (1998). A probabilistic approach to aircraft design emphasizing guidance and stability and control uncertainties. PhD thesis of Georgia Institute of Technology.

[21] Li L. (2008). Structural design of composite rotor blades with consideration of manufacturability, durability, and manufacturing uncertainties. PhD thesis of Georgia Institute of Technology.

[22] Zhang T., Yan X., Huang W., Che X., Wang Z. (2021). MDO of a wide speed range vehicle with waveride airframe and RBCC engine. Energy, 235: 121386.

[23] Wang S., Yang M., Niu W., Wang Y., Yang S., Zhang L., Deng J. (2021). MDO of underwater glider for improving endurance. Structural and Multidisciplinary Optimization, 63: 2835–2851.

[24] Helton J. C., Johnson J. D., Sallaberry C. J., Storlie C. B. (2006). Survey of sampling-based methods for uncertainty and sensitivity analysis. Reliability Engineering & System Safety, 91(10–11): 1175–1209.

[25] Iman R. L., Helton J. C. (1988). An investigation of uncertainty and sensitivity analysis techniques for computer models. Risk Analysis, 8(1): 71–90.

[26] Cacuci D. G., Ionescu-Bujor M. (2004). A comparative review of sensitivity and uncertainty analysis of large-scale systems—II: Statistical methods. Nuclear Science and Engineering, 147(3): 204–217.

[27] Liu H., Chen W., Sudjianto A. (2004). Probabilistic sensitivity analysis methods for design under uncertainty. 10th AIAA/ISSMO Multidisciplinary Analysis and Optimization Conference, 4589.

[28] Morris M. D. (1991). Factorial sampling plans for preliminary computational experiments. Technometrics, 33(2): 161–174.

[29] Campolongo F., Cariboni J., Saltelli A. (2007). An effective screening design for sensitivity analysis of large models. Environmental Modelling & Software, 22(10): 1509–1518.

[30] Helton J. C., Johnson J. D., Oberkampf W. L., Sallaberry C. J. (2006). Sensitivity analysis in conjunction with evidence theory representations of epistemic uncertainty. Reliability Engineering & System Safety, 91(10–11): 1414–1434.

[31] Oberguggenberger M., King J., Schmelzer B. (2009). Classical and imprecise probability methods for sensitivity analysis in engineering: A case study. International Journal of Approximate Reasoning, 50(4): 680–693.

[32] Simpson T. W., Poplinski J. D., Koch P. N., Allen J. K. (2001). Metamodels for computer-based engineering design: Survey and recommendations. Engineering with Computers, 17: 129–150.

[33] Powell J. D. (1987). Radial basis function approximations to polynomials. Proceedings of 12th Biennial Numerical Analysis Conference, 223–241.

[34] Broomhead D. S., Lowe D., Casadevall A. (1988). Multivariable functional interpolation and adaptive networks. Complex Systems, 2: 312–355.

[35] Park J., Sandberg I. W. (1991). Universal approximation using radial-basis-function networks. Neural Computation, 3(2): 246–257.

[36] Haykin S. (1998). Neural Networks: A Comprehensive Foundation. Upper Saddle Rivcr, NJ: Prentice Hall PTR.

[37] Verleysen M., Hlavackova-Schindler K. (1996). Learning in RBF networks. International Conference on Neural Networks, 199–204.

[38] Du X., Chen W. (2004). Sequential optimization and reliability assessment method for efficient probabilistic design. Journal of Mechanical Design, 126(2): 225–233.

[39] Chen X., Hasselman T., Neill D., Chen X., Hasselman T., Neill D. (1997). Reliability based structural design optimization for practical applications. 38th Structures, Structural Dynamics, and Materials Conference, 1403.

[40] Agarwal H., Renaud J., Lee J., Watson L. (2004). A unilevel method for reliability based design optimization. 45th AIAA/ASME/ASCE/AHS/ASC Structures, Structural Dynamics & Materials Conference, 2029.

[41] Bryson D. E., Rumpfkeil M. P. (2017). All-at-once approach to multifidelity polynomial chaos expansion surrogate modeling. Aerospace Science and Technology, 70: 121–136.

[42] Calado J. M. F., da Costa J. S., Bartys M., Korbicz J. (2006). FDI approach to the DAMADICS benchmark problem based on qualitative reasoning coupled with fuzzy neural networks. Control Engineering Practice, 14(6): 685–698.

[43] Vroemen B., Serrarens A., Veldpaus F. (2001). Hierarchical control of the zero inertia powertrain. JSAE Review, 22(4): 519–526.

[44] Kodiyalam S. (1998). Evaluation of Methods for Multidisciplinary Design Optimization (MDO), Phase I. Hampton, VA: National Aeronautics and Space Administration, Langley Research Center.

[45] Hao W. J. (2014). Research on key technology of large mining excavator design. PhD thesis of Jilin University (in Chinese), Changchun, China.

[46] Gu X., Renaud J. E., Penninger C. L. (2006). Implicit uncertainty propagation for robust collaborative optimization. Journal of Mechanical Design, 128(4): 1001–1013.

[47] Agarwal H., Renaud J., Lee J., Watson L. (2004). A unilevel method for reliability based design optimization. 45th AIAA/ASME/ASCE/AHS/ASC Structures, Structural Dynamics & Materials Conference, 2029.

[48] Ahn J., Kwon J. H. (2006). An efficient strategy for reliability-based MDO using BLISS. Structural and Multidisciplinary Optimization, 31: 363–372.

[49] Kokkolaras M., Mourelatos Z. P., Papalambros P. Y. (2004). Design optimization of hierarchically decomposed multilevel systems under uncertainty. International Design Engineering Technical Conferences and Computers and Information in Engineering Conference, 46946: 613–624.

[50] Chiralaksanakul A., Mahadevan S. (2007). Decoupled approach to MDO under uncertainty. Optimization and Engineering, 8: 21–42.

[51] Yao W., Chen X., Ouyang Q., Van Tooren M. (2013). A reliability-based MDO procedure based on combined probability and evidence theory. Structural and Multidisciplinary Optimization, 48: 339–354.

[52] Fu C., Liu J., Xu W. (2021). A decoupling strategy for reliability analysis of multidisciplinary system with aleatory and epistemic uncertainties. Applied Sciences, 11(15): 7008.

[53] Man H. H., Deng J. T. (2007). Uncertainty modeling in MDO. Journal of Beihang University, 33(01): 115–118.

[54] Acar E., Bayrak G., Jung Y., Lee I., Ramu P., Ravichandran S. S. (2021). Modeling, analysis, and optimization under uncertainties: A review. Structural and Multidisciplinary Optimization, 64(5): 2909–2945.

[55] Walsh J., Weston R., Samareh J., Mason B., Green L., Biedron R. (2000). Multidisciplinary high-fidelity analysis and optimization of aerospace

vehicles. II-Preliminary results. 38th Aerospace Sciences Meeting and Exhibit, 419.
[56] Li Y., Zhang F. P., Yan Y., Zhou J. H., Li Y. F. (2019). Multi-source uncertainty considered assembly process quality control based on surrogate model and information entropy. Structural and Multidisciplinary Optimization, 59: 1685–1701.
[57] Xie T. W. (2020). MDO of turbine rotor considering multi-source uncertainty. Master thesis of University of Electronic Science and Technology of China (in Chinese), Chengdu, China.
[58] Liu C. W., Jin X. X., Li L. S. (2014). A web services-based MDO framework for complex engineering systems with uncertainties. Computers in Industry, 65(4): 585–597.
[59] Yue G. D. (2009). Research on MDO of bulldozer working device. Master thesis of Zhejiang University (in Chinese), Hangzhou, China.

Chapter 2

Basic knowledge of UBMDO

2.1 MDO THEORY

The field of aerospace is where Multidisciplinary Design Optimization (MDO) theory first emerged. Conventional aircraft designs use a serial architecture, and different key disciplines are selected in different design stages to design and optimize the aircraft. This artificially separates the various disciplines that affect the aircraft's performance and does not fully use the synergistic effects that may occur due to mutual coupling between the disciplines. Therefore, the design obtained by this method is most likely not the overall optimal solution of the system, thereby reducing the aircraft's overall performance. The emergence of MDO solves the defects of traditional aircraft design methods. Since its proposal, it has received full attention from countries and industries.

Aiming at the whole process of complex system (product) design, people integrate the MDO theory with the relevant knowledge of various disciplines (systems), which applies effective design optimization strategies and distributed computer network systems to organize and manage the product design process. By entirely using the synergistic effect generated by the interaction between various disciplines (systems), the overall optimal solution or engineering satisfaction of the system is obtained. MDO technology makes up for the inherent defects of traditional optimization design methods. From a holistic system design perspective, achieving the overall optimal system performance is more attainable by taking into account the synergistic effects of each subsystem and leveraging the latest advancements in various disciplines. The basic idea of MDO was first proposed by Sobieszczanski-Sobieski in 1974, and then he formally proposed the concept and core idea of MDO in 1982. Thus, MDO officially entered the field of vision of researchers and has been studied and developed by scholars and researchers worldwide for more than 40 years. So far, MDO theory has made much progress and rich achievements and has been widely used in engineering. Briefly, the development history of MDO is shown in Figure 2.1.

DOI: 10.1201/9781003464792-2

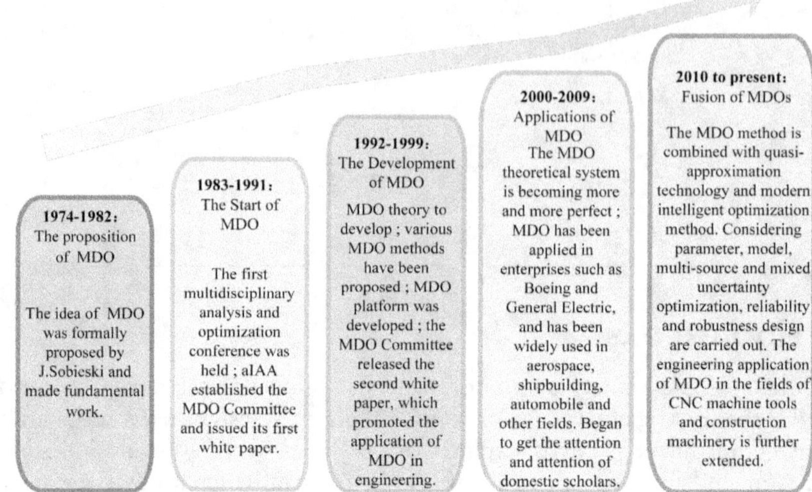

Figure 2.1 The development history of MDO.

The first stage was from 1974 to 1982, when the concept of MDO was formally proposed. In 1974, NASA senior researcher Sobieszczanski-Sobieski first proposed the basic idea of MDO, then formally proposed the concept in 1982, and successively described the main research contents, key technologies, and frameworks of MDO. Moreover, in the later research, Concurrent Subspace Optimization (CSSO) and Bi-level Integrated System Synthesis (BLISS) were proposed successively, laying a foundation for MDO theory.

The second stage was in 1983–1991, MDO began to start. At this stage, the American Institute of Aeronautics and Astronautics (AIAA) established the MDO committee, and at the proposal of the US Air Force, NASA, and AIAA, the first multidisciplinary analysis and optimization conference was held at the Langley Research Center of NASA. The conference has evolved into a prominent international academic event in MDO and structural optimization, garnering increased attention from the global scientific community. AIAA also released the first white paper on MDO, which describes the research content and key technologies of MDO in detail. At the same time, the European Society for the International Society for Structural and Multidisciplinary Optimization (ISSMO) journals 'Structure and MDO (SMDO)' and 'World Congress on Structure and MDO (WCSMDO)' attach great importance to MDO and greatly promote the development and engineering application of MDO theory.

In the third stage, in 1992–1999, MDO technology was revitalized and vigorously developed. The research on the MDO method is very active. So far, the most commonly used MDO methods have been proposed: Multidisciplinary Feasible (MDF, also known as the All-In-One method, AIO), Simultaneous Analysis and Design (SAND), also known as the All-At-Once method (AAO), Individual Discipline Feasible (IDF), CSSO, Collaborative Optimization (CO), BLISS, and Analytical Target Cascading (ATC). At the same time, various multidisciplinary design platforms have been developed, such as Insight, Model Center, Optimus, mode FRONTIER, and other commercial multidisciplinary integrated design software. At the same time, some scientific research institutions have developed their dedicated platforms. At the same time, AIAA conducted a survey of General Electric and Boeing in the United States, investigated the application of MDO technology, and released the second white paper on MDO, summarized the application of MDO technology in engineering, and summarized the specific application experience, which strongly encouraged the use of MDO in engineering.

In the fourth stage, in 2000–2009, the United States incorporated MDO technology into the national key technology development plan. Researchers and engineers around the world have widely used MDO technology. The MDO theoretical system is becoming more and more complete, and the application technology is becoming more and more mature. At the same time, domestic researchers have also attached great importance to MDO technology and have made many achievements. The applications of MDO technology in aviation, aerospace, missile design, ship, manned submersible, marine engineering equipment, automobile, engineering machinery, and CNC machine tools are discussed, respectively.

The fifth stage is 2010–present. The combination of methods and approximation techniques in MDO has dramatically improved the solution efficiency and promoted the engineering application of MDO technology. Combined with modern intelligent optimization algorithms, people proposed various MDO methods for specific problems, and the research on multi-objective MDO problems is also very active. At the same time, since engineering problems often contain various uncertainties, the research of Uncertainty-based MDO (UBMDO) has become very active. Considering the uncertainty of parameters, models, multi-source, and time-varying, uncertainty theories such as probability theory, fuzzy theory, convex model, interval theory, evidence theory, and generalized probability theory are used to carry out uncertainty optimization design. Combined with the deterministic MDO method, the corresponding UMDO method is proposed, significantly expanding the theoretical framework of MDO technology and its application in engineering product design. At the same time, the Asian Society for Structural and Multidisciplinary Optimization (ASSMO) and its Asia Symposium on Structural and Multidisciplinary Optimization

(ACSMO) have played a significant role in promoting the scientific research and engineering application of MDO in Asia, including China.

2.1.1 Basic concepts of MDO

2.1.1.1 Definition and characteristics of MDO

Definition 2.1 Discipline: The basic module of the system, which is relatively independent and has data exchange relationships with each other.

In MDO, the discipline is also called a subsystem, and the so-called MDO is based on NASA's definition of MDO, which is the methodology of designing complex systems and subsystems by fully exploring and utilizing the synergistic mechanism of interaction in the system.

Taking the non-hierarchical system of the three disciplines in Figure 2.2 as an example, we explained the basic concepts of MDO.

Definition 2.2 Design variables: A set of independent controllable variables used to describe the system's characteristics during the design process.

The design variables can be divided into system design variables and local design variables, such as X in Figure 2.2, which play a role in the whole system, so it is called system design variables; at the same time, X_1, X_2, X_3, and so on only play a role within the discipline, so they are called local design variables or discipline variables, and some books are also called sub-space design variables.

Definition 2.3 State variable: The response calculated in the analysis model or computer model in the discipline to describe the performance or characteristics of the engineering system.

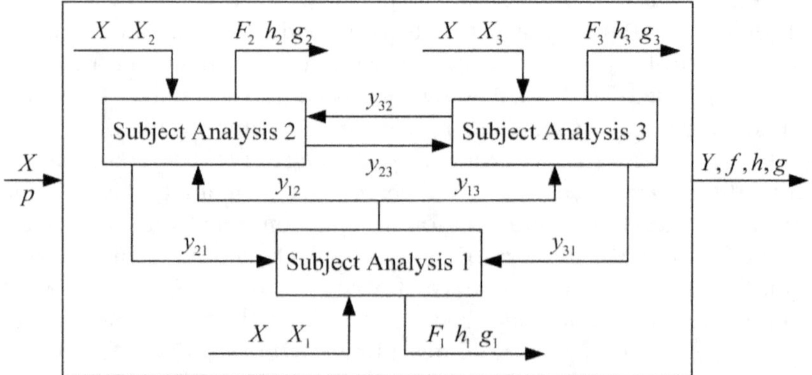

Figure 2.2 Non-hierarchical system of three disciplines.

The state variables need to be obtained through model analysis and calculation. According to the described object, it can be divided into system state variables, discipline state variables, and coupling state variables. As shown in Figure 2.2, Y describes the characteristics of the whole system, which is the system state variable. y_1, y_2, and y_3 describe the characteristics of the corresponding disciplines, which are called discipline variables or subspace state variables. The coupling state variable is also known as the nonlocal state variable. It refers to the variable that the state variables of other disciplines are input as design variables into a certain discipline for discipline analysis at the same level, which is represented by y_{ij}, where i refers to the discipline to which the state variable belongs and j refers to the discipline using the input design variable to be the state variable. The y_{12}, y_{21}, and so on in Figure 2.2 are the above coupling state variables.

Definition 2.4 Constraints: The system must meet conditions in the design process.

Constraints can be separated into constraints on inequality and equality, which are represented by h and g , respectively in Figure 2.2. According to the constraint object, the constraint conditions can be divided into system and discipline constraints. The system constraints refer to the constraints received in the whole system layer, such as h and g in Figure 2.2, and the discipline constraints refer to the constraints to be received in the scope of each discipline, such as h_1 and g_1 in Figure 2.2.

Definition 2.5 System Parameter: A set of fixed-value parameters describe the system characteristics during the design process, like p in Figure 2.2.

Definition 2.6 Contributing Analysis: This process involves adhering to the physical laws of the discipline to derive the discipline's unique characteristics. It entails utilizing the discipline's design variables, coupled state variables from other disciplines, and system parameters as inputs.

The process of discipline analysis solves the discipline state equation. It is assumed that the state equation of discipline i is as follows:

$$G_i\left(y_i;X,X_i,y_{ji}\right)=0 \tag{2.1}$$

where y_{ji} denotes the coupling state variable from discipline j to discipline i, and $i \neq j$, ';' denotes that only y_i is unknown and needs to be solved. X represents system design variables, and X_i denotes the design variable of discipline i.

According to the above definition, disciplinary analysis is the process of solving the disciplinary state equation, that is,

$$y_i = CA_i\left(X, X_i, y_{ji}\right) = G_i^{-1}\left(0; X, X_i, y_{ji}\right) \tag{2.2}$$

Definition 2.7 System Analysis: In the whole system, a set of design variables X is given as the initial value, and then the system's state variables are obtained by solving the state equation of the system.

For a system consisting of n disciplines, the system analysis process can be represented by the following formula:

$$y = SA\left(X, X_1, X_2, \ldots, X_n\right) \tag{2.3}$$

If it is a complex engineering system, its system analysis involves multi-disciplinary analysis. For the non-hierarchical system shown in Figure 2.2, due to the coupling between disciplines, system analysis requires multiple iterations to complete.

In addition, there are couplings between disciplines and conflicting relationships. System analysis does not guarantee immediate outcomes. Therefore, there are the following definitions:

Definition 2.8 Consistent Design: In the system design process, a design scheme comprises design variables and corresponding system state variables that satisfy the system state equation.

Definition 2.9 Feasible Design: A consistency design that meets the design requirements under constraints.

Definition 2.10 Optimal Design: The feasible design scheme that best meets the design goal.

2.1.1.2 Mathematical expression of MDO

In general, the deterministic mathematical model of the MDO problem can be expressed as:

$$\begin{aligned} &\text{Find } x \\ &\min \ F(x,y) \\ &\text{s.t.} \ G(x,y) \le 0 \\ &\qquad H(x,y) = 0 \\ &\quad x^{\min} \le x \le x^{\max} \end{aligned} \tag{2.4}$$

where x is the design variable vector, $F(x, y)$ is the objective function, y is the coupling state variable vector, $G(x, y)$ is the inequality constraint vector, and $H(x, y)$ is the equality constraint vector. Calculating y in Eq. 2.4 is often the most complex in solving the whole MDO problem. This process is called multidisciplinary or system analysis, which can be solved by repeated iterations. The Gauss-Seidel iteration method can be used to solve the linear equation, and the steepest descent method can be used to solve the nonlinear equation. The MDO problem frequently involves multiple interconnected subsystems, necessitating iterative solutions.

2.1.2 Research content of MDO

The research content of MDO mainly includes the research of optimization methods, the establishment of a multidisciplinary optimization (MDO) model, the establishment of an integrated design environment, and so on. The detailed explanation of the research content is as follows [1].

2.1.2.1 Multidisciplinary optimization method

The MDO method is also called the multidisciplinary strategy. This field primarily focuses on decomposing intricate MDO problems into more manageable design optimization problems across different disciplines. It seeks to enhance the coordination of the design processes in various domains and integrate the resulting designs effectively. The commonly used deterministic MDO methods currently include MDF, AAO, IDF, CSSO, CO, BLISS, and ATC. The following briefly overviews these seven methods' basic principles and characteristics.

2.1.2.1.1 Multidisciplinary feasible

The MDF method is the most traditional and original method for solving MDO problems, and it is also the most basic method in MDO methods. MDF is a single-stage MDO optimization method. In line with the system's initial framework, the subsystem is organized, and complex multidisciplinary analysis is carried out. All objective functions, design variables, and constraints are solved in a solver. The coupling state variables between disciplines are solved by repeated multidisciplinary analysis. Figure 2.3 shows a block diagram of the MDF method for the MDO problem involving two disciplines. The multidisciplinary analysis process is iterated repeatedly using the MDF method to solve the MDO problem. The MDF method attains an optimal solution ensuring multidisciplinary feasibility upon convergence.

The MDF method has a simple structure, is relatively easy to implement, and is more suitable for solving problems with fewer subsystems and tight coupling. For example, it performs well in solving aeroelastic problems in aviation design. However, the MDF method requires time-consuming

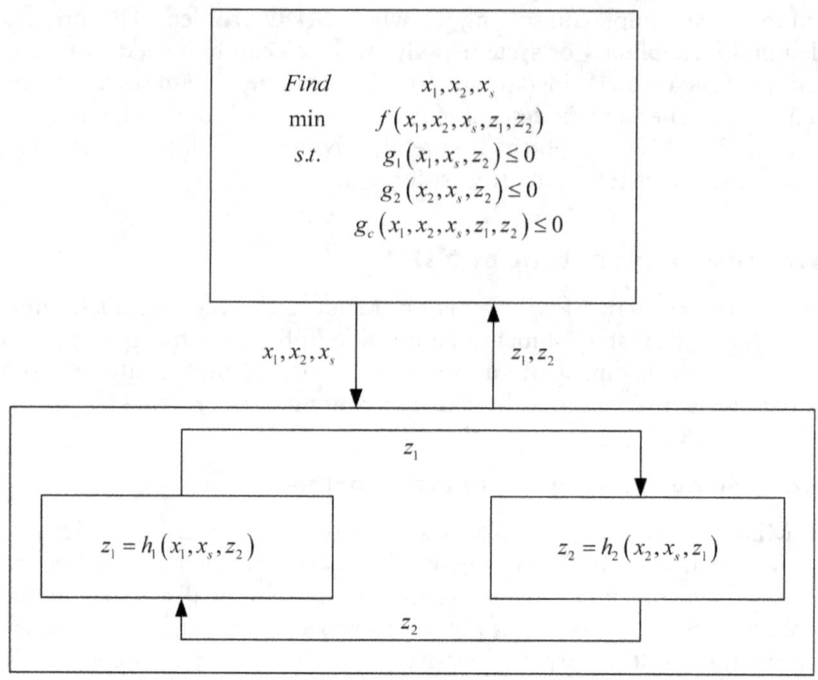

Figure 2.3 The block diagram of the MDF method for MDO problems in two disciplines.

multidisciplinary analysis, and the subsystem balance iteration is sensitive to the choice of the coupling variable's starting value.

2.1.2.1.2 Simultaneous analysis and design

At the same time, the Simultaneous analysis and optimization method is also called the AAO method, which is a single-stage MDO method. The main design concept of the SAND method is to change the analysis process into the solution process and to optimize the solution and subsystem analysis in the exact solver. The basic idea of the SAND method is to use two auxiliary variables: Coupling auxiliary design variables and state coupling design variables. Each sub-discipline remains independent of the other, and a consistency constraint condition is added at the system level to minimize the error between the auxiliary variables and the coupling variables so that they can be solved uniformly in the system-level optimizer. The structural block diagram of the SAND method for the MDO problem, including two disciplines, is shown in Figure 2.4. When the optimization process is terminated, it satisfies both single-disciplinary feasibility and multidisciplinary feasibility.

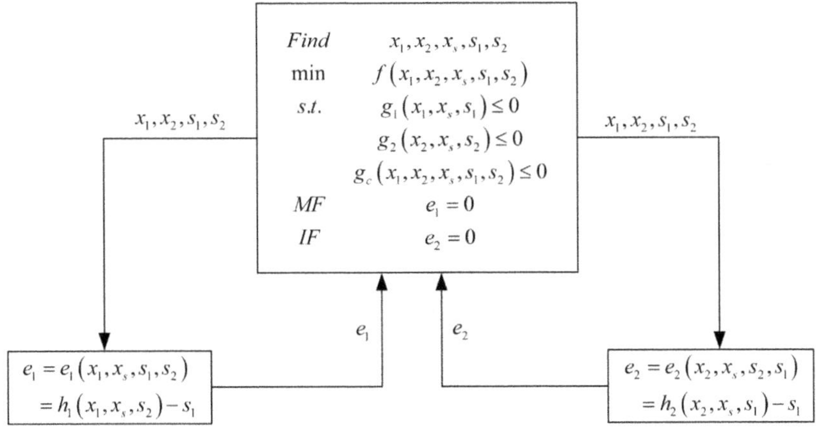

Figure 2.4 Structure diagram of the SAND method for MDO problems in two disciplines.

In solving the SAND method, the design variables contain all the auxiliary variables, and the coupling relationship between the subsystems is completely decoupled. It is suitable for the MDO problem, where the coupling between the subsystems is relatively loose, and the subsystem analysis is time-consuming. However, the SAND method introduces new auxiliary design variables, which increases the design space dimension. At the same time, the consistency constraint conditions introduced also increase the difficulty of the constraint conditions, which will increase the difficulty of solving the optimal solution. As such, the problem of more coupling variables and state variables is not suited for the SAND method.

2.1.2.1.3 Individual discipline feasible

The single-disciplinary feasibility method is a single-stage MDO method. The difference between the IDF method and the SAND method is that the IDF method only introduces the coupling auxiliary design variable and does not introduce the state auxiliary design variable. The direct relationship between the total system and the subsystem is retained, and the coupling auxiliary design variable is used to decouple each subsystem. For computing constraints and objective functions, the state variables of every subsystem are directly communicated to the entire system. Figure 2.5 displays the structure diagram of the IDF method for the MDO problem, which incorporates two disciplines. The IDF method does not need to perform complex multidisciplinary analysis processes. Each discipline calculates independently and adds auxiliary variables and consistency constraints. When the equation constraints are satisfied, the optimal solution is MDF.

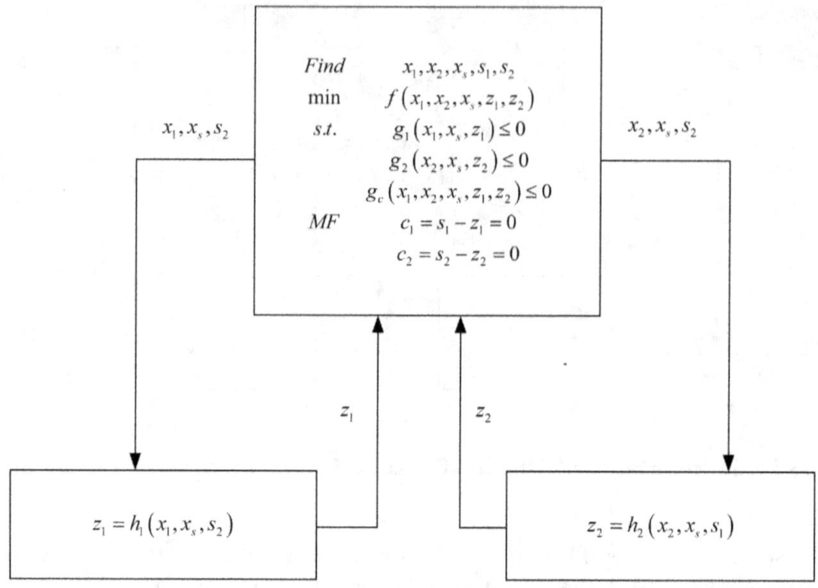

Figure 2.5 Structure diagram of IDF method for MDO problems in two disciplines.

The IDF method is very similar to the SAND method. One could consider the IDF method to be an incomplete SAND method. Its characteristics are similar to those SAND method. It is suitable for solving the MDO problem with loose coupling between systems and time-consuming subsystem analysis. It is not appropriate for addressing the issue of subsystems that are closely coupled. The IDF method's design space dimension for the identical MDO problem is less than that of the SAND method. However, the problem is expected to be that there are more coupling variables, and the number of state variables is not high enough. Currently, the reduction of the design space dimension of the IDF method is not very obvious compared with the SAND method.

2.1.2.1.4 Concurrent subspace optimization

The original CSSO method is a non-hierarchical two-level MDO method. The basic idea of CSSO is to create an approximate relationship between the design variables and the coupling variables using the global sensitivity equation by utilizing the first-order Taylor series expansion. In the subsystem optimization, the global sensitivity equation provides the approximate values of the state variables and the coupling variables. Figure 2.6 shows the flowchart of the CSSO for the MDO problem involving two disciplines.

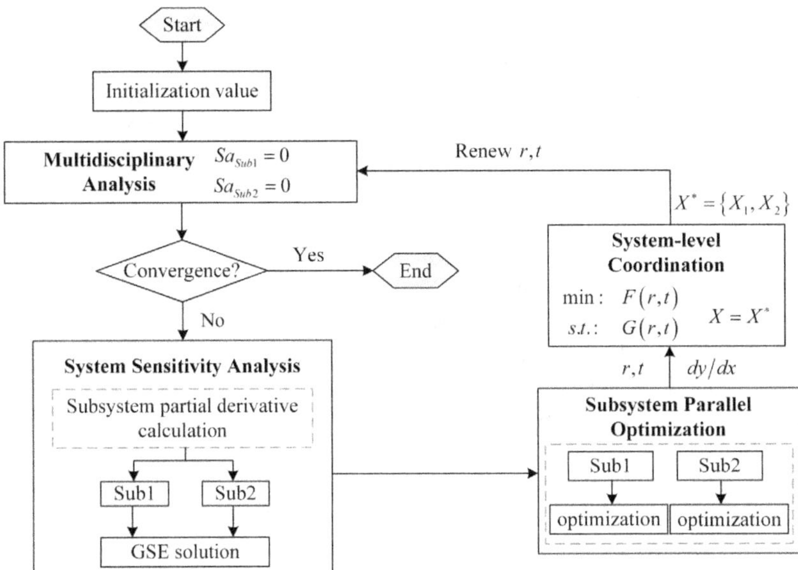

Figure 2.6 Flowchart of CSSO method for MDO problems in two disciplines.

First, the multidisciplinary analysis is carried out to obtain the initial design value. Then, the sensitivity analysis of each subsystem is carried out to calculate the partial derivative of each subsystem, and the global sensitivity equation is solved. Each subsystem's gradient information is used to optimize each subsystem in parallel, and the obtained optimization value is transmitted to the system level for coordination. After optimization, the coordination coefficient of each subsystem is obtained, and the convergence judgment is carried out. The optimization is completed, and the optimal solution is obtained. If unsatisfied, the next iteration is continued until convergence to obtain the optimal solution.

The CSSO method maintains multidisciplinary consistency through the global sensitivity equation and realizes the decoupling between subsystems through approximation technology. Each subsystem can be independently analyzed and optimized, and parallel computing and optimization can be realized. However, the CSSO method needs gradient calculation, which has a large amount of calculation and cannot deal with the mixed case of continuous and discrete variables. At the same time, it needs multidisciplinary analysis, which is unsuitable for large-scale engineering MDO problem optimization with many design variables. Of course, the solution's efficiency has been dramatically improved in various CSSO methods, especially the CSSO method based on approximation technology.

2.1.2.1.5 Collaborative optimization

The Collaborative Optimization (CO) method is the first hierarchical two-level MDO method. The standard CO method divides the MDO problem into upper and lower layers. The upper layer is system-level optimization, and the lower layer is subsystem optimization. Like the IDF and SAND methods, the fundamental idea behind the standard CO method is to decouple subsystems at the system level by introducing auxiliary design variables and consistency constraints. This allows each subsystem to be independent, autonomous, capable of carrying out parallel design, not carrying out complex multidisciplinary analysis processes, and only analyzing at the subsystem level. However, different from the two single-stage MDO methods, the IDF and SAND methods, the system and subsystem levels of the CO method are optimized. The framework of the standard CO method for the MDO problem involving two disciplines is shown in Figure 2.7.

According to the structural framework diagram of the standard CO method, the basic solution process can be obtained: The expected target value is assigned at the system level. For the coupling variables between the subsystems and transmits it to each subsystem. Under the constraints of the subject, each subsystem aims to minimize the discrepancy between the target value assigned at the system level and the coupling variables' values across disciplines. After analysis and optimization, the results are returned to the system level. The system-level takes the minimization of the objective

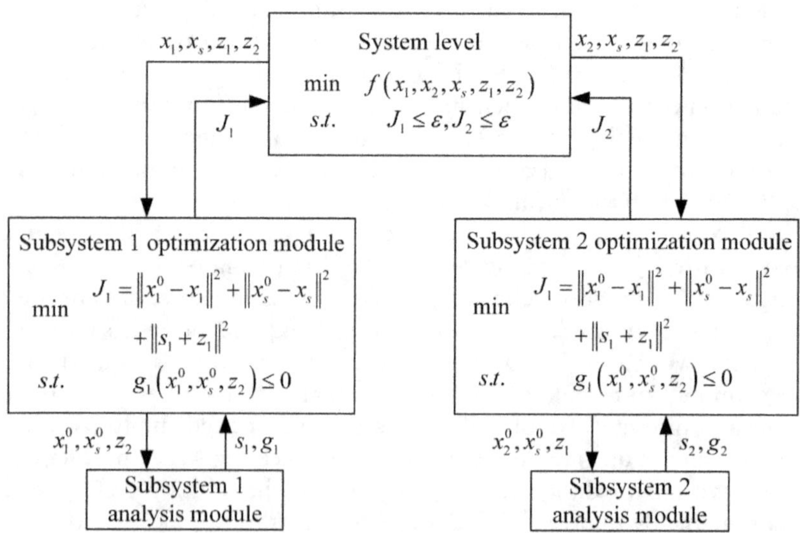

Figure 2.7 Standard CO method framework for MDO problems in two disciplines.

function of the original MDO problem as the optimization goal and relies on satisfying the consistency constraint condition to decouple and coordinate the consistency of the coupling variables between disciplines and then redistributes the new target value obtained after the system-level optimization to each subsystem. Iterate repeatedly until the termination condition is satisfied and an optimal design scheme for the MDO problem is obtained.

The CO method is consistent with distributed design and the division of labor in large-scale engineering today. Each subsystem maintains independent autonomy, and the implementation is relatively simple. It is convenient for management and organization. It can better solve the two major problems faced by solving MDO problems: Calculation complexity and organization. Large-scale engineering product distributed design is a better fit for it. Therefore, it has become one of the most widely used MDO methods. However, because additional design variables were added, the CO method increases the dimension of the design space, which is unsuitable for solving the MDO problem, which is closely coupled between various disciplines. Moreover, the consistency constraint conditions often make it difficult for the system-level optimization problem to satisfy the Karush-Kuhn-Tucker (KKT) condition so that it cannot converge to the optimal solution. Furthermore, the nonlinearity of the linear MDO problem will increase with its solution, adding to its difficulty. However, many scholars have made many improvements to the shortcomings of the CO method. The convergence of the CO method has been improved, and the scope of application has been expanded.

2.1.2.1.6 BI-level integrated system synthesis

The BLISS method is a two-stage MDO method. The original BLISS method was first proposed by Sobieski in 1998. This paper is denoted as BLISS98. The fundamental idea is to use the first-order Taylor approximation to extend the system-level objective function to the subsystem design variables. However, due to the need for many gradient calculations, it is unsuitable for complex engineering practical problems. In 2000, Sobieski proposed an improved BLISS method based on the response surface, called BLISS2000 in this paper. The basic principle is to create response surfaces in each subsystem based on satisfying the constraints of each subsystem. System-level optimization is performed to achieve the optimal design of the entire system by utilizing each subsystem's shared design variables and coupling variables to satisfy the consistency requirements. The process of the BLISS2000 method for the MDO problem with two disciplines is shown in Figure 2.8.

The diagram shows that the BLISS2000 method includes four parts: Primary system analysis, subsystem optimization, optimal response surface model construction of each subsystem, and system optimization. The weight coefficient connects the subsystem optimization and system optimization.

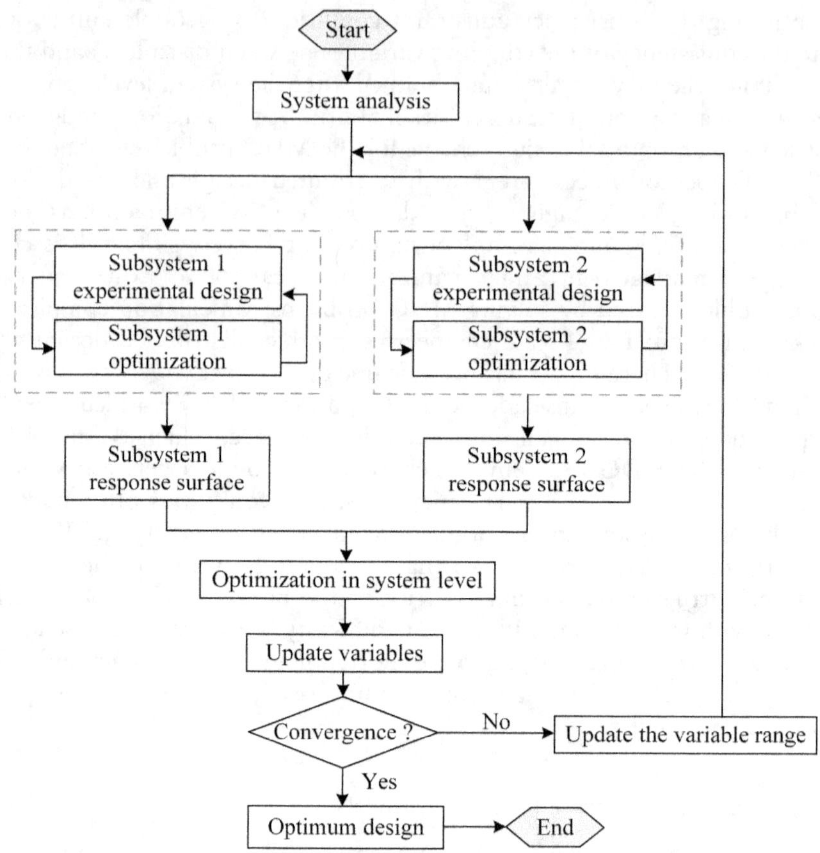

Figure 2.8 Flowchart of BLISS2000 method for MDO problems in two disciplines.

The basic flow of the BLISS2000 method is as follows: Subsystem-level optimization's objective function is defined as the weighted sum of the coupling state variables that are produced by each discipline in the optimization process. The DOE method is used to sample the system-level design space, and the extracted sample points are used to optimize the subsystem. The obtained optimal value of the coupling variable is used as the input of the system-level optimization, and the response surface proxy model is constructed. Finally, the system-level optimization is performed to meet the termination condition. That is, the optimal design is obtained. If not, the design space range released by the system level to the subsystem level is updated, and the next iteration is entered until the optimal design is solved.

Compared with the CSSO method, the BLISS method is more suitable for optimizing multivariable problems. The BLISS method suits the MDO problem with variable severe coupling more than the CO method. Moreover, the BLISS method only needs one multidisciplinary analysis to ensure multidisciplinary feasibility, reduce the time consumption of multidisciplinary analysis, and avoid the complex and time-consuming sensitivity calculation, which saves much time. Therefore, the BLISS2000 method is considered to be an auspicious MDO method. However, the BLISS method introduces weighting coefficients into the system design variables. The dimension of the system-level design variables is excessively large for the optimization problem with numerous state variables for each subsystem, which increases the difficulty of the optimization search and reduces the efficiency of the solution. In addition, it is necessary to divide the system-level design variables properly. Otherwise, the accuracy of the results will not be high.

2.1.2.1.7 Analytical target cascading

ATC is a model-based multi-level MDO method. The ATC method was initially developed for layered product design and is suitable for solving hierarchical systems. The basic idea of the ATC method is to optimize each subsystem of each level independently through target-level analysis and finally obtain a solution that meets the requirements of system consistency. The basic process is as follows: The ATC method decomposes the hierarchical product design into the subsystems of the next level, forming a hierarchical structure design of multiple subsystems; the design index of the product is transmitted in the hierarchical structure, and the system index is decomposed into subsystem index, and then the subsystem index is decomposed into a component index to continue to decompose according to the hierarchy until the decomposition is completed. The optimization design results are continuously fed back to the corresponding upper level at each level, and this process of alternating between the upper and lower levels of optimization continues until the convergence conditions are met. The schematic diagram and decomposition coordination diagram of the three-layer decomposition using the ATC method and the data flow diagram of the optimization process correspond to Figure 2.9, Figure 2.10, and Figure 2.11, respectively.

2.1.2.2 Establishment of multidisciplinary optimization model

In MDO, the form and format of information exchange between different modules are very complex, so it is necessary to conduct modeling research on the MDO analysis model. Modeling is the basis of multidisciplinary integrated design and an essential means and premise of the research system. The so-called modeling establishes the system model, that is, modeling the actual system. Modeling uses models, such as mathematical, geometric (entity), and structural models, to describe the system's causal or mutual

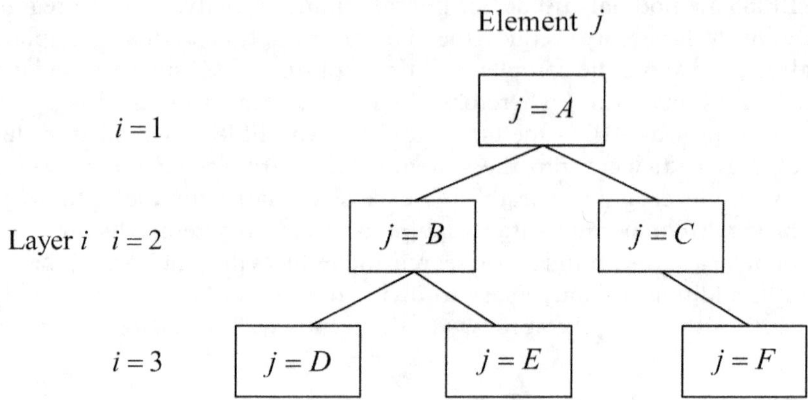

Figure 2.9 Hierarchical decomposition diagram using the ATC method.

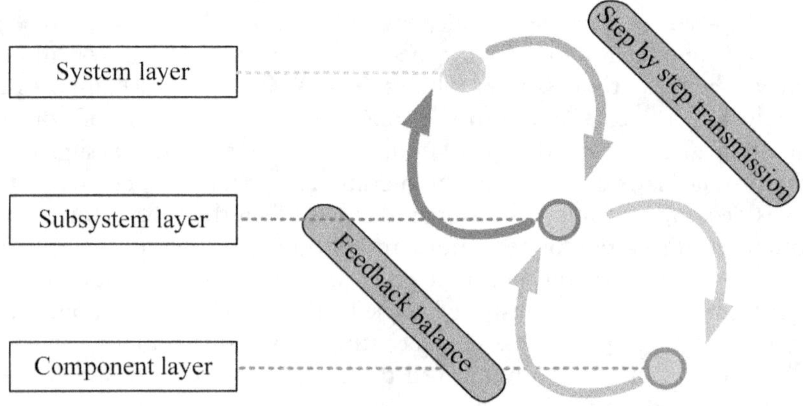

Figure 2.10 The diagram of decomposition and coordination of the ATC method.

relationship. Establishing an integrated design model that can fully reflect the characteristics of the system and can quickly and accurately optimize the design is a vital piece of multidisciplinary integrated design technology. Multidisciplinary integrated design modeling mainly includes integrated parametric models, discipline analysis models, and optimization model research.

2.1.2.2.1 Parametric geometric model

The parametric geometric model is the basis of multidisciplinary integrated design, and its role is to provide a unified geometric model for the analysis

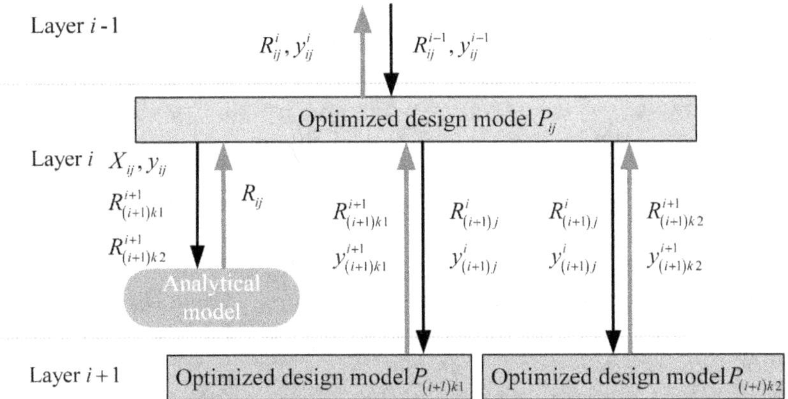

Figure 2.11 Data flow diagram in the ATC method optimization process.

and optimization of various disciplines. The parameters of the parametric model should include the shape parameters of the model, the design variables, and the state variables of each discipline, which can truly describe the characteristics of each discipline/system and the relationship between them. For example, accurately describing the aircraft's shape with fewer parameters is a key problem for complex aircraft shapes. There are relatively mature methods for parametric description of regular aircraft shapes, such as wings and fuselages. Still there are few studies on the establishment of parametric models for new or complex shapes.

2.1.2.2.2 Subject analysis model

The discipline analysis model is based on the parametric geometric model of the aircraft, which can automatically generate an aerodynamic analysis model, structural analysis model, mass center calculation model, etc. Its essence is to automatically prepare data files for the calculation program (software) of each discipline, establish the analysis model of each discipline, and carry out design analysis with the help of various computer-aided design and analysis tools. The discipline analysis model necessitates not only a precise representation of the design object's physical attributes within this field but also the ability to swiftly evaluate the discipline's performance.

2.1.2.2.3 Surrogate model

Multidisciplinary integration design requires high-precision numerical analysis models for various disciplines to achieve good design results. It will be very time-consuming and laborious for some complex systems to establish numerical analysis models for various disciplines, carry out corresponding discipline analysis, establish inter-disciplinary relationship models, and realize inter-disciplinary information exchange. To solve the rapid response problem of numerical analysis models, the research on surrogate models has developed rapidly in recent years, which plays a vital role in promoting the application of MDO methods in practical engineering design. For MDO of some complex systems or engineering, the original high-precision analysis model can be replaced by the surrogate model. There are three steps to construct a surrogate model:

First, the sample points of the design variables are generated. The methods of generating sample points include full factorial design, central composite design, Latin hypercube, uniform design, and other experimental design methods.

Second, the sample points are analyzed to obtain a set of input/output data;

Finally, the fitting method is used to fit the sample data, construct the approximate model, and evaluate the credibility. The main methods of constructing approximate models include the polynomial response surface method, artificial neural network, Kriging model, and radial basis function.

2.1.2.3 Establishment of integrated design environment

Building an integrated design environment means establishing a computer environment that supports integrated design. In this environment, we can integrate the stages of product design, various processes, and various disciplines involved and realize the communication. Aircraft design optimization is complex and computationally intensive work. Interaction and communication between disciplines and systems are required in design optimization, which requires an integrated design environment. Designers can easily manage and optimize disciplines in an integrated design environment [1,2].

2.1.3 Modeling of MDO

The main work of system modeling is to parameterize the geometric model and transform the optimization problem into a mathematical model. The purpose of establishing a mathematical model is to facilitate the transformation of engineering problems into computable mathematical functions. Accurate mathematical models inevitably bring high computational costs, and vice versa. Therefore, in establishing the function model, different precision models should be selected according to different problems, and the balance between the accuracy of the mathematical model and the calculation cost should be grasped (Figure 2.12). When developing a model, it is crucial to determine the approach for achieving design balance. This approach should not only fulfill the practical requirements of engineering design, such

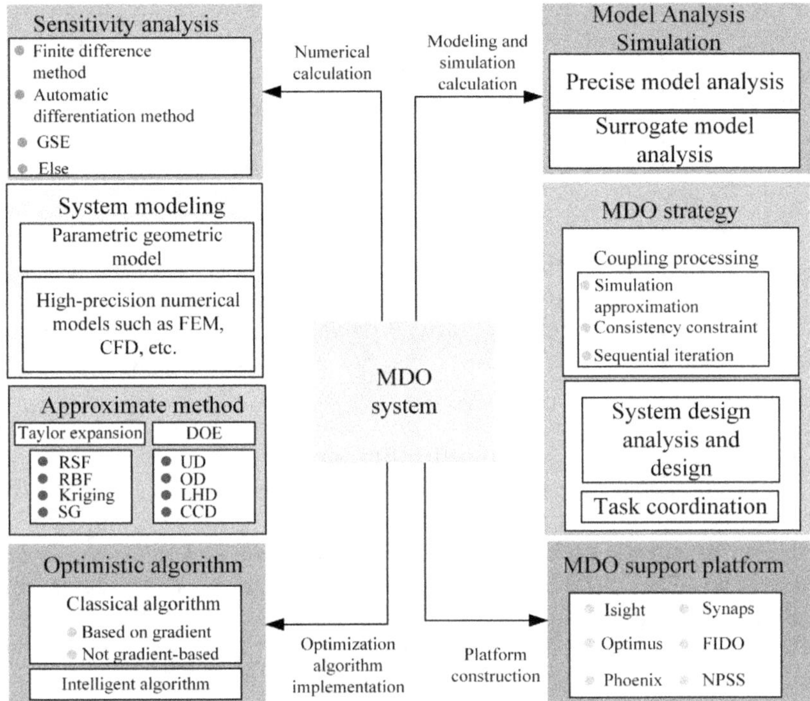

Figure 2.12 The MDO system diagram.

as ensuring that the specified nodes adhere to the design requirements, but also meet the precision demands of engineering design. It is essential to strike a balance between completing the design on schedule and upholding the quality of engineering design [3].

MDO, as the name implies, is the design optimization of the cross-combination of many disciplines. Therefore, each discipline of multidiscipline can find its original discipline analysis module, which is only integrated into MDO. The mathematical models of MDO's sub-disciplines are all derived from the mathematical models of its original discipline analysis module, which are simplified, deformed, and then added to MDO. After understanding the principle, the appropriate model can be selected according to the actual needs of the problem. For example, in MDO design, for important disciplines, we need to find the disciplines that play a decisive role in the optimal solution and select complex discipline analysis to improve the accuracy of the optimization results; for those disciplines that have less impact on the results, simple models can be selected to reduce the cost of optimization.

The model is an idealized abstract or simplified representation for studying and solving practical system problems. The model employs various elements, including text, charts, symbols, relationships, and solid models, to elucidate the system objects under investigation. It effectively represents the key constituents of the system and the interplay among each element. There are many classification methods for the model. According to a typical classification method, the model can be divided into the following categories:

1. Knowledge model. The knowledge model mainly uses the methods and techniques of artificial intelligence and knowledge engineering to build the model. It can use people's qualitative and empirical knowledge about things, conduct qualitative analysis and logical reasoning, and express and solve related problems. For example, an expert system is a knowledge model
2. Mathematical model. Mathematical models are generally established using control theory, system identification, operational research, and other mathematical methods and techniques. It can quantitatively describe the static characteristics or dynamic process of things, which is convenient for quantitative analysis and numerical calculation of the problem. This model is most used in optimization design, such as the differential equation model
3. Relationship model. Relational models are generally established using graph theory, logic, and other methods. This model is mainly used to qualitatively or quantitatively describe the relationship between the internal and external of the research object, such as the organizational structure model, process model, and so on. It is also called the structural model, and it is used for structural analysis and structural synthesis of the system

In some large-scale system modeling, various models can be flexibly applied according to the needs of specific tasks and environmental conditions. In recent years, based on the above three models, people have used computer software to propose an integrated model, that is, a generalized model.

The generalized model is an integrated model that combines the knowledge model, mathematical model, and relational model. This model can comprehensively (qualitatively, quantitatively, statically, dynamically) describe the system's structure, parameters, functions, and characteristics. Based on the generalized model, a generalized optimization design method has been developed to obtain the optimal design of the whole system, the whole performance, and the whole process. The generalized optimization model has been successfully applied in complex mechanical design.

In MDO problems, designers have the flexibility to select an appropriate modeling approach based on the specific characteristics of the research object. This adaptability allows for a nuanced understanding of the research object from various perspectives. However, it is worth noting that research in the field of MDO modeling is relatively limited, primarily focusing on the articulation of specific models. An example of this is the establishment of multidisciplinary product models through hierarchical tree structures within mechanical systems. The following mainly analyses the characteristics of MDO problem modeling and general modeling steps, and then introduces several modeling problems thar are often encountered in MDO research.

2.1.3.1 Characteristics of the MDO problem modeling

Analyzing and designing some very complex engineering systems is difficult to carry out directly. Generally, the more effective method is to decompose these systems into several subsystems according to components or disciplines (or other principles). According to the relationship between subsystems, complex systems can be divided into two categories: Hierarchical and non-hierarchical systems. The subsystems in the hierarchical system have no coupling relationship, and each subsystem is only directly related to the upper or lower-level subsystem, as shown in Figure 2.13(a). The information flow between subsystems in the non-hierarchical system has a coupling relationship, so the non-hierarchical system is also called a coupling system, as shown in Figure 2.13(b). In practice, some complex engineering systems may also be hybrid systems that include hierarchical and non-hierarchical systems.

In the MDO of some complex engineering systems mentioned above, because many factors need to be considered in the design, analysis, and optimization models of systems and subsystems, the coupling between disciplines will make modeling MDO problems very difficult. Some common problems can be divided into the following categories:

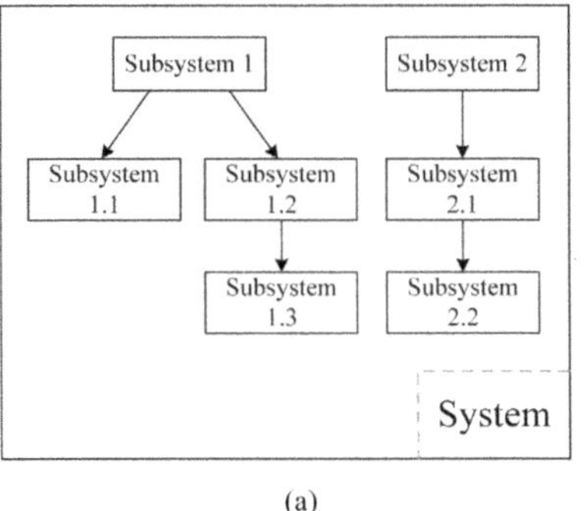

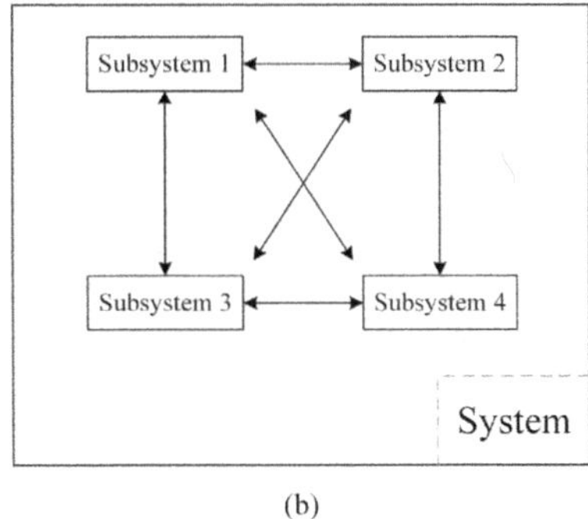

Figure 2.13 Classification of complex systems: (a) Hierarchical systems; (b) Non-hierarchical system.

1. Physical modeling. Physical modeling is the process of abstracting practical problems for analysis and research. Physical modeling generally uses simplification, assumptions, and other methods to abstract practical problems, focusing on the factors of concern. Physical modeling must be reasonable and accurate, and different levels of physical models are suitable for different analysis and design methods. For the MDO problem, it is challenging to clarify the design objectives and determine the coupling relationship of each part
2. Mathematical modeling. Mathematical modeling is to select design variables, establish mathematical equations, etc. It mathematicalizes the physical model so that appropriate analysis and optimization methods can process it. The mathematical model established should also consider the problems implemented on the computer
3. Model solving. In the MDO problem, some disciplines can adopt simple models, while others may adopt more complex models. The complexity of the system model increases due to the coupling between various disciplines, resulting in a complex nonlinear model. This complexity adds to the challenges in analyzing and solving the model. Moreover, handling extensive information exchange between disciplines and aligning design objectives across multiple fields further complicates the solution process of MDO problems. It can be seen that it is unrealistic to find a general modeling method suitable for MDO problem modeling under the complexity and diversity of practical problems. However, some basic modeling principles and steps can be proposed to model MDO problems

2.1.3.2 General steps and principles of MDO problem modeling

To abstract the actual problem into an optimization model, it is necessary to have a deep understanding of the research object. Choosing the appropriate expression method to describe is more complicated. The following basic principles can be used to establish the model:

1. Accuracy. The established model should be able to sufficiently reflect the design object's authenticity. In order to reflect the essence of the design with a certain accuracy, the results predicted by the model should be correct
2. Practicality. The established model should be as simple as possible to make it easy to solve. That is to say, the establishment of the model should consider the calculation cost factor
3. Applicability. The established model should reflect the facts as much as possible, and do not want the adaptation of the model to be too narrow. The adaptation of the model is also reflected in its inheritability and versatility

Although these principles are not specific modeling methods, they have universal guiding significance for modeling. For MDO modeling, it is also necessary to emphasize:

1. Rationality of system decomposition. The system is decomposed into subsystems according to disciplines or components because the solution of the subsystems is easy to carry out. Therefore, system decomposition should be conducive to enhancing the autonomy of subsystems. That is, system decomposition should minimize the coupling between subsystems
2. The accuracy of the coupling relationship between disciplines. Determining the coupling relationship between disciplines and measuring the degree of coupling will affect the design optimization of disciplines and systems
3. The coordination between the system design goal and discipline design goal

Balancing the discipline design goals and system design goals will affect the optimality of the overall system design results.

In general, the process of modeling is a process of gradual refinement from macro modeling to micro modeling. From the perspective of macro modeling, it is to separate the system from the environment, clarify the research content, consider the relevant influencing factors, and abstract the actual system through appropriate assumptions to obtain the overall concept of the system. The microscopic modeling concretizes the determined object and obtains a mathematical model that can be designed, analyzed, and optimized. There is no clear boundary between macro modeling and micro modeling. For example, the process of system decomposition is not only modeling at the macro level but also the micro level. Figure 2.14 shows the general steps of MDO problem modeling.

2.1.4 Meaning of MDO

The main idea of MDO is to decompose a complex system into several simple disciplines or subsystems according to disciplines or components and design and optimize each discipline or subsystem separately. At the same time, the coupling effect between disciplines is fully considered, and the effective design optimization solution strategy and distributed parallel computer network system are used to organize and manage the design process of the entire complex system. By making full use of the synergistic effect generated by the coupling of various disciplines, the overall optimal solution of the system is obtained [4].

Compared with the traditional engineering product design optimization method, MDO has the following characteristics:

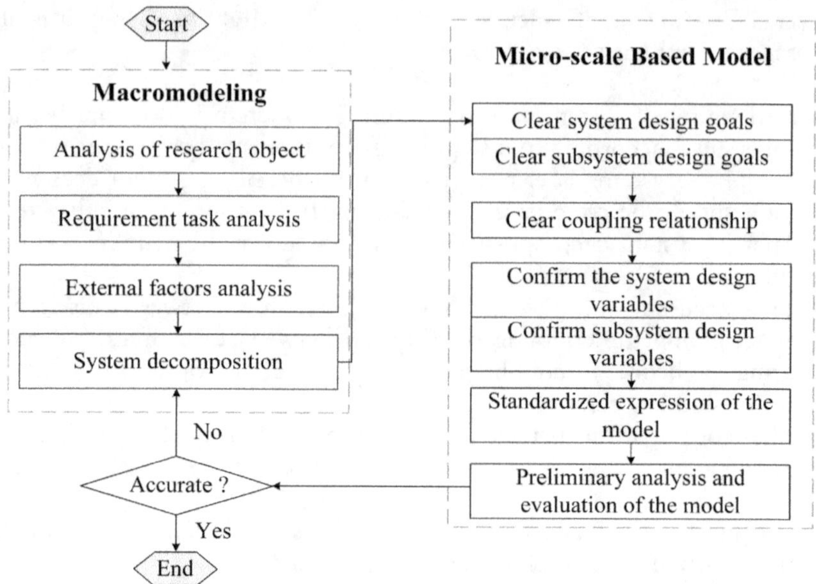

Figure 2.14 General steps of MDO problem modeling.

1. By entirely using the coupling mechanism between different disciplines, the design optimization of the whole system or engineering products is carried out to obtain the optimal design scheme of the overall performance of the engineering products and improve the comprehensive performance of the engineering products
2. Engineering product design is decomposed into the design of different disciplines or subsystems through the system decomposition technology. Through the computer network, the computing modules and designers are scattered in different regions, and design departments are organized to achieve concurrent design and shorten the development cycle of engineering products
3. Applying an approximate model reduces computational complexity, the number of system analyses is reduced, and the efficiency of design optimization is improved
4. The efficient solution strategy effectively resolves interdisciplinary inconsistencies by organizing data and information exchange between computing modules and engineering product designers across various regions and design departments. This streamlined approach simplifies organizational complexity. The idea of decomposing the MDO problem fully plays into the technical advantages of experts in various disciplines. It obtains the global optimal solution of the system by realizing

parallel design optimization. The emergence of MDO has dramatically reduced the time and cost of design and created a new discipline, which has also played a certain role in interdisciplinary communication and talent training

2.2 UNCERTAINTY MODELING THEORY

Uncertainty design is an essential direction in the development of design theory. In traditional engineering design, the uncertainty in the system is usually assumed to be a deterministic quantity, which is essentially represented by the variable's mean value. Because of ignoring the inherent uncertainties of the system, the optimal solution pursued by this design optimization only has relative or mathematical significance or only exists in a very narrow range. However, there is no real optimal solution in the complex real world. First of all, this is because the indicators to evaluate the merits of the scheme are vague and subjective, as well as the randomness that changes with time and conditions. Second, various information and models as the basis of optimization are uncertain. In order to overcome the shortcomings of deterministic design, various uncertainty analysis and design methods have developed rapidly since the 1950s and have achieved remarkable results in vehicle engineering design.

2.2.1 Sources and classification of uncertainty

Uncertainty pervades various aspects of engineering design, stemming from factors like variations in material properties, component dimensions, operating conditions, and assumptions made during mathematical model development. These sources of uncertainty primarily result from two key factors during the modeling process.

1. The system response of the model is not unique. This is due to the uncertainty of the model input parameters
2. The design system and design environment have many optional mathematical models. From the point of view of the input and output of the system, since the simulation tool is determined, the output of the system input after passing through the simulation tool should also be determined. This results in different outputs for the same input due to different simulation tools (mathematical models) [5]

Generally speaking, there are three categories of uncertainty: Aleatory uncertainty (also called accidental, irreducible, inherent, or variability), epistemic uncertainty (also called reducible, subjective, model, or simple uncertainty), and numerical uncertainty (error). To estimate all uncertainties and identify all potential sources of variability, uncertainty, and error in mathematical models and simulation tools, Oberkampf et al. [6] have put

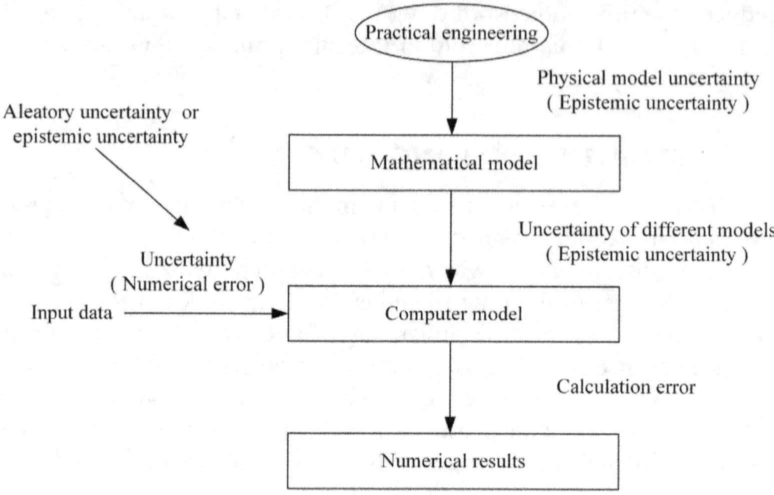

Figure 2.15 Sources of uncertainty.

forth some methods. Computer software, such as Finite Element Analysis (FEA) software, is widely adopted by designers. This simulation software has many sources of uncertainty, as shown in Figure 2.15.

First, when the physical model of the actual engineering system is converted into a mathematical model, epistemic uncertainty is generated because all the nonlinearities in the physical model cannot be accurately converted into mathematical equations; second, there are uncertainties in the data input of the system (which may be aleatory uncertainty and epistemic uncertainty); third, a variety of techniques can solve these mathematical equations, and the results calculated by these different methods are slightly different. Therefore, we will use different fidelity models to analyze the project and get different calculation results. This is usually called model form uncertainty. In short, there are uncertainties in the models used by computers. Finally, it should be pointed out that numerical errors will inevitably occur because some data will be rounded in the calculation process.

According to the above, there are three categories of uncertainty: aleatory, epistemic, and numerical (error). These uncertainties will be described in detail below.

2.2.1.1 Aleatory uncertainty

The term aleatory uncertainty refers to the internal variations of the physical system that are typically brought about by the random environment of the input data. These variations can be uncontrollable variations in the external environment or process tolerance. This uncertainty is often defined as a random phenomenon expressed as a probability distribution. Usually, the relative frequency of events is used to construct the probability distribution, which requires much information. However, this information generally does not exist, and designers often assume the characteristics of random phenomena (mean, variance, correlation coefficient, etc.) that cause internal changes in the system.

The following two cases are given to illustrate:

1. The boundary of deterministic change is known, but the probability density function (PDF) or the distribution of management change is unknown. For example, the maximum and minimum temperatures at which the aircraft can operate normally are known, but the potential aleatory process characteristics are unknown
2. The individual PDF or design variable of deterministic parameter change is known, but their relationship is unknown

Although using the probability mean to represent the random uncertainty in the design variables and parameters is reasonable in most engineering applications, this representation method has been questioned in some cases. In certain engineering examples, the system's distribution, mean, and variance are unknown, but the system's input data interval is known. Discrete data can occasionally be acquired from earlier experiments. It is difficult to estimate the uncertainty of the system output and use it for design when these input data are a part of the engineering system. The convex set model, interval method, and possibility theory can represent this uncertainty caused by insufficient information.

2.2.1.2 *Epistemic uncertainty*

Carelessness, ignorance, or insufficient information are the sources of epistemic uncertainty. Parameter or model-based uncertainty is an example of epistemic uncertainty in engineering systems. Furthermore, making decisions can also lead to epistemic uncertainty.

2.2.1.2.1 *Uncertainty of parameters and uncertainty of tools*

The majority of epistemic uncertainty in engineering systems is model-based or parameter-based uncertainty. The model is uncertain due to a lack of physical knowledge of the system, and the uncertain parameters stem from a lack of information that is currently available. Certain engineering applications utilize models or tools that are excessively conservative or frequently overestimate or underestimate deterministic features. Applying the consistent mass matrix, for instance, in structural dynamics overestimates the natural frequency of the structure. However, the mass matrix won't make such an error if the application is concentrated. Model uncertainty is typically less than that of high-fidelity analysis tools.

Using various mathematical models to simulate different states is also associated with model uncertainty. One of the reasons is the absence of unified modeling technology due to the lack of state information or changes in the system state. For instance, in a typical fluid mechanics application, laminar and turbulent flows are represented by different models.

2.2.1.2.2 Decision-related uncertainties

The uncertainty related to decision-making is the ambiguity or inaccuracy in decision-making problems, which are usually a multi-objective optimization problem. However, constraints and conflicting objectives make it almost impossible to minimize all objectives. In this environment, designers usually make decisions based on what goals to weigh or how to implement the trade-off ratio. Traditional multi-objective optimization methods (such as fuzzy multi-objective optimization and preference design) can implement design optimization based on this situation. A fuzzy set is a typical modeling method for this multi-objective problem.

2.2.1.3 Numerical uncertainty (error)

An error called numerical uncertainty is generally related to the numerical model used for modeling and simulation. This category of errors encompasses tolerance, rounding, truncation errors in convergence during coupled system analysis, and errors associated with solving ordinary and partial differential equations.

2.2.2 Probabilistic modeling theory

In the case of sufficient available data, probability theory is the most commonly used uncertainty modeling theory. It expresses uncertainty as a random variable.

Variables can generally be either discrete or continuous. In this section, we only consider continuous random variables. Lowercase letters (e.g., x) represent the random variable, and uppercase letters (e.g., X) represent the random variable vector. The PDF $f_X(x)$ expresses the information about probability and randomness.

If it is known that X has a value between x_1 and x_2, then the probability formula of X is:

$$P(x_1 < X < x_2) = \int_{x_1}^{x_2} f_X(x)dx \tag{2.5}$$

In order to facilitate the calculation, we use $F_X(x)$ to represent $P(X \le x)$, which is also called Cumulative Distribution Function (CDF) or Distribution Function (DF). In the calculation, it is necessary to integrate all X values less than or equal to x, and calculate the area of the corresponding area under PDF, that is, to calculate the integral from $-\infty$ to x in theory, as shown in the following formula:

$$P(X < x) = F_X(x) \int_{-\infty}^{x} f_X(x) dx \tag{2.6}$$

The CDF directly gives the probability of random variables less than or equal to a certain value. The PDF is the first derivative of the CDF, expressed as:

$$f_X(x) = \frac{dF_X(x)}{dx} \tag{2.7}$$

The mean value of X is represented by μ, the standard deviation is represented by σ, and the correlation between random variables X_1 and X_2 is represented by ρ_{12}, which is called correlation coefficient. The formulas of the above quantities are as follows:

$$\mu = \int_{-\infty}^{\infty} x f_X(x) dx \tag{2.8}$$

$$\sigma^2 = \int_{-\infty}^{\infty} (x - \mu)^2 f_X(x) dx \tag{2.9}$$

$$\rho_{12}\sigma_1\sigma_2 = \int_{-\infty}^{\infty}\int_{-\infty}^{\infty} (x_1 - \mu_1)(x_2 - \mu_2) f_{X_1,X_2}(x_1, x_2) dx_1 dx_2 \tag{2.10}$$

where $f_{X_1,X_2}(x_1, x_2)$ is the joint PDF of X_1 and X_2. These statistical properties can often be obtained from sampled data.

For example, there is a scalar function $g(X_1, X_2, \ldots, X_n)$, then the mean value of g can be calculated by the following formula:

$$\mu_g = \int_{-\infty}^{\infty} g(x_1, x_2, \ldots, x_n) f_{X_1,X_2,\ldots,X_n}(x_1, x_2, \ldots, x_n) dx_1 dx_2 \cdots dx_n \tag{2.11}$$

Similarly, the distribution of g and other statistical properties can be obtained. However, sometimes it is not the case, for example, when g is obtained with a relatively large computational analysis tool. By the way, approximation techniques can achieve this goal, but this has nothing to do with the content of this section, so do not elaborate on it.

2.2.3 Non-probabilistic modeling theory

In addition to probabilistic modeling theory, some uncertainty theory courses are used to quantify uncertainty, such as Dempster-Shafer theory (D-S evidence theory), convex set model, possibility theory, and fuzzy set theory.

2.2.3.1 D-S evidence theory

Evidence theory, also known as Dempster-Shafer theory (D-S theory), is a mathematical theory that measures uncertainty by determining credibility and plausibility based on the known evidence (information) of a proposition. These metrics can define the lower and upper bounds of probabilities (interval ranges) rather than assigning precise probabilities to propositions. This is very useful in research when there is less knowledge of uncertainty. The information or evidence that measures credibility and plausibility come from various sources, such as experimental data, theoretical evidence, experts' views on the credibility of parameters or the occurrence of events, and evidence that can be combined through combination rules.

Let Ω be a general set of all possible states of the system under consideration, and the elements of its power set 2^{Ω} can be used to represent propositions related to the actual state of the system. Evidence theory assigns the m function to each element of the set through the Basic Probability Assignment (BPA) function m: $2^{\Omega} \rightarrow [0,1]$, which has the following three attributes:

$$m(A) \geq 0, \forall A \in 2^{\Omega} \tag{2.12}$$

$$m(\phi) = 0 \tag{2.13}$$

$$\sum_{A \in 2^{\Omega}} m(A) = 1 \tag{2.14}$$

$m(A)$ is called the basic credibility number of A, where the set A of $m(A) \geq 0$ is called the focal element of m. The basic credibility number reflects the degree of evidence's support for an element of Ω belonging to the set A, which is similar to the PDF in the probability method. When there are multiple interval variables, the combination rules can aggregate multiple BPA structures, similar to the joint probability in probability theory. For example, if there are interval variables Z_1 and Z_2, then Eq. 2.15 can be used to calculate the joint BPA

$$S_Z(A_Z)\begin{cases} S_{Z_1}(A_1)S_{Z_2}(A_2), A_Z = A_1 \times A_2 \\ 0, Else \end{cases} \tag{2.15}$$

2.2.3.2 Convex model

In some engineering examples, uncertain events can be modeled by the convex set model [7]. These convex models include intervals, ellipses, or any convex sets. Compared with the uncertainty probability model, the uncertainty convex set model only needs less detailed information to characterize the uncertainty. In design applications, uncertain convex set models often require worst-case analysis. Local or global optimization techniques will be applied based on the performance function; the interval analysis technique is used when the convex set model is an interval.

The convex model is divided into the interval and ellipsoid convex models. These two models are different in mathematical expressions, but in the geometric sense, there is a certain correlation between the two models.

2.2.3.2.1 Interval model

2.2.3.2.1.1 DEFINITION OF INTERVAL VARIABLES

The parameters of existing engineering structures (such as elastic modulus, material strength, geometric size, etc.) are uncertain. Assuming that the uncertain parameter x of the structure changes in an interval, and its upper and lower bounds are X^u and X^l, respectively, then $x \in X\left[X^l, X^u\right]$ is called an interval variable. Let,

$$x^c = \frac{x^u + x^l}{2}, \quad x^r = \frac{x^u - x^l}{2} \tag{2.16}$$

There is,

$$x^l = x^c - x^r, \quad x^u = x^c + x^l \tag{2.17}$$

And X and x can be expressed as:

$$X = \left[x^l, x^u\right] = x^c + x^r \xi, \quad x = x^c + x^r \delta \tag{2.18}$$

where $\xi = [-1,1]$ is called the standardized interval, and $\delta \in \xi$ is called the standardized interval variable. It can be seen that for any real-valued interval X and interval variable x, as long as the two parameters x^c and x^r can be known, any real-valued interval X and x interval variables can be determined. The absolute value $|X|$ of interval X is defined as:

$$|X| = \max\left(\left|x^l\right|, \left|x^u\right|\right) \tag{2.19}$$

Therefore, for any variable $x \in X$, there is,

$$|x| \leq |X| \tag{2.20}$$

We call n-ordered interval arrays $X = (X_1, X_2, \ldots, X_n)$ as interval variables. Geometrically, the interval vector is an n-dimensional cuboid, and its width is defined as:

$$w_x = \max(w(X_1), w(X_2), \ldots, w(X_n)) \tag{2.21}$$

Its norm is defined as:

$$\|X\| \max(|X_1|, |X_2|, \ldots, |X_n|) \tag{2.22}$$

And the midpoint:

$$X_c = (m(X_1), m(X_2), \ldots, m(X_n)) \tag{2.23}$$

2.2.3.2.1.2 BASIC RULES OF INTERVAL OPERATION

For a bounded closed interval in the real number set R, it can be expressed as:

$$Y = [y^l, y^u] = \{y \in R | y^l \leq y \leq y^u\} \tag{2.24}$$

where $y^l, y^u \in R$, and $y^l \leq y^u$. We denote the set of such intervals as IR. Denote IR^+ for all interval sets satisfying $y^l \geq 0$ and IR^- for all interval sets satisfying $y^l \leq 0$.

Suppose that there are two interval numbers, $X = [x^l, x^u]$ and $Y = [y^l, y^u]$, and there are the following four operations.

$$\begin{aligned} X + Y &= [x^l, x^u] + [y^l, y^u] = [x^l + y^l, x^u + y^u] \\ X + Y &= [x^l, x^u] + [y^l, y^u] = [x^l + y^l, x^u + y^u] \\ X \times Y &= [x^l, x^u] \times [y^l, y^u] \\ &= [\min(x^l y^l, x^u y^u, x^l y^u, x^u y^l), \max(x^l y^l, x^u y^u, x^l y^u, x^u y^l)] \\ X \div Y &= [x^l, x^u] \div [y^l, y^u] = [x^l, x^u] \times \left[\frac{1}{y^u}, \frac{1}{y^l}\right], 0 \notin Y \end{aligned} \tag{2.25}$$

Therefore, suppose there exists $X, Y \in IR^+$, then there is,

$$X \times Y = [x^l y^l, x^u y^u] \tag{2.26}$$

The operation result of the interval is related to the expression of the operation.

Assuming that there is a constant λ, it can be considered that λ is an interval variable with equal upper and lower bounds, and its mean value is λ and the deviation is 0. Then there is,

$$\lambda X=\begin{cases}\lambda\left[x^{l},x^{u}\right],\lambda>0\\ \lambda\left[x^{u},x^{l}\right],\lambda<0\end{cases} \tag{2.27}$$

$$A=\lambda_x X+\lambda_y X\begin{cases}\left[\lambda_x x^{l}+\lambda_y y^{l},\lambda_x x^{u}+\lambda_y y^{u}\right],\lambda_x>0,\lambda_y>0\\ \left[\lambda_x x^{u}+\lambda_y y^{l},\lambda_x x^{l}+\lambda_y y^{u}\right],\lambda_x<0,\lambda_y>0\\ \left[\lambda_x x^{l}+\lambda_y y^{u},\lambda_x x^{u}+\lambda_y y^{l}\right],\lambda_x>0,\lambda_y<0\\ \left[\lambda_x x^{u}+\lambda_y y^{u},\lambda_x x^{l}+\lambda_y y^{l}\right],\lambda_x<0,\lambda_y<0\end{cases} \tag{2.28}$$

$$A^{u}+A^{l}=\lambda_x\left(x^{u}+x^{l}\right)+\lambda_y\left(y^{u}+y^{l}\right) \tag{2.29}$$

$$\begin{aligned}A^{u}-A^{l}&\begin{cases}\lambda_x\left(x^{u}-x^{l}\right)+\lambda_y\left(y^{u}-y^{l}\right),\lambda_x>0,\lambda_y>0\\ -\lambda_x\left(x^{u}-x^{l}\right)+\lambda_y\left(y^{u}-y^{l}\right),\lambda_x<0,\lambda_y>0\\ \lambda_x\left(x^{u}-x^{l}\right)-\lambda_y\left(y^{u}-y^{l}\right),\lambda_x>0,\lambda_y<0\\ \lambda_x\left(x^{u}-x^{l}\right)+\lambda_y\left(y^{u}-y^{l}\right),\lambda_x<0,\lambda_y<0\end{cases}\\ &=\left|\lambda_x\right|\left(x^{u}-x^{l}\right)+\left|\lambda_y\right|\left(y^{u}-y^{l}\right)\end{aligned} \tag{2.30}$$

The following criteria are weak criteria obtained from the real number calculation criteria:

Distribution law:

$$A\times(B\pm C)\subseteq A\times B\pm A\times C \tag{2.31}$$

$$(A+B)\times C\subseteq A\times C+B\times C \tag{2.32}$$

Cancellation law:

$$A-B\subseteq(A+C)-(B+C) \tag{2.33}$$

$$A/B\subseteq (A\times C)/(B+C) \tag{2.34}$$

These simple calculation methods can effectively carry out the operation between interval variables. However, their applicability has limitations, particularly when dealing with numerous interval variables or when these intervals have significant width, especially in cases involving intercorrelations among them. Using the above calculation methods can easily lead to interval expansion results.

2.2.3.2.1.3 INTERVAL COMBINATION METHOD

The interval combination method is proposed to suppress the problem of interval expansion. The specific process is as follows.

For interval-valued functions:

$$F(X) = F(x_1, x_2, \dots, x_n) \tag{2.35}$$

Assuming that $F(X)$ is an increasing function of the uncertain interval parameter $x_1, x_2, x_3, \dots, x_k$ and a decreasing function of the uncertain interval parameter $x_{k+1}, x_{k+2}, x_{k+3}, \dots, x_n$, two combinations of X_i should be considered to solve the upper and lower bounds of the function $F(X)$, which are

$$F^{1}(X) = F(x_1^l, \dots, x_k^l, x_{k+1}^u, \dots, x_n^u) \tag{2.36}$$

$$F^{u}(X) = F(x_1^u, \dots, x_k^u, x_{k+1}^l, \dots, x_n^l) \tag{2.37}$$

The above two formulas, which are the upper and lower bounds, can be used to obtain the minimum and maximum values of the interval-valued function.

Ensuring the function's monotonicity is highly challenging in most cases. The function appears in the case of piecewise monotonicity in the range of its independent variable interval, and the interval range of the function is obtained by combining the interval segments. Let $X_i\,(i = 1, 2, \dots, n)$ be divided into n_i subintervals $X_{ij}\,(i = 1, 2, \dots, n; j = 1, 2, \dots n)$, then

$$F^{l}(X) = \min_{i=1,2,\dots,n_1; j=1,2,\dots,n_2; \cdots; k=1,2,\dots,n_n} F(x_{1i}^o, x_{2j}^p, \dots, x_{nk}^q) \tag{2.38}$$

$$F^{u}(X) = \max_{i=1,2,\dots,n_1; j=1,2,\dots,n_2; \cdots; k=1,2,\dots,n_n} F(x_{1i}^o, x_{2j}^p, \dots, x_{nk}^q) \tag{2.39}$$

where $o, p, \dots, q$ represents l or u, respectively, that is, the lower bound and upper bound of the response interval.

Example: For function $f(x) = x(x-1)$, the function graph is shown in Figure 2.16.

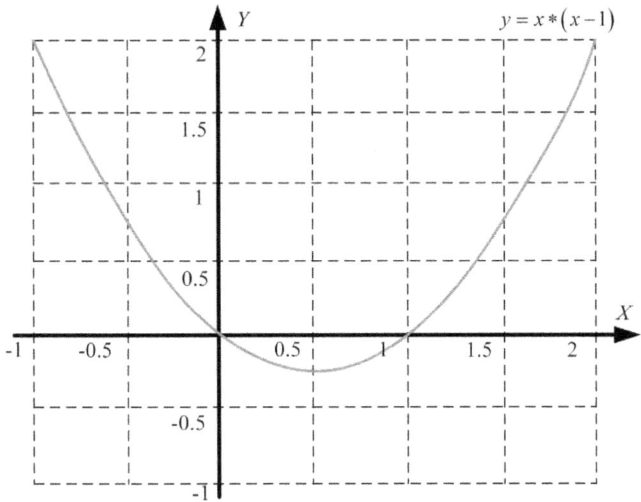

Figure 2.16 The curve of function $f(x) = x(x - 1)$.

It can be seen from Figure 2.16 that function $f(x) = x(x-1)$ is a decreasing function on $(-\infty, 0.5)$ and an increasing function on $(0.5, +\infty)$. Suppose that the domain $x \in [0,1]$, the true range of a function $f(x)$ is $[-0.25, 0]$; the interval algorithm is directly applied to obtain

$$f(x) = [0,1] \times ([0,1] - 1) = [0,1] \times [-1,0] = [-1,0] \tag{2.40}$$

The interval algorithm is directly applied to obtain the interval of function $f(x)$ is $[-1,0]$, which is contrary to the actual value range interval $[-0.25,0]$; now the interval combination method is applied to divide the interval $[0,1]$ into two sub-intervals $[0,0.5]$ and $[0.5,1]$, and then the upper and lower bounds are solved.

$$f^{l}(x) = \min_{x_1=0, x_2=0.5} = F(x_1, x_2) = -0.25 \tag{2.41}$$

$$f^{u}(x) = \min_{x_1=0.5, x_2=1} = F(x_1, x_2) = 0 \tag{2.42}$$

The value range of the function $f(x)$ using the interval combination method is $[-0.25, 0]$, which is equal to its true value range.

2.2.3.2.2 Ellipsoid model

The expression of the ellipsoid convex set model is:

$$U(x,\alpha)=\{x\colon x^TWx\le\alpha\}=\left\{x\colon \sum_{i=1}^{n}\frac{(x_i-\overline{x}_i)^2}{e_i^2}\le\alpha^2\right\}\tag{2.43}$$

where e_i is the radius axis of the ellipsoid, $i=1,2,\ldots,n$;
α is the radius of the ellipsoid, $\alpha>0$;
$\overline{x}_i$ is the theoretical value of the uncertainty parameter;
e_i and α represent the uncertainty or change interval of structural parameters. When $\alpha\neq 0$, the ellipsoid model can be transformed into the ellipsoid of $\alpha=1$. That is,

$$U(x,\alpha)=\left\{x\colon \sum_{i=1}^{n}\frac{(x_i-\overline{x}_i)^2}{e_i^2}\le 1\right\}\tag{2.44}$$

At this time, the uncertainty or variation range of the parameters will be described by the radius axis e_i of the ellipsoid.

2.2.3.2.3 The relationship between the interval model and ellipsoid model

The ellipsoid model and interval model are interrelated in a geometric sense. The interval model can be regarded as a special convex model, and the ellipsoid model is based on it. The interval model is obtained by expanding the interval domain of uncertain parameters, as shown in Figure 2.17. The ellipsoid model can be regarded as an extension of the interval model. The calculation expressions between the interval model and the ellipsoid model are η_1 and η_2, respectively.

$$\eta_1=\frac{R^c-S^c}{R^r+S^r}\tag{2.45}$$

$$\eta_2=\frac{R^c-S^c}{\sqrt{(R^r)^2+(S^r)^2}}\tag{2.46}$$

It can be seen from Figure 2.17 that the failure domain of the ellipsoid model is smaller than the failure domain of the interval model (the shadow part in the graph). The interval model yields result in failure cases, while the

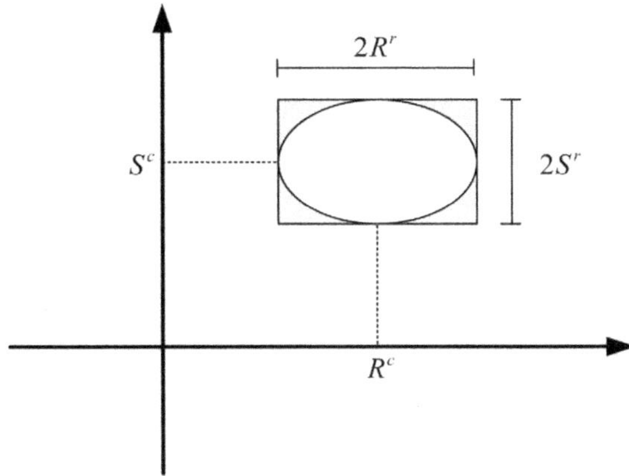

Figure 2.17 Correlation between interval model and ellipsoid model.

ellipsoid model may not necessarily do so. That is, when evaluating the reliability of the bridge, the results calculated by the ellipsoid model are more accurate. Therefore, this paper uses the ellipsoid model to evaluate the real bridge.

2.2.3.3 Method based on possibility theory and fuzzy set theory

2.2.3.3.1 Possibility theory and fuzzy variables

The possibility theory was first proposed by Zadeh [8], who proposed the theory of dealing with fuzzy propositions based on fuzzy sets. It is an effective mathematical tool for dealing with incomplete information. The theory of possibility is based on the possibility of measuring space. A complete possibility measure space can be represented by (Ω, Γ, Π) triples, where Ω is the sample space, Γ is the Ample domain, and Π is the possibility measure defined in Γ. The possibility theory uses the possibility measure Π and the necessity measure N to measure the uncertainty.

For the possibility measure Π and the necessity measure N of an event, it satisfies the following property:

$$\begin{cases} N(A) = 1 - \Pi(\overline{A}) \\ N(A) + N(\overline{A}) \le 1 \\ \Pi(A) + \Pi(\overline{A}) \ge 1 \end{cases} \tag{2.47}$$

For events A and B, the possibility measure Π and the necessity measure N satisfy the axiom:

$$\begin{cases} \Pi(A \cup B) = \Pi(A) \vee \Pi(B) = \max\{\Pi(A), \Pi(B)\} \\ N(A \cap B) = \Pi(A) \wedge \Pi(B) = \min\{\Pi(A), \Pi(B)\} \end{cases} \tag{2.48}$$

The fuzzy variable X is a measurable function from the energy measure space (Ω, Γ, Π) to the real number set [9]. The fuzzy variable X is characterized by the membership function $\mu_X(x)$ corresponding to different points, and these different points belong to the real number field $\Re r$.

2.2.3.3.2 Determination of membership function

When the amount of data of input uncertainty variables is small, especially in the following two cases, it is necessary to model these uncertainties as fuzzy variables.

1. When the uncertainty of the input is random and has the distribution type of the experiment, but the data used to describe the distribution type of these experiments is not entirely credible
2. When the input uncertainty is stochastic, the dataset may not be sufficiently extensive even with some available data. It is not enough to allocate a reasonable probability for the basic event

When using fuzzy theory to model input variables as fuzzy variables, the most critical work using existing data to determine variables is the Fuzzy Membership Function (FMF). The determination of FMF has been studied in the relevant literature. Considering that the fuzzy variables with normal attributes FMF have strong practicability, this paper uses the method of determining the membership function in the literature to establish the FMF of fuzzy uncertain variables. The established FMF has normal, continuous, and bounded attributes.

First, according to the form of the existing data, that is, whether the data form belongs to case (1) or case (2), a temporary PDF is established by using the parametric method (corresponding to case (1)) or the nonparametric method (corresponding to case (2)). Then, FMF is obtained based on the generated temporary PDF:

$$\mu_X(x) = 1 - |2F(x) - 1| = \begin{cases} 2F(x) & x \in \{x : F(x) \leq 0.5\} \\ 2 - 2F(x) & x \in \{x : F(x) > 0.5\} \end{cases} \tag{2.49}$$

where $F(x)$ is the CDF. It is worth noting that the obtained FMF is symmetric about the temporary CDF and satisfies the probability-possibility consistency principle and the most conservative principle.

After the FMF of the fuzzy variable is obtained by the above method, the fuzzy uncertainty is modeled as a fuzzy variable denoted as x_f, and the corresponding membership function is μ_{x_f}.

2.3 UBMDO THEORY

2.3.1 Uncertainty in multidisciplinary system

In a multidisciplinary system, disciplines are not independent but interrelated and interact. One scientific output may be the input of another discipline. Especially in large-scale complex systems, there is a strong coupling between disciplines, and a scientific change may cause changes in the entire system. In the research of multidisciplinary systems with uncertainty, multidisciplinary uncertainty analysis is the key link of the whole design optimization. Due to the characteristics of multidisciplinary coupling, the uncertainty in multidisciplinary is more complex than that in a single discipline. The system's uncertainty between disciplines will spread with the input and output between disciplines. Moreover, due to the uncertainty of each discipline itself, the uncertainty of the final output of the system will become larger and larger. The magnitude of uncertainty plays a crucial role in a designer's decision-making process. As a result, uncertainty analysis is of paramount importance in multidisciplinary systems. [10].

2.3.1.1 Sources of uncertainty in a multidisciplinary system

In the multidisciplinary system, the sources of uncertainty mainly include the following three aspects:

1. Uncertainty of input parameters. It mainly refers to each discipline's design variables and design parameters and the shared input design variables and parameters of the discipline
2. Uncertainty of model parameters. It mainly refers to the uncertainty of the parameters in the multidisciplinary system model when calculating their values due to insufficient information
3. The uncertainty of the model itself. It mainly refers to the uncertainty of the model established in modeling multidisciplinary systems and the modeling process of the discipline itself

2.3.1.2 The propagation of uncertainty in multidisciplinary systems

In a multidisciplinary system, the sub-disciplines are coupled, and the uncertainty constantly spreads in various disciplines. The uncertainty of one discipline may therefore permeate other disciplines through coupling variables, which is a major issue in multidisciplinary design. The uncertainty of a single discipline will accumulate in the final output of a multidisciplinary system. Figure 2.18 briefly introduces the spread of uncertainty between disciplines in multidisciplinary systems.

Figure 2.18 represents the design evaluation simulation program of the discipline. x_s is a shared variable for all disciplines. $x_i\ (i = 1, 2, \ldots, n)$ is the input variable of subject i. $y_{ij}\ (i \neq j)$ is a coupling variable, which is the function output calculated by discipline i, and is also the input of discipline j. $y_i = \left\{ y_{ij} \mid j = 1, 2, \ldots, n; j \neq i \right\}$ represents the output of discipline i and the set of coupling variables as the input of other disciplines, $y_i' = \left\{ y_1, \ldots, y_{i-1}, y_{i+1}, \ldots, y_n \right\}$ represents the output of other disciplines except discipline i and the set of coupling variables as the input of discipline i. For discipline i, according to the subsystem simulation model $F_{yi}(\bullet)$ and

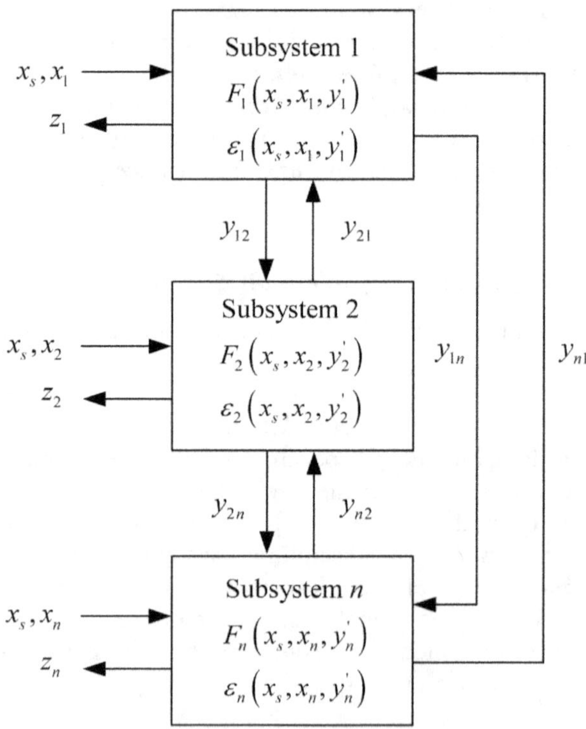

Figure 2.18 The propagation of uncertainty in the multidisciplinary coupling system.

the corresponding model error $\varepsilon_{yi}(\bullet)$, the coupling variables can be derived as follows:

$$y_i = F_{yi}\left(x_s, x_i, y_i'\right) + \varepsilon_{yi}\left(x_s, x_i, y_i'\right), \quad i = 1, 2, \ldots, n \tag{2.50}$$

Similarly, the total output of the subsystem i is derived as follows:

$$z_i = F_{zi}\left(x_s, x_i, y_i'\right) + \varepsilon_{zi}\left(x_s, x_i, y_i'\right), \quad i = 1, 2, \ldots, n \tag{2.51}$$

2.3.2 MDO modeling considering multi-source uncertainty

For some more complex engineering systems, the primary work of MDO is to complete the construction of optimization models. The MDO model needs to follow some criteria, such as the accuracy, practicability, and applicability of the model. Therefore, in the process of MDO modeling, the system decomposition should be reasonable, the judgment of the discipline coupling relationship should be accurate, and the system and discipline design objectives should be coordinated.

MDO modeling is mainly divided into two aspects. One is to solve the problems of system confirmation, analysis, system layer design optimization, and coordination in system layer modeling. The second challenge involves resolving the issue of defining precise tasks for each subsystem when modeling the subsystem layer. An appropriate model should be established according to the discipline's characteristics and the system's design requirements. In addition, there are many modeling methods for multidisciplinary design problems, such as process modeling, variable complexity modeling, and parametric modeling. The modeling process under different uncertainty conditions will be introduced separately in the following sections.

2.3.2.1 UBMDO model under aleatory uncertainty

2.3.2.1.1 Mathematical model of UBMDO

Under the condition that the reliability of the design constraint is greater than the given value, the optimization design is generally constructed as a minimization problem of the total cost or weight of the product. Reliability optimization design can ensure that the reliability of the designed engineering system or product meets the given reliability requirements and reduces its failure probability to an acceptable level. A typical single-disciplinary reliability design optimization model is shown as follows:

$$\begin{cases} \min f\left(d_s, d_i, x_s, x_i, p\right) \\ s.t. \Pr\left(g_i\left(d_s, d_i, x_s, x_i, p, y_{.i}\right) \geq 0\right) \geq R_i \\ g_i\left(d_s, d_i, y_{.i}\right) \geq 0 \\ d_s^L \leq d_s \leq d_s^U, d_i^L \leq d_i \leq d_i^U \\ x_s^L \leq x_s \leq x_s^U, x_i^L \leq x_i \leq x_i^U \\ i = 1,2,3 \end{cases} \tag{2.52}$$

where d_s is a shared deterministic design variable, which is a deterministic design variable shared by all disciplines, d_s^U and d_s^L are the upper and lower limits of d_s, respectively; d_i is the local deterministic variable of discipline i, d_i^U, and d_i^L are the upper and lower limits of d_i, respectively. x_s is a random input design variable shared by all disciplines, x_s^U and x_s^L are the upper and lower limits of x_s, respectively; x_i is the local random design variable of discipline i, x_i^U, and x_i^L are the upper and lower limits of x_i, respectively. p is the design parameter; R_i is the allowable reliability of discipline i; $g_i(\cdot)$ is the deterministic design constraint condition of discipline i; $\Pr\left(g_i(\cdot) \geq 0\right)$ is the probabilistic reliability non-failure model of discipline i, $\Pr\left(g_i(\cdot) \geq 0\right) \geq R_i$ is the probabilistic reliability design constraint of discipline i; $y_{.i}$ is the input state variable of discipline i, which is

$$y_{.i} = y_{.i}\left(d_s, d_i, x_s, x_i, y_{ji}\right) \quad (i, j = 1,2,3, i \neq j) \tag{2.53}$$

where y_{ji} denotes the coupling state variable from the output of discipline j to discipline i.

2.3.2.1.2 UBMDO with a multidisciplinary feasible method

Multidisciplinary system analysis includes integrated sub-disciplinary analysis with coupled inputs and outputs. Therefore, MDO is much more complex than single-disciplinary optimization, and the computational cost is also higher. The coupling between multidisciplinary systems is reflected by state variables, which refer to the variables that the output of one discipline can be used as the input of other disciplines. The state variables must remain consistent across these disciplines to ensure interdisciplinary compatibility. In order to make the coupling equation represent a complete multidisciplinary analysis, the state variables between all subsystems in the multidisciplinary analysis should be consistent. Figure 2.19 shows a typical three-disciplinary system. d_1 is the shared variable of subsystem 1, subsystem 2, and subsystem 3, d_2 is the local design variable of subsystem 2. Variables y_1 and y_2 are state variables, for example, y_{ij} is the output of subsystem i to subsystem j; g_i is the output of subsystem i.

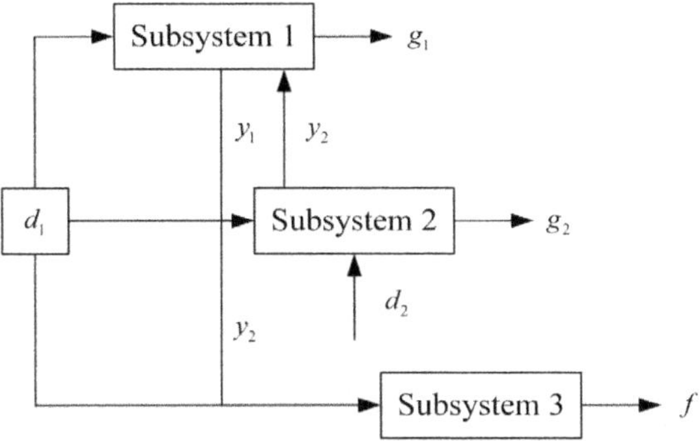

Figure 2.19 Typical multidisciplinary problems of the three subsystems.

The optimization formula is as follows:

$$\begin{cases} \min_{d} f\left[d, y(d)\right] \\ s.t. g\left(d, y(d)\right) \le 0 \end{cases} \tag{2.54}$$

Equation f represents the target, $g(d)$ is the constraint vector. In Eq. 2.54, only the independent variable d is the design variable, and the state variable $y(d)$ is updated in each iteration of system optimization.

In order to improve efficiency, UBMDO can make use of the existing MDO strategy. MDF optimization strategy is one of the most basic MDO strategies in reliability design optimization. The mathematical model of UBMDO under the MDF optimization strategy can be expressed as follows:

$$\begin{cases} \min_{d} f\left[d, y(d)\right] \\ s.t.\ \Pr\left\{g\left[x, y(x)\right] \ge 0\right\} \le \Phi\left(-\beta_t\right) \end{cases} \tag{2.55}$$

2.3.2.2 UBMDO mathematical model based on probability theory and convex model

2.3.2.2.1 Reliability evaluation index based on convex model

In engineering practice, some design parameters cannot be expressed and quantified by probability theory, such as development cycle, economic conditions, test equipment, and other design parameters with epistemic uncertainty. However, the magnitude or boundary of these epistemic uncertainties is easy to determine, and the convex model can be used to deal with such uncertainties.

The fundamental concept of reliability based on the convex set model is that a system is considered reliable if it can withstand substantial uncertainties without experiencing failure. Otherwise, if the system can only tolerate minor uncertainties, the system is unreliable. The system's reliability can be measured by the maximum degree of uncertainty that the system can tolerate.

In multidisciplinary reliability analysis, the limit state function of each discipline is generally described by $g(x)$, $g(x)=0$ is called the limit state equation of the discipline, which is an essential basis for structural reliability analysis. After standardized transformation, the limit state equation divides the standard space V into two parts: The security domain $\left(g(v)>0\right)$ and failure domain $\left(g(v)<0\right)$.

In the case of a single convex model in the two-dimensional space V shown above, the curve in Figure 2.20 is the limit state curve, the unit circle is the normalized convex domain corresponding to the uncertain variable, and δ is the shortest distance from the origin to the limit state function curve. It can be seen from Figure 2.20 that if $\delta=1$, the failure domain is just tangent to the convex model and is in the state of 'critical failure'. When $\delta>1$, the variation of uncertain variables is in the security domain, and the structure is reliable. The larger the δ value is, the greater the variation of uncertain parameters that the structure can tolerate and the more reliable the structure is. Therefore, δ can be selected as the measurement index of structural non-probabilistic reliability.

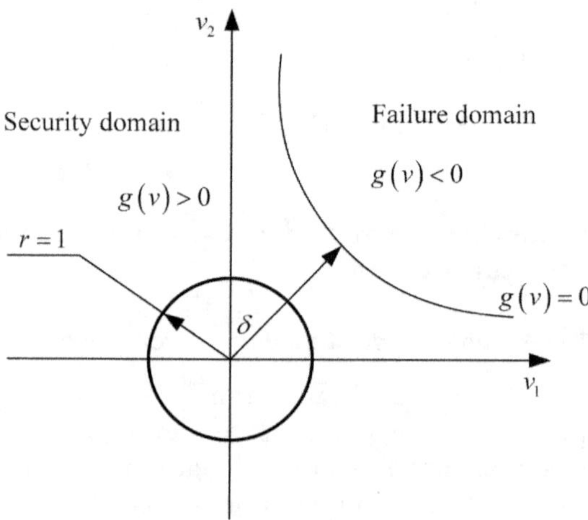

Figure 2.20 Reliability index based on convex model.

2.3.2.2.2 Reliability comprehensive evaluation index based on probability theory and convex model

The reliability based on probability theory and the reliability based on the convex set model are proposed under the condition of considering single uncertainty. However, there are often many multi-source and multi-type uncertainties in the actual complex engineering design. The traditional reliability considering single uncertainty can no longer meet the reliability design requirements of current complex engineering systems. Therefore, the multidisciplinary reliability comprehensive evaluation index considering both aleatory and epistemic uncertainty has gradually entered people's field of vision.

When the aleatory and epistemic uncertainty are fully considered, the original single aleatory uncertainty design parameter variable and the coupling state variable are replaced with the aleatory uncertainty parameter variable in the interdisciplinary reliability design optimization model's limit state function, the epistemic uncertainty parameter variable and the coupling state variable. It is precisely because of epistemic uncertainty design parameter variables that the value of the limit state function is no longer a single value (no longer a single failure surface). However, a series of values between the minimum and maximum values of the limit state function (there is a series of failure surface families) and the reliability value at this time becomes an interval. This is another kind of uncertainty, which is caused by uncertainty.

In order to ensure the high reliability of the design, the minimum value of the limit state function can be used as the standard to measure the reliability, and the difference of the limit state function value is used to characterize the influence of the corresponding epistemic uncertainty on the reliability. The greater the interval difference, the greater the variation of epistemic uncertainty, and vice versa. The value range of the limit state function also directly prompts the designer. To improve the reliability of the design, it is necessary to do more experiments on the design parameters of these epistemic uncertainties or obtain more data and knowledge through other channels to reduce the influence of the epistemic uncertainty of the design parameters on the limit state function, and then improve the reliability of the design of complex engineering systems.

2.3.2.3 UBMDO modeling considering aleatory-fuzzy-interval mixed uncertainty

The main task of uncertainty mathematical modeling is to mathematically describe various types of uncertainty inputs in the complex design process. Based on different theoretical tools, the aleatory uncertainty input, fuzzy uncertainty input, and interval uncertainty input in the design can be described as different mathematical variables or parameters. Probability theory models aleatory uncertainty, while evidence theory models interval uncertainty. The reliability evaluation based on probability can be used to consider various random uncertainties in the design. The probability theory calculates the probability of failure caused by these random uncertainties, and the degree of system safety is evaluated according to the calculated value.

First, reliability evaluation is based on possibility. If the uncertainty in the design is fuzzy, the uncertain variables will be modeled as fuzzy variables. Due to the influence of the fuzziness of the input variables, the output function is also fuzzy. Therefore, whether the system is safe and reliable is no longer based on a certain probability value as the standard but on the possibility of failure events.

According to the possibility theory, for an impossible event A, the possibility degree $\Pi(A)$ is 0. When the definition of performance function g_i to meet the requirements of the form of $g_i \geq 0$, if $\Pi(g_i < 0) = 0$, then our current design must always meet the requirements. Therefore, in the possibility-based reliability evaluation, $\Pi(g_i < 0)$ is used to evaluate the reliability of the current design. The calculation formula of $\Pi(g_i < 0)$ is

$$\Pi(g_i < 0) = \sup_{g_i \in (g_i < 0)} \mu_{g_i}(g_i) \tag{2.56}$$

The solved membership value indicates the possibility of $g_i < 0$ event, and the reliability of the design can be judged accordingly.

Second, reliability evaluation is based on evidence theory. When there is interval uncertainty in the design process, the interval uncertainty is modeled as an evidence variable. Since several possible intervals describe the uncertainty variable and the basic probability distribution corresponding to the interval, the performance function g is under the interval uncertainty of multiple joint basic probability distributions. According to the evidence theory, when the performance function g is under the interval uncertainty of multiple joint basic probability distributions, the true failure probability $\Pr(F)$ is between the confidence $Bel(F)$ and the plausibility $Pl(F)$. At the same time, if $Pl(F)$ does not exceed the required failure probability, then it is sure that $Bel(F)$ will not exceed the required failure probability. Therefore, the reliability evaluation based on evidence theory is based on the $Pl(F)$ of failure events.

The solution of Pl (F) is to add all the BPAs corresponding to the possible failure interval.

$$Pl(F) = \sum_{g_{\min}^k < 0} m(k) \tag{2.57}$$

where $m(k)$ represents the basic probability distribution corresponding to the k joint interval.

Third, combining the reliability evaluation of probability, possibility theory, and evidence theories. The easiest way to deal with reliability evaluation with various uncertain variables is to assume a distribution for these non-random variables, such as a uniform distribution. These non-random variables are transformed into random variables, and then the reliability is evaluated based on probability. However, Du et al. [11] pointed out that, in this case, it is best not to make any assumptions to use the information obtained fully. Therefore, the reliability evaluation with aleatory, fuzzy, and interval uncertainty needs to be based on probability and evidence theory.

Under the influence of mixed uncertainty, the probability of the failure event of g is first in the form of membership degree, and then the probability events under different membership degrees are presented in the form of confidence and plausibility. Therefore, the reliability under the three uncertainties is evaluated by the confidence $Bel(F)$ and plausibility $Pl(F)$ of the failure probability under the specified membership degree.

According to the unified reliability analysis model URAM and its analysis results, the failure probability of performance function g under three uncertainties is shown in Figure 2.21.

It can be seen from the figure that if $Pl^{\alpha}(F)$ under the membership degree α does not fail, then failure will not occur under this membership degree. This is because $Pl^{\alpha}(F)$ is the greatest possibility of failure under this membership degree. Therefore, the reliability evaluation content when aleatory

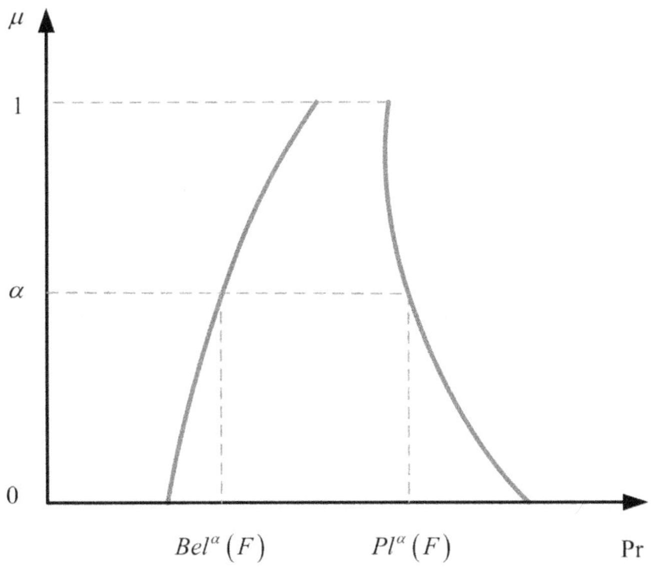

Figure 2.21 Membership degree of failure probability under aleatory-fuzzy-interval uncertainty.

uncertainty, fuzzy uncertainty and interval uncertainty coexist is whether the plausibility $Pl^{\alpha}(F)$ of the maximum failure probability under the allowable possibility α_t satisfies the plausibility $Pl^{\alpha}(F)$ of the allowable failure probability.

Using the *FORM-α-URA* method proposed above, the plausibility degree $Pl^{\alpha}(F)$ under the allowable possibility degree α_t can be obtained, and the reliability can be evaluated according to the value of $Pl^{\alpha}(F)$.

2.3.3 MDO under multi-source uncertainties

In order to meet the high reliability and safety design requirements of complex engineering systems, the multidisciplinary reliability design optimization method considering uncertainty has attracted more and more attention. It has become one of the focuses of current MDO research.

The traditional UBMDO is a typical three-layer nested loop optimization process: The first layer is to perform MDO in a deterministic space to search for a deterministic optimization solution of a multidisciplinary system; the second layer is the multidisciplinary probabilistic reliability analysis in the probability space, which searches out the Most Probable Point (MPP) of each probability constraint. The third layer is the multidisciplinary analysis in the innermost layer, which provides the value of the coupling state variable for the deterministic MDO and multidisciplinary reliability analysis. Research indicates that [12] within UBMDO, multidisciplinary reliability analysis significantly influences overall computational efficiency. Practical engineering design often involves numerous highly nonlinear probability constraints, intensifying the complexity of the problem. Research shows that Multidisciplinary reliability analysis dominates the computational efficiency of the entire UBMDO, and there are many highly nonlinear probability constraints in practical engineering design, which makes the problem more prominent. The reliability constraints of multidisciplinary reliability design optimization under multi-source uncertainties include both aleatory and epistemic uncertainties. It is not only required to deal with aleatory uncertainties according to probabilistic reliability analysis methods but also to use reasonable mathematical methods to perform non-probabilistic multidisciplinary reliability analysis on epistemic uncertainties.

2.3.3.1 Reliability-based MDO

As shown in Figure 2.22, the traditional UBMDO method integrates deterministic MDO and reliability analysis to form a typical three-layer nested loop optimization problem. From the UBMDO optimization process, it can be seen that the outermost layer is the overall optimization of the system, that is, the MDO in the deterministic space, which is responsible for

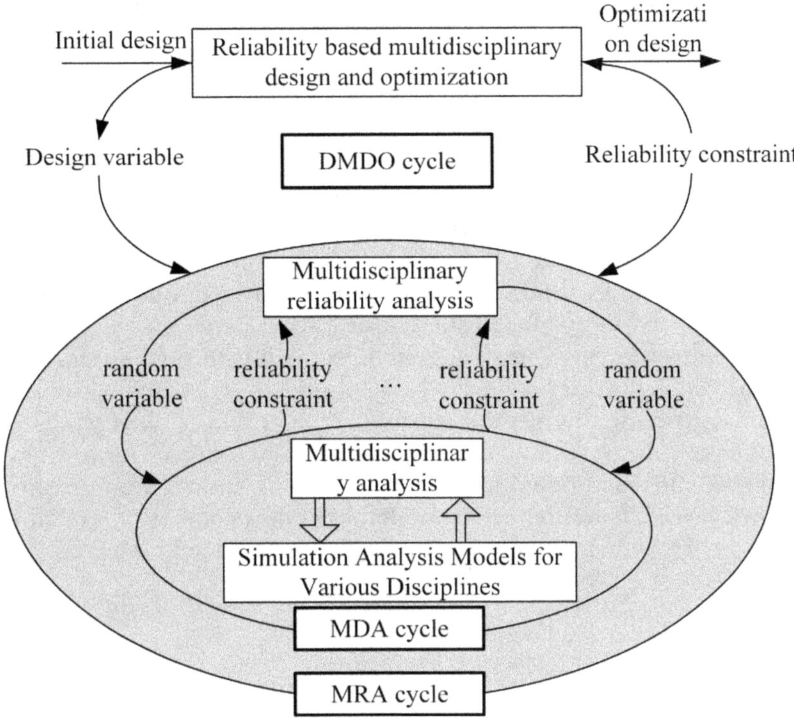

Figure 2.22 **UBMDO process diagram under aleatory uncertainty.**

searching for the overall optimal solution of the system (only the optimal solution in the theoretical sense). The middle layer is a multidisciplinary reliability analysis cycle, mainly used to calculate the MPP of each probability constraint and verify whether it meets the reliability design requirements. The innermost layer is a multidisciplinary system analysis that calls for both deterministic MDO and multidisciplinary reliability analysis, and each multidisciplinary system analysis contains a large number of disciplinary analyses, so the total number of disciplinary analyses is vast, especially when large-scale and coupled, the problem is particularly prominent, and the computational efficiency is very low.

The uncertainties in practical engineering design are rising due to the growing complexity of multidisciplinary systems. Consequently, addressing multiple sources of uncertainty complicates the process of multidisciplinary reliability design optimization. The UBMDO process will change from three-layer nesting to four-layer nested loop optimization. The computational cost and efficiency problem have become the biggest obstacle to realizing UBMDO engineering applications.

The solution of the UBMDO model involves a three-layer nested loop optimization problem. In this regard, we can mainly study from the following two aspects.

2.3.3.1.1 Decoupling strategy of optimization process

In order to improve the computational efficiency of UBMDO, Du et al. [13] from Northwestern University proposed a Sequence Optimization and Reliability Assessment (SORA) strategy based on serialization. This method decouples the deterministic MDO and multidisciplinary reliability analysis. First, the deterministic optimization cycle is performed, and then the reliability analysis of the probabilistic constraints is performed at the deterministic optimal design point. The reliability analysis provides information for reconstructing the deterministic MDO model to repeat the optimization until convergence. In this method, the number of multidisciplinary reliability analyses equals the number of optimization cycle iterations. Due to its rapid convergence, this method typically requires only a limited number of optimization cycles. Therefore, the computational efficiency of this method is much higher than that of the traditional method of directly integrating reliability analysis with MDO.

2.3.3.1.2 KKT equivalent replacement strategy

Agarwal et al. [14] proposed a Uni-level reliability design optimization strategy using KKT conditions instead of probabilistic constraints. Under this strategy, the multidisciplinary reliability analysis cycle in the inner layer is replaced by the consistency constraint in the outer layer. This can avoid the expensive reliability analysis and form a similar deterministic MDO problem. However, the additional number of design variables in the KKT equivalent substitution method greatly influences computational efficiency, especially when there are many design variables and design constraints. To improve the computational efficiency and the robustness of the algorithm, Liang et al. improved the method and proposed a Single-Loop Algorithm (SLA). However, the SLA will cause non-convergence when dealing with nonlinear or uncertain limit state functions.

In summary, the UBMDO optimization based on the KKT equivalent replacement strategy has problems low computational efficiency, complex convergence, and instability caused by using approximate methods due to adding many additional design variables. These problems greatly limit the application of this method in large-scale, multi-coupling, and highly nonlinear complex engineering systems. The results demonstrate that the decoupling strategy excels in computational efficiency, robustness, sensitivity to uncertainty, and managing nonlinear limit state functions.

Regarding the Sequential Approximate Programming (SAP) algorithm and the SORA method, the SAP algorithm uses the first-order linear Taylor expansion to approximate the reliability index at the current design point, which has poor robustness and relies heavily on the starting point selection. However, most of the current complex engineering systems are nonlinear. Therefore, the SAP algorithm is suitable for dealing with simple structural reliability design optimization and is not suitable for dealing with multidisciplinary reliability design optimization problems of complex engineering systems. The SORA method can effectively solve the UBMDO problem.

2.3.3.2 Sequence optimization and reliability evaluation strategy

2.3.3.2.1 Sequence optimization and reliability evaluation strategy

The SORA approach is an efficient decoupling technique for the reliability design optimization problem. In contrast to the conventional approach, the SORA method is predicated on the decoupling and serialization of the traditional nested loop's reliability design optimization process, resulting in a deterministic design optimization and sequential reliability analysis execution, which forms a recursive optimization loop. As shown in Figure 2.23, the main idea of the SORA method is to transform the UBMDO problem into an approximate MDO problem by using equivalent constraints and then use the deterministic MDO method to solve it. The SORA method can make the equivalent constraint gradually 'shift' in the direction of the probability constraint and quickly obtain the optimal solution.

The SORA method uses SLA for continuous deterministic optimization cycles and reliability analysis. In each cycle, optimization and reliability analysis do not interfere with each other. Reliability analysis is used to verify the feasibility of probabilistic constraints after optimization. The key of this method is to use the results of reliability analysis to constantly modify the constraints in the optimization so that it is close to the expected probability constraints, to achieve the optimal design as soon as possible, reduce the number of optimizations, and then reduce the number of reliability analysis.

2.3.3.3.2 UBMDO based on SORA and CO

2.3.3.3.2.1 OPTIMIZATION MODEL

Taking a system with two coupled disciplines as an example, people established a UBMDO model by integrating the CO algorithm and Performance Measure Approach (PMA). The model is as follows:

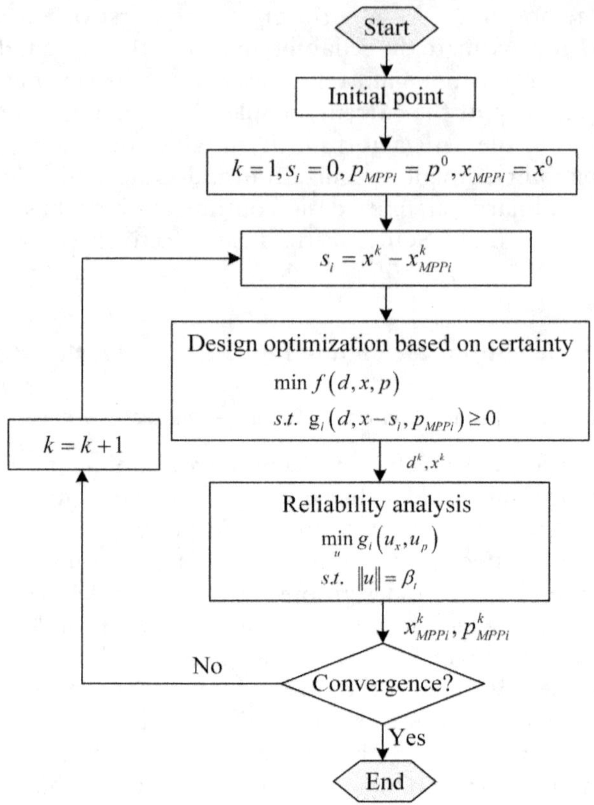

Figure 2.23 Flowchart of the SORA method.

$$\begin{cases} \min\limits_{(d_s,d_i,x_s,x_i)} f(d_s,d_i,x_s,x_i) \\ s.t\quad J_1(d_s,d_i,x_s,x_i) \le \varepsilon \\ \qquad J_2(d_s,d_i,x_s,x_i) \le \varepsilon \\ \qquad \Pr\{giJ_1(d_s,d_i,x_s,x_i) \le 0\} \ge R_i \\ \qquad d_S^L \le d_s \le d_S^U, d_i^L \le d_i \le d_i^U \\ \qquad x_S^L \le x_s \le x_S^U, x_i^L \le x_i \le x_i^U \\ \qquad i = 1,2 \end{cases} \tag{2.58}$$

where d_s and x_s represent the system 's shared design output and random input vector, respectively, that is, the common variables of each discipline; d_i is the local design variable of discipline i; x_i is the local random variable of discipline i; J_1 and J_2 represent the relaxation constraints of subsystem 1 and subsystem 2, respectively. $\Pr\{\bullet\}$ is the failure probability of each discipline constraint condition; g_i and R_i are the reliability constraints and allowable reliability of the discipline i, respectively. ε is a very small positive number that changes dynamically. Here, the consistency constraint relaxation method is used to improve the convergence performance of the CO algorithm.

2.3.3.3.2.2 OPTIMIZATION PROCESS AND STEPS

This method decouples multidisciplinary reliability analysis from the three-layer nested loop of traditional UBMDO, thus serializing deterministic MDO and multidisciplinary reliability analysis and forming a single-layer recursive loop optimization process. Multidisciplinary reliability analysis is performed after deterministic MDO. It is unnecessary to call the MRA loop to perform reliability analysis on all probability constraints after each iteration of deterministic MDO, which can significantly reduce the number of systems analyses and reliability analyses and improve the computational efficiency of UBMDO. The specific steps are as follows:

Step 1. Deterministic MDO based on CO algorithm. In the first cycle of this step, the MDO method does not consider the influence of uncertainty factors on the optimization. That is, all random variables are characterized by their mean values, and the traditional MDO method is used to obtain the optimal solution of the system only in the mathematical sense. The optimization results will be used as the initial data for the multidisciplinary reliability analysis. Considering the convergence difficulty or slow convergence speed caused by the consistency constraints of CO, the CO algorithm is improved. The adaptive relaxation constraints are constructed according to the difference information between disciplines, which overcomes the convergence difficulty caused by the internal definition defects of the CO algorithm. In addition, to improve the optimization efficiency of the CO algorithm and expand the scope of solving MDO problems, the genetic algorithm is used as the system-level optimizer of the CO algorithm to verify the adaptability of the proposed optimization strategy to various algorithms.

Step 2. Multidisciplinary collaborative reliability analysis based on PMA. Aiming at the optimization points obtained in **Step 1**, people used the multidisciplinary collaborative reliability analysis method based on PMA to analyze the reliability of all probability constraints, that is, to search the MPP of each probability constraint and verify whether the probability constraints meet the reliability design requirements. The specific process is as follows.

Step 2.1. Allocation of design variables and state variables. The system level transfers the corresponding design and state variables to each subsystem and solidifies them in optimization.

Step 2.2. Subsystem parallel optimization. The subsystems are optimized in parallel collaboratively manner to obtain their respective optimal solutions. The optimal solution replaces the initial value from the system level to each subsystem, and the system-level constraints are updated to facilitate the maintenance of inter-disciplinary consistency constraints.

Step 2.3. System-level optimization. Based on the optimization results of each subsystem, the system-level automatic optimization is performed to obtain the optimal solution, which is assigned to each subsystem as a new target value as the objective function of the next cycle generation optimization.

Step 2.4. Convergence test. The iterative optimization is carried out, and the system level judges whether to converge according to the convergence criterion and constraint condition verification algorithm. If it converges, the calculation is completed; otherwise, the initial value of **Step 1** is updated using the obtained analysis results, and the analysis is re-analyzed.

Step 3. Convergence verification. If all reliability requirements are met, and the system objective optimization function tends to be stable, the entire optimization process ends; otherwise, turn to **Step 4**.

Step 4. Reconstruct the deterministic MDO model. The deterministic MDO model is reconstructed using the mobile strategy based on the MPP information obtained in **Step 3**. Suppose the probability constraint condition does not meet the reliability requirement. In that case, the reliability analysis result of the probability condition (MPP information) is used to find the moving vector s from the MPP to the deterministic optimization point. Then, the probability constraint condition is moved along the vector to the security domain. After moving, a new deterministic constraint condition is generated to construct a new MDO model. In the reliability design optimization cycle, all the MPPs with probability constraints can be moved to the reliable area through multiple 'move-optimization.'

2.3.3.3.2 UBMDO based on SORA and BLISCO

2.3.3.3.2.1 DETERMINISTIC MDO BASED ON BLISCO

The deterministic MDO is carried out based on the BLISCO algorithm. The core idea is to use the BLISS algorithm to divide the design variables into system-level and subsystem-level design variables while retaining the coordination mechanism. The weighted sum of the subsystems' coupling output is used instead of the consistency constraint as the objective function of the subsystem optimization. System-level optimization coordinates the differences between subsystems, and subsystem-level optimization minimizes the comprehensive impact of its coupling state variables on the system objective function.

The BLISCO algorithm integrates the advantages of the BLISS and CO algorithms. The BLISCO algorithm completely abandons the consistency constraint in the CO algorithm subsystem. It takes the comprehensive influence of the subsystem on the system-level objective function as the objective function of the subsystem, which improves the convergence performance. The algorithm integrates the coordination mechanism of the CO algorithm. After the system passes the optimization goal to the subsystem, it can realize the distributed parallel independent design of the subsystem, which is suitable for the current organization form of large-scale and complex coupling system engineering design. The algorithm integrates the advantages of the BLISS algorithm, which divides the design variables into system-level and subsystem-level design variables. The subsystem regards the system-level design variables as constants in the optimization process. Reduce the scale of subsystem optimization; the algorithm embeds subsystem-level optimization into system-level optimization and abandons the complicated system iteration process of the BLISS algorithm. It does not need the complex system analysis of BLISS-98 and the approximate modeling of BLISS-2000, which can significantly reduce the complexity of the optimization algorithm and the computational cost of MDO.

The mathematical model of system-level MDO based on BLISCO is:

$$\begin{cases} s_s^{(i)} = x_s^{M,k-1} - x_{MPPs}^{i,k-1} \\ s_i^{(i)} = x_i^{M,k-1} - x_{MPPs}^{i,k-1} x_i \\ i = 1,2,\ldots,n \end{cases} \tag{2.59}$$

where $\left(z_s, z_y\right)$ is the system-level design volume; $y*$ is the system-level coupling design variable after subsystem optimization; z_y is the subsystem-level blending state variable; C is the consistency constraint between the coupling design variables and the coupling state variables. In order to improve the convergence performance of the algorithm, the relaxation factor ε (ε is a small positive number) is introduced into the consistency constraint. g_s and h_s are system-level inequality constraints and equality constraints, respectively. z_s^U and z_s^L are the upper and lower limits of Z_s, respectively; z_y^U and z_y^L are the upper and lower limits of z_y, respectively.

In order not to lose generality, the optimization model of the subsystem i based on BLISCO is expressed as:

$$\begin{cases} \min J_i = \sum_j D\left(f, z^0_{y_{i,j}}\right) y_{i,j} \\ s.t.\ \ g_i = \left(x_i, z_i, y_i\left(x_i, z_i\right)\right) \leq 0 \\ \qquad h_i = \left(x_i, z_i, y_i\left(x_i, z_i\right)\right) = 0 \\ \qquad j = 1, 2, \ldots, m \end{cases} \tag{2.60}$$

where $D(\bullet)$ represents the derivative information of function f to $z^0_{y_{i,j}}$; $z_i = \left(z_{s,i}, z_{y,i}\right)$ is the target value of the system-level design variables passed from the system-level to the i subsystem; x_i and y_i are the subsystem-level design variables and state variables corresponding to the shared design variable $z_{s,i}$ and the coupled design variable $z_{y,i}$; g_i and h_i are the inequality constraints and equality constraints of the ith subsystem, respectively.

2.3.3.3.2.1 UBMDO OPTIMIZATION PROCESS BASED ON SORA AND BLISCO

Based on serialization, the UBMDO optimization process of three-layer nested loop optimization is decoupled to form a deterministic MDO based on BLISCO and a single-loop optimization process of sequential execution of multidisciplinary reliability analysis based on PMA. The specific steps are as follows.

Step 1. Initialize the deterministic and random design parameters and perform deterministic MDO based on BLISCO. The deterministic MDO in the first loop iterative optimization does not consider the influence of uncertain factors. That is, all random variables are characterized by their mean values. The coupling state variables are represented by their standard values, and the BLISCO algorithm is used for deterministic optimization to obtain the theoretically optimal solution. The optimization results will be used as the initial data of SMRA.

Step 2. Multidisciplinary reliability analysis using SMRA. For the theoretical optimal solution obtained in **Step 1**, the system and sensitivity analyses are carried out based on the SMRA method. The PMA method based on the angle update strategy is used to analyze the reliability of all probability constraints, and the MPP of all probability constraints are searched to verify whether the probability constraints meet the reliability design requirements.

Step 3. Convergence verification. If all reliability requirements are met, and the system objective optimization function tends to be stable, the optimization process ends. Otherwise, the steering **Step 4**.

Step 4. Based on the MPP obtained in **Step 3**, the deterministic MDO model is reconstructed to continue the process optimization. The mobile strategy is used to reconstruct the mobile vector.

$$\begin{cases} s_s^{(i)} = x_s^{M,k-1} - x_{MPPs}^{i,k-1} \\ s_i^{(i)} = x_i^{M,k-1} - x_{MPPi}^{i,k-1}\ k-1 \\ i = 1,2,\ldots,n \end{cases} \tag{2.61}$$

where $s_s^{(i)}$ and $s_i^{(i)}$ are the moving vectors of the shared design variable x_s and the subject design variable x_i, respectively; $x_{MPPs}^{i,k-1}$ and $x_{MPPi}^{i,k-1}$ are the most likely failure points of the $(k-1)$th iteration optimization of the shared design variable x_s and the discipline design variable x_i, respectively. Substituting Eq. 2.61 into Eq. 2.60, a new deterministic MDO model can be reconstructed.

REFERENCES

[1] Yi Y. S. (2019). Research on MDO method based on collaborative approximation and set strategy. PhD thesis of Huazhong University of Science and Technology (in Chinese), Wuhan, China.

[2] Li Z. Z. (2018). Overall MDO of winged reentry vehicle considering handling and stability characteristics. PhD thesis of Nanjing University of Aeronautics and Astronautics (in Chinese), Nanjing, China.

[3] Wang X. (2018). Research on modeling and solving methods of MDO considering uncertainty factors. Master thesis of University of Electronic Science and Technology of China (in Chinese), Chengdu, China.

[4] Xiao M. (2012). Research on approximate model and solving strategy in MDO. PhD thesis of Huazhong University of Science and Technology (in Chinese), Wuhan, China.

[5] Oberkampf W. L., Helton J. C., Sentz K. (2001). Mathematical representation of uncertainty. 42nd AIAA/ASME/ASCE/AHS/ASC Structures, Structural Dynamics and Materials Conference & Exhibit, number AIAA 2001–1645, 16–19.

[6] Oberkampf W. L., Diegert K. V., Alvin K. F., et al. (1998). Variability, uncertainty, and error in computa-tional simulation. 7th AIAA/ASME Joint Thermophysics and Heat Transfer Conference, 15–18, 2: 259–272.

[7] Ben-Haim Y., Elishakoff I. (2013). Convex models of uncertainty in applied mechanics. Amsterdam, The Netherlands: Elsevier Science.

[8] Zadeh L. A. (1978). Fuzzy sets as a basis for a theory of possibility. Fuzzy Sets and Systems, 1(1): 3–28.

[9] Di Nola A., Pedrycz W., Sessa S., Sanchez E. (1991). Fuzzy relation equations theory as a basis of fuzzy modelling: An overview. Fuzzy Sets and Systems, 40(3): 415–429.

[10] Zhou S., Zhang J., Zhang Q., Wen M. (2022). A new chance reliability-based design optimization approach considering aleatory and epistemic uncertainties. Structural and Multidisciplinary Optimization, 65(8): 233.

[11] Du X., Sudjianto A., Huang B. (2005). Reliability-based design with the mixture of random and interval variables. Journal of Mechanical Design, 127(6): 1068–1076.
[12] Du X., Guo J., Beeram H. (2008). Sequential optimization and reliability assessment for multidisciplinary systems design. Structural and Multidisciplinary Optimization, 35: 117–130.
[13] Du X., Sudjianto A., Chen W. (2004). An integrated framework for optimization under uncertainty using inverse reliability strategy. Journal of Mechanical Design, 126(4): 562–570.
[14] Agarwal H., Renaud J., Lee J., Watson L. (2004). A unilevel method for reliability based design optimization. 45th AIAA/ASME/ASCE/AHS/ASC Structures, Structural Dynamics & Materials Conference, 2029.

Chapter 3

Approximate method

The approximate method is a crucial calculation method. This mainly refers to the mathematical calculation method with an approximate number as the calculation object, which approximately represents the true value of a certain amount. In uncertainty analysis, many system analysis models need to be called. If high-precision model analysis is directly used, it will generate immeasurable computational costs. Therefore, constructing an approximate model using an approximate method is essential to compromise costs. This can improve the solution's efficiency while ensuring accuracy meets the requirements.

3.1 AN OVERVIEW OF APPROXIMATION TECHNIQUE AND EXPERIMENTAL DESIGN

3.1.1 Basic concepts of approximation technology

The approximation method could contribute to MDO problems in uncertainty analysis and realize discipline decoupling to support discipline autonomy and parallel optimization. It is a key technology that can solve the computational and organizational complexity of UBMDO problems. Approximation techniques encompass model approximation, represented by surrogate models, and function approximation, involving display expressions. The model approximation mainly starts from the perspective of simplifying the system model and reducing the scale of the optimization problem. It improves efficiency of the design optimization by reducing the number of design variables and constraints. Design variable chaining and reduction basis, envelope function, and constraint function reduction method are commonly used [1]. The main goal of model approximation is to approximate the complex original model by establishing a simple model and retaining the key features and properties of the original model as much as possible. In some cases, the goal of model approximation is not only to understand the behavior of a single model but also to understand a set of

DOI: 10.1201/9781003464792-3

models, which requires processing a large amount of data. In this case, a model approximation technique called the response surface method is often used. This method uses a small number of representative samples to fit the complex model linearly or nonlinearly, resulting in a simple and easy-to-handle approximation model.

The surrogate model is an essential part of model approximation. The most common surrogate models include the polynomial, radial basis function, and Kriging models. When employing model approximation techniques, it is essential to emphasize the associated verification and testing. This guarantees that the approximated model adheres to precise accuracy requirements while retaining its capability to address the intended problem. Among them, cross-validation is the most commonly used verification method.

However, model approximation techniques are not omnipotent. We should select appropriate models and approximation techniques in practical applications according to specific problems and scenarios. At the same time, understanding the uncertainty in the results and how this uncertainty affects the accuracy and stability of the solution is the key to effective uncertainty analysis.

According to the scope of the design space that the approximation function can simulate, the modeling method can be divided into local, medium-range, and global approximation methods. The local approximation method is a numerical calculation method used to approximate the value of the function near a certain point. Based on the concept of Taylor series expansion, the value of the function at a given point is estimated locally by approximating the function. The approximation function is primarily effective in the vicinity of the design point, often referred to as a single-point approximation. The mid-range approximation method uses the data of multiple points to construct an approximate display function, which is also called a multi-point approximation method. Its effective range is between the local and global approximation methods. The global approximation method is mainly used to construct a global model describing the design space's objective function or constraint condition. This model aims to succinctly depict the overall behavior of objective functions and constraints, including essential characteristics like potential energy surfaces, nonlinearity, multimodality, and more. Its wide design space makes it suitable for parallel computing and is most widely used in UBMDO.

3.1.2 Basic concepts of experimental design

The experimental design was born in the 1920s. Its main research content is based on mathematical statistics theory, which is used to study obtaining data information reasonably and effectively. Experimental design plays a vital role in uncertainty analysis. In the face of complex problems, especially when the factors affecting the results are numerous and interrelated, a reasonable experimental design can help us more effectively understand the relationship between dependent variables and various influencing factors and evaluate the uncertainty of these factors. In uncertainty quantification, a good experimental design can achieve more precise stage division so that researchers can analyze each stage's influencing factors and uncertainties and better understand and evaluate the whole system. To this end, a multivariate test scenario should be designed to facilitate observing the impact of different variables on the results rather than just focusing on a single dependent variable.

Full factorial design, fractional factorial design, and Central Composite Design (CCD) are commonly used experimental design methods. Further, the experimental design can also introduce statistical methods, such as random seeds and Monte Carlo methods, to increase the rigor and accuracy of the test. Efficient experimental design techniques, such as orthogonal and Latin hypercube designs, can effectively address large-scale and high-dimensional scenarios, enhancing resource utilization and experimental efficiency. A brief overview of each design method is introduced in the next part.

3.1.2.1 Full factorial design method

All m_i of the design variable x_i of each dimension are horizontally combined to obtain $k = \prod_{i}^{n} m_i$ test schemes. The full factorial design sampling diagram is as follows in the three-dimensional design space, as shown in Figure 3.1.

This design method can fully reflect the influence of the interaction between the design variables and their interaction on the response value. However, this method will experience a sharp increase in the number of trials as the number of design variables and levels increase. Therefore, the full factorial design method is mainly used to design the experimental design with few variables and levels.

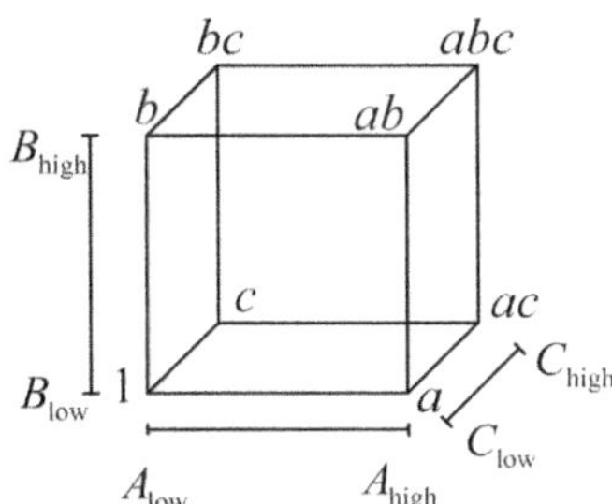

Figure 3.1 Full factorial design sampling diagram.

3.1.2.2 Partial factorial design method

Since the full factorial design method is limited by the number of design variables and test levels, the partial factorial design method is developed. This method is mainly used to deal with the problem of too complex or too many factors. Partial factor design simplifies the model by focusing on a subset of factors and their interaction effects. It is a cost-effective and practical approach when considering all possible factor combinations that are infeasible or prohibitively expensive. Partial factorial design methods include hierarchical design, Plackett-Burman design, Taguchi design, and so on. These partial factor design methods usually only consider the main and low-order interaction effects to maximize the acquisition of adequate information when resources are limited.

3.1.2.3 Central composition design

The CCD method is mainly used to establish and optimize the response surface model, which can be used to predict and optimize the processing parameters and conditions, as well as uncertainty analysis. The method consists of a two-level full factorial design, a central point, and two additional test points along each dimension direction. The corresponding number of tests is $2^n + 2n + 1$. For the two-dimensional design space, the sampling diagram of the CCD method is shown in Figure 3.2.

In uncertainty analysis, a CCD offers richer data by estimating a wide range of effects, encompassing linear, quadratic, and interaction effects. As a result, it can provide more comprehensive and accurate information for uncertainty modeling and analysis. At the same time, CCD also has a relatively strong effect in predicting complex nonlinear phenomena, especially when trying to solve optimization problems or find the minimum or maximum value of the response function.

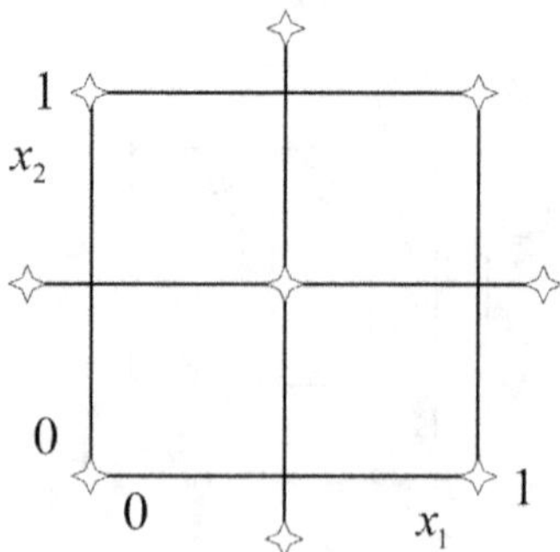

Figure 3.2 Partial factorial design sampling diagram.

3.1.2.4 Monte Carlo method

This method randomly selects test points in the design space to simulate the information of the accurate model. The selected points are shown in Figure 3.3.

The basic idea of the Monte Carlo method is straightforward. By constructing a model and randomly selecting values from the input parameters of the model, we run multiple (up to several million times) calculations based on this model, simulate the possible results, and evaluate the uncertainty of the model. Each time the model is run, a random value is generated for each input parameter, which usually comes from a predefined probability distribution. However, due to the randomness of its selection, the test points can be concentrated in a certain area. For this purpose, a stratified Monte Carlo sampling method is developed, as shown in Figure 3.4. This method first divides the design space into several subspaces with equal probability distribution along the direction of each dimension design variable. It then randomly selects points, as shown in the following figure. This can ensure that each area has a test point and improve the uniformity of the point.

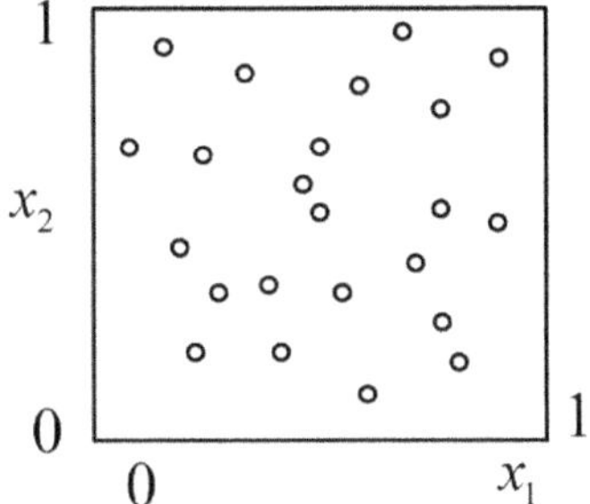

Figure 3.3 Monte Carlo sampling diagram.

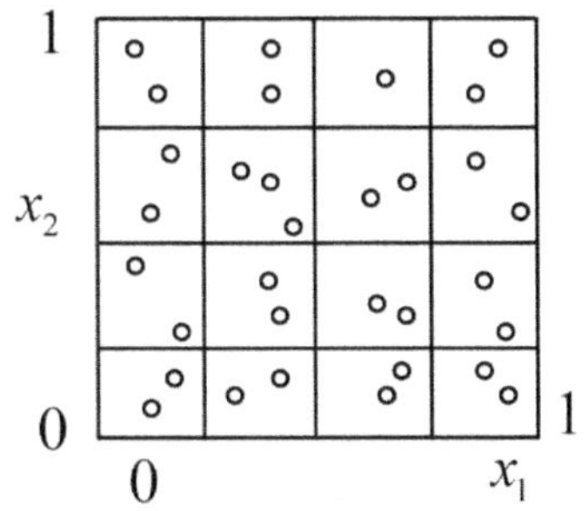

Figure 3.4 Stratified Monte Carlo sampling diagram.

3.1.2.5 Orthogonal experimental design method

The orthogonal experimental design method is based on an orthogonal array and table to reduce the number of experiments and obtain the optimal combination of multi-factor and multi-level combinations. The comprehensive combination of test factors and their levels is classified and sampled in the orthogonal experimental design. That is, a specific combination appears only once in all possible combinations, which ensures the independence of each factor, so it is called orthogonal. This design method can effectively reduce the number of tests, obtain as much helpful information as possible in the limited number of tests, and then find the key factors affecting the results to optimize the product or process. A three-dimensional, second-level, four-time orthogonal experimental design is shown in the following Figure 3.5.

3.1.2.6 Latin hypercube design method

Latin Hypercube Sampling (LHS) is a widely used sampling strategy in experimental design that aims to achieve uniform sampling of high-dimensional parameter space. In the traditional Monte Carlo sampling method, there may be a problem of insufficient data sampling in a certain interval. Especially when dealing with multi-dimensional parameter space, the efficiency becomes very low. The Latin hypercube design method uniformly samples each dimension with stratified sampling, effectively avoiding parameter correlations. This method greatly enhances sampling efficiency in a concise and academic manner. Specifically, LHS divides each dimension equally and randomly selects a point in each division as a sampling point. The advantage of this is that it can ensure uniform sample coverage in each dimension while avoiding the value judgment of setting the bias in advance

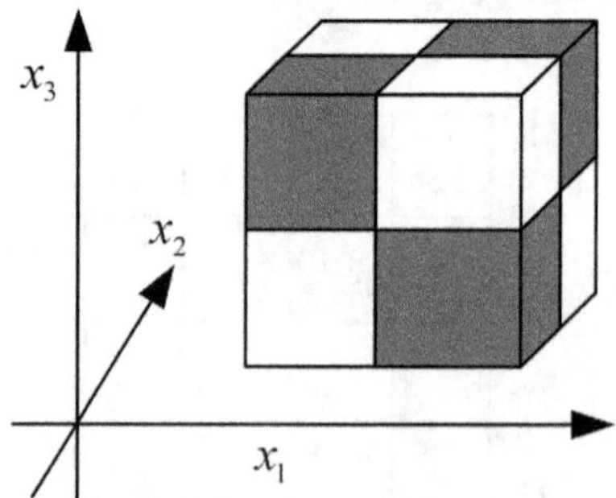

Figure 3.5 Orthogonal experimental design diagram.

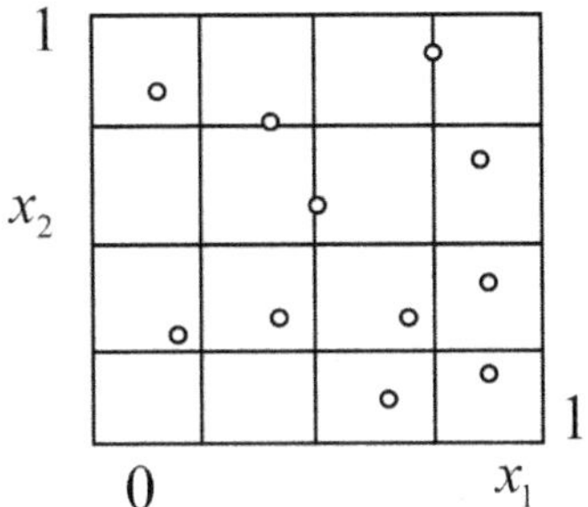

Figure 3.6 Latin hypercube design sampling diagram.

in a specific dimension. A two-dimensional Latin hypercube design method takes the points shown in Figure 3.6.

In the above methods, the full factorial design method, the partial factorial design method, the CCD method, and the Monte Carlo method all have the problem that the number of test points increases sharply with the increase of the number of design variables and levels. This will cause the test cost to be high and uncontrollable, and the scope of application is relatively small.

The Monte Carlo method is mainly used as a comparison method for developing new methods to test their effectiveness because it has the highest model accuracy and is closest to the actual effect. The orthogonal experimental design and Latin hypercube design methods efficiently yield well-dispersed and highly representative design points from the design space at a minimal cost. This also makes these two methods more effective in obtaining accurate model information. However, the orthogonal test design method needs to be arranged based on the orthogonal table, and the user's flexibility for the operation of the number of test points is not strong. Therefore, the Latin hypercube design method has more robust flexibility and wide application.

3.2 APPROXIMATE MODEL AND MODELING METHOD

This section will introduce several commonly used approximate models, including the polynomial model, Kriging model, Support Vector Machine (SVM), and Radial Basis Function Neural Network (RBFNN) method.

3.2.1 Polynomial model

The polynomial model uses polynomial expansion to approximate the function, which is used to simulate or predict the behavior of a complex

Table 3.1 Polynomial regression function model

Order	*Polynomial function row vector* $F^T(x)$	*The number of components of* β
0	$[1]$	1
1	$[1, x_1, x_2, \ldots, x_n]$	$n+1$
2	$[1, x_1, x_2, \ldots, x_n, x_1^2, x_1x_2, \ldots x_1x_n, \ldots, x_n^2]$	$\frac{(n+1)(n+2)}{2}$

system. Its basic form is polynomial equations because polynomial equations can describe the development trend of complex curves or models.

Define the real function is $y(x)$, $x = [x_1, x_2, \ldots, x_n]^T$ is the independent variable vector. Then by sampling in the design space, N_T training sample points are obtained, where $T = \{(x_k, y_k) : y_k = y(x_k)\}_{k=1}^{N_T}$, The ith component of sample point x_k is x_{ki}. Define the approximate function is $\hat{y}(x)$.

$$\hat{y}(x) = F^T(x) \cdot \beta \tag{3.1}$$

where $F^T(x)$ is the row vector of the polynomial regression basis function, and β is the regression coefficient sequence vector. A can be calculated by Eq. 3.2

$$\min J(x) : J(x) = \sum_{i=1}^{N_T} \left[F^T(x_i) \cdot \beta - y_i \right]^2 \tag{3.2}$$

Due to the existence of the Runge phenomenon, the basis function is generally not more than three times. Table 3.1 shows its basic model.

In this model, the response variable is a linear combination of the powers of the input variables. The power of all input variables is considered, and the maximum number is usually a user-defined parameter. The advantage of the polynomial model is that it can contain nonlinear relationships and has high flexibility, making it adaptable to many different data shapes. In addition, it only needs basic mathematical operations, and the computational efficiency is also good. However, the polynomial model also has limitations. For example, data over-fitting may occur, especially when a high degree of polynomial is used. The chosen model structure and parameters can influence the polynomial model's accuracy. Incorrect selections may result in inaccurate outcomes.

3.2.2 Kriging model

In recent years, the Kriging model has received extensive attention on uncertainty analysis and reliability. This model is an unbiased estimation model with the smallest estimation variance. The basic idea is to express the real unknown function as a global approximation function and a correction function, which is constructed as follows:

$$\hat{y}(\boldsymbol{x}) = \hat{f}(\boldsymbol{x}) + z(\boldsymbol{x}) \tag{3.3}$$

where $\hat{f}(\boldsymbol{x})$ is the mean value of the Gaussian process, which reflects the prediction trend of the Kriging model and can be expressed as:

$$\hat{f}(\boldsymbol{x}) = \mathrm{F}^{\mathrm{T}}(x) \cdot \beta \tag{3.4}$$

where $\mathrm{F}^{\mathrm{T}}(x)$ represents the polynomial regression function vector. β represents the corresponding regression coefficient sequence vector. The error term $z(\boldsymbol{x})$ represents the steady-state Gaussian process and the error term between any two points of $z(\boldsymbol{x}) \sim N(0, \sigma^2)$ is correlated. The covariance is:

$$\mathrm{cov}\left[z(x_i), z(x_j)\right] = \sigma^2 R_\theta (x_i, x_j) \tag{3.5}$$

σ^2 is the process variance. $R_\theta (x_i, x_j)$ represents the correlation function between any two points in the variable space, and its size is related to the distance between two points.

The Kriging model can provide the optimal prediction of unknown values and quantify the prediction uncertainty. This is particularly crucial when addressing uncertainty in the form of probability. Specifically, in situations where the complete system response is either unknown or uncertain, the Kriging model proves valuable in forecasting the system's response and assessing the associated prediction uncertainties. This is achieved using observed data and spatial or temporal relationships to establish a random process model. Therefore, the Kriging model provides a powerful method to deal with complex systems with uncertain information, which can generate accurate predictions and quantify the prediction uncertainty.

3.2.3 Support vector machine

SVM is a powerful and flexible supervised learning model for classification and regression analysis. The main goal is to find a hyperplane to classify data. In two-dimensional space, this decision boundary is a line. The hyperplane of SVM can maximize the distance between the nearest training samples, also called support vectors. SVM includes two categories: Support Vector Classification (SVC) machine for classification problems and Support Vector Regression (SVR) machine for regression problems.

3.2.3.1 SVC

In the case of linear separability, SVC is developed from the optimal classification surface, and its basic classification idea is shown in Figure 3.7.

In this classification surface, the equations of the two dotted lines and the solid line are shown in Eq. 3.6

$$\begin{cases} w^T x + b = 1, \text{upper dotted line} \\ w^T x + b = 0, \text{ solid line} \\ w^T x + b = -1, \text{ lower dotted line} \end{cases} \tag{3.6}$$

The solid line is the classification line, and the distance from the two dotted lines is equal. The gap between the two dashed lines is referred to as the margin. Following the principle of structural risk minimization, the optimal classification line in SVC is the one that maximizes this classification margin. It can be expressed as the following constrained optimization problem

$$\begin{cases} \min\limits_{w,b} \ J_p(w) = \dfrac{1}{2} w^T w \\ s.t. \quad y_k\left(w^T x_k + b\right) \geq 1, k = 1, \ldots, N \end{cases} \tag{3.7}$$

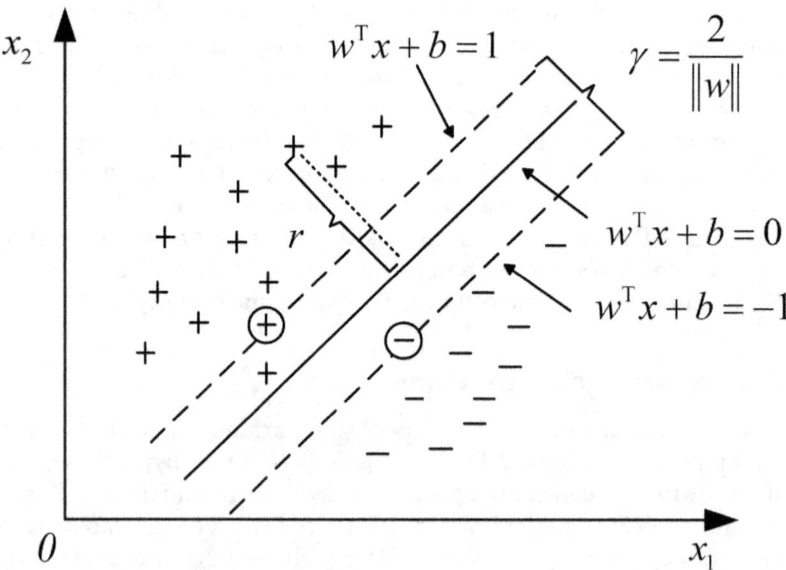

Figure 3.7 Optimal classification surface schematic diagram.

where N is the number of sample points. For the problem of high position, the classification function is its optimal classification surface.

When the data sample does not satisfy the current separability, the optimal classification line mentioned above will fail. At this time, we can introduce $\xi_k \geq 0$ and allow the existence of misclassified samples, then the most classification problem can be expressed in the following form

$$\begin{cases} \underset{w,b}{min} \ J_p(w) = \frac{1}{2} w^T w + C\sum_{k=1}^{N} \xi_k \\ s.t. \ \ y_k\left(w^T x_k + b\right) \geq 1, k = 1,\ldots,N \end{cases} \tag{3.8}$$

where C is a constant greater than 0, as a penalty factor. Assuming the optimal solution obtained by the above equation is $\{w_1, b_1, \xi_1\}$, then the optimal classification surface equation is $w_1 x + b_1 = 0$. For a given sample point, it is only necessary to substitute this formula to determine the category according to the symbol.

This method can also be used to solve regression problems. The regression problem of function is mainly to find an approximate function $y = \hat{f}(x)$ to predict the true value of the function under the condition of given sample data. The difference between the regression problem and the classification problem is that the y value range of the output is different. In the classification problem, $y = \pm 1$, it only plays a classification role; in the regression problem, y can take any value.

3.2.3.2 SVR

In practical engineering scenarios, numerous classification problems exhibit nonlinearly separable characteristics. Consequently, the application of linear SVMs solutions may lack practical significance due to the introduction of significant empirical risk. Therefore, it is necessary to introduce a nonlinear SVM to deal with practical problems better. Using the kernel technique in the kernel function, the problems that cannot be solved in the low dimension can be reflected in the high dimensional space. The nonlinear mapping problem is shown in Figure 3.8.

As shown in Figure 3.8, if we want to make the problem linearly separable, we must introduce a suitable kernel function through the original two-dimensional space. Then, the problem is mapped to a suitable three-dimensional space to find a suitable partition hyperplane. The key is to transform the problem to be solved into a high-dimensional space and introduce the internal basis function to simplify the calculation process.

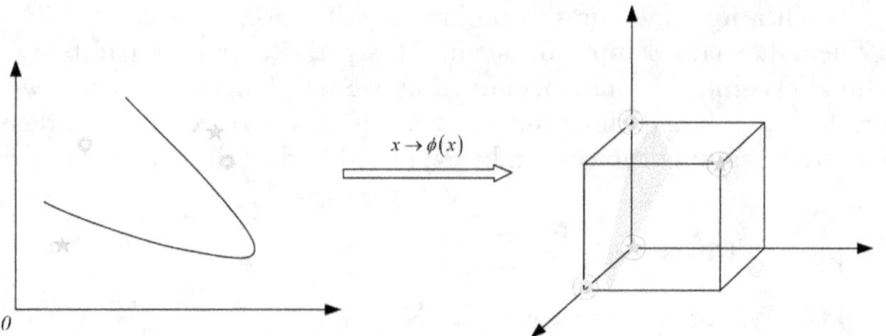

Figure 3.8 Nonlinear mapping problem graph.

Table 3.2 Kernel function expression

Name-kernel function	*Expression*	*Constraint condition*
Linear	$x_i^T x_j$	–
Polynomial	$\left(x_i^T x_j\right)^d$	$d \geq 1$
Sigmoid	$\tanh\left(\beta x_i^T x_j + \theta\right)^d$	$\beta > 0, \theta < 0$
Gaussian	$\exp\left(-\left\lVert x_i - x_j \right\rVert^2 / 2\sigma^2\right)$	$\sigma > 0$
Laplacian	$\exp\left(-\left\lVert x_i - x_j \right\rVert^2 / \sigma\right)^d$	$\sigma > 0$

Given an input space χ, the symmetric function defined on $\chi \times \chi$ is $\kappa(\cdot,\cdot)$. Only if for any data $D = \{x_1, x_2, \cdots, x_m\}$, the kernel matrix K is always positive semidefinite, then, $\kappa(\cdot,\cdot)$ is called a kernel function.

$$K \begin{bmatrix} \kappa(x_1, x_1) & \cdots & \kappa(x_1, x_j) & \cdots & \kappa(x_1, x_m) \\ & \ddots & \vdots & & \\ \kappa(x_i, x_1) & \cdots & \kappa(x_i, x_j) & \cdots & \kappa(x_i, x_m) \\ & & \vdots & \ddots & \\ \kappa(x_m, x_1) & \cdots & \kappa(x_m, x_j) & \cdots & \kappa(x_m, x_m) \end{bmatrix} \tag{3.9}$$

As shown in Table 3.2, there are many kernel functions, including linear kernel function, polynomial kernel function, Gaussian kernel function, Laplace kernel function, and so on. The selection depends on the size of the data. Polynomials can be added to correct the overfitting phenomenon that is prone to occur for the selection of simple models with small data sets. The extensive dataset necessitates the utilization of a sophisticated model to mitigate overfitting issues efficiently.

3.2.4 Radial basis function neural networks

The artificial neural network is a model that imitates the human brain to explore knowledge, store information, deal with problems, and process information by adjusting the connection mode of internal nodes. It can well approximate complex nonlinear problems. Inside the neural network, the information is qualitatively stored in the nodes. Even in the event of a node failure, the overall functionality of the artificial neural network remains unaffected. This attribute enhances artificial neural networks' fault tolerance and robustness, resulting in superior operational speed and adaptability to uncertain systems. Powell proposed the RBFNN in 1985. It is a real-valued function whose value depends only on the distance from the origin, that is, the function satisfies $\phi(x) = \phi(\|x\|)$. In addition to the origin, we can also choose any point a as the center point, that is, the function satisfies $\phi(x,a) = \phi(\|x - a\|)$.

The Gaussian kernel function mentioned above is an RBFNN. The RBFNN is a three-layer feedforward neural network, including the input, hidden, and output layers. The input layer to the hidden layer in the region is nonlinearly transformed, and the hidden layer to the output layer is linearly transformed. The structure diagram is shown in Figure 3.9.

The RBFNN interpolation model accurately passes through the given N_T sample point, and its output $\hat{f}(\boldsymbol{x})$ is

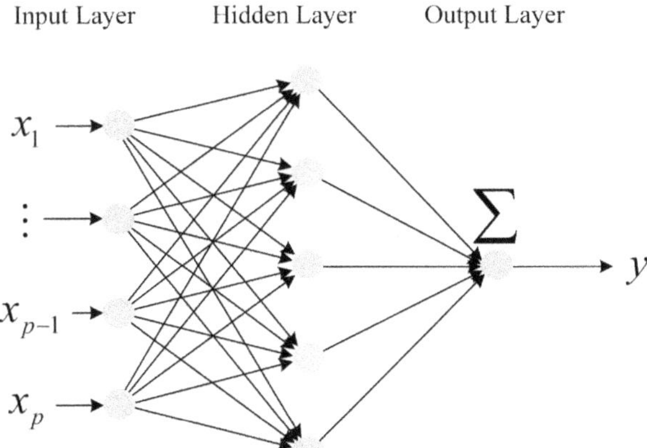

Figure 3.9 The structure diagram of RBFNN.

$$\hat{f}(c_k) = f(c_k), 1 \le k \le N^T \tag{3.10}$$

where $c_k \in R^n$ is the center point of the radial basis function of the kth neuron.

The relationship between the linear sum of all sample points and a matrix form can be used to express the training sample points.

$$\Phi w = F_T = \left[f(c_1), f(c_2), \cdots, f(c_{N^T})\right]^T$$
$$\Phi = \begin{bmatrix} \phi_1(c_1, c_1) & \phi_2(c_1, c_2) & \cdots & \phi_{N^T}(c_1, c_{N^T}) \\ \phi_1(c_2, c_1) & \phi_2(c_2, c_2) & \cdots & \phi_{N^T}(c_2, c_{N^T}) \\ \vdots & \vdots & \vdots & \vdots \\ \phi_1(c_{N^T}, c_1) & \phi_2(c_{N^T}, c_2) & \cdots & \phi_{N^T}(c_{N^T}, c_{N^T}) \end{bmatrix} \tag{3.11}$$

All the sample points give a set of fixed values for the shape parameter σ_k, then the above equation can be a linear equation group. Suppose the center point of the radial basis function of each neuron is different. In that case, the matrix Φ is positive definite and reversible, and the above formula can directly solve the weight vector w. Therefore, the adjustable parameters of the RBFNN interpolation model include only the shape parameter σ_k. The parameter optimization of the approximate model is a shape parameter optimization problem.

Given the validation sample point set A, the widely used RBFNN approximation modeling criterion is to minimize the sum of squared regular errors as follows:

$$\text{obj} = (1 - \lambda) e^T e + \lambda w^T w \tag{3.12}$$

where λ is the regular parameter and e is the error vector.

In Eq. 3.10, $\lambda w^T w$ is mainly used to avoid overfitting, and the model can be made smoother and more accurate in approximations by adjusting the size of λ [2, 3]. For the approximate modeling of complex systems, a large number of sample points are usually required for training to obtain an approximate model that meets the accuracy requirements, and the number of RBFNN interpolation model neurons is the same as the number of training sample points. If the shape parameters of each neuron radial basis function are optimized as independent variables, the complexity of optimization will be too high, and the computational cost will be unbearable. To simplify the optimization process, one can reduce the number of variables. Considering that the adjacent sample points generally have similar spatial distribution characteristics and the nonlinear distribution characteristics of the exact function response value, the sample points can be clustered to achieve a compromise between approximation accuracy and modeling efficiency.

The training sample point set is divided into non-overlapping N_S clusters each cluster is denoted by:

$$TS_i = \left\{c_p\right\}, i = 1,2,\ldots,N_S, p \in I_{TS_i} \tag{3.13}$$

where I_{TS_i} is the number set of sample points in cluster TS_i, which contains N_{TS_i} elements.

Reducing the total distances between each sample point and the cluster center is the clustering criterion, which the *K*-means clustering algorithm can achieve [4]. The dimension coordinate of the center point of each cluster is the average value of all sample points in this cluster in this dimension coordinate. The number set of all clusters can form a clustering scheme, that is, $I_{TS_i} = \left\{I_{TS_i}\right\}_{i=1}^{N_S}$.

For the above clustering scheme, the shape parameter σ_{Si} of the sample points in the same cluster is the same. It can be seen that

$$\forall k \in I_{TS_i}, \sigma_k = \sigma_{Si}, i = 1,2,\ldots,n \tag{3.14}$$

Then, the shape parameter optimization problem only contains N_S independent optimization variables, which greatly reduces the complexity of optimization. For a given training sample set *T*, validation sample set *V* and clustering scheme I_{TS}, the RBFNN interpolation model parameter optimization problem can be expressed as Eq. 3.15.

$$\begin{cases} \text{find} & \boldsymbol{\sigma}_S = \left[\sigma_{S1}, \sigma_{S2}, \cdots, \sigma_{SN_S}\right] \\ \min & \text{obj} = (1-\lambda)\boldsymbol{e}^{\mathrm{T}}\boldsymbol{e} + \lambda\boldsymbol{w}^{\mathrm{T}}\boldsymbol{w} \\ s.t. & \boldsymbol{\sigma}_S \in \Omega_S,\ \boldsymbol{\sigma} = \left[\sigma_1, \sigma_2, \cdots, \sigma_{N_T}\right] \\ & \forall k \in I_{TS_i}, \sigma_k = \sigma_{Si}, i = 1,2,\cdots,n \\ & \Phi(\sigma)\boldsymbol{w} = F_T, \boldsymbol{e} = F_V - \hat{F}_V(\boldsymbol{\sigma}, \boldsymbol{w}) \end{cases} \tag{3.15}$$

where Ω_S is the domain of $\boldsymbol{\sigma}_S$. For the Gaussian function, obj is an even function of the optimization variable $\boldsymbol{\sigma}_S$, so only the non-negative region of $\boldsymbol{\sigma}_S$ needs to be considered. The setting principle of the upper limit of $\boldsymbol{\sigma}_S$ is to avoid the serious ill-condition of the matrix Φ, which can ensure the stability of the weight coefficient.

One of the main characteristics of RBFNN is that it has good interpolation and approximation ability for new data. Because of its strong nonlinear mapping ability, it can deal with the problem of complex nonlinear relationships. However, determining the appropriate RBFNN center and width is a major challenge to use this network. For extensive datasets, a substantial number of neurons may be necessary, leading to complex computations.

3.3 PREDICTION ACCURACY EVALUATION OF APPROXIMATE MODEL

The approximate model is based on the simulation calculation model of the actual situation, so there is a certain error with the actual situation. Its accuracy and error must be considered when using the approximate model method. Different practical problems have different accuracy requirements. Selecting the suitable model minimizes computational expenses while fulfilling accuracy criteria. Evaluating the prediction accuracy of the approximate model can ensure its effectiveness.

3.3.1 Error analysis methods

The construction of the approximate model based on sample points requires its accuracy to meet the requirements so that the approximate model can be used to replace the high-precision model for approximate analysis. Otherwise, more sample information is needed to adjust the approximate model parameters or replace more reasonable approximation methods. The error analysis method is crucial to evaluate the approximate model, which quantitatively analyses the model's accuracy through the error index. Next, the commonly used error indicators are introduced below.

3.3.1.1 Maximal Error (ME)

$$\mathrm{ME} = \max\left\{\left|f_i - \hat{f}_i\right|\right\}, i = 1,2,\ldots,N_V \tag{3.16}$$

where f_i is the accurate model response value of the ith sample point, $\hat{f}_i$ is the approximate model estimation value of the ith sample point, N_V is the number of sample points used to verify the model's accuracy. The same variable represents the same meaning in the following.

3.3.1.2 Sum of the Square Error (SSE)

$$\mathrm{SSE} = \sum_{i=1}^{N_V}\left(f_i - \hat{f}_i\right)^2 \tag{3.17}$$

3.3.1.3 Root Mean Square Error (RMSE)

$$\mathrm{RMSE} = \sqrt{\frac{\mathrm{SSE}}{N_V}} = \sqrt{\frac{\sum_{i=1}^{N_V}\left(f_i - \hat{f}_i\right)^2}{N_V}} \tag{3.18}$$

3.3.1.4 Mean Absolute Error (MAE)

$$\mathrm{MAE} = \frac{\sum_{i=1}^{N_V}\left|f_i - \hat{f}_i\right|}{N_V} \tag{3.19}$$

3.3.1.5 Mean Relative Error (MRE)

$$\mathrm{MRE} = \frac{1}{N_V}\sum_{i=1}^{N_V}\left|\frac{f_i - \hat{f}_i}{f_i}\right| \tag{3.20}$$

3.3.1.6 Square Sum of Total Departure (SSTD)

$$\mathrm{SSTD} = \sum_{i=1}^{N_V}\left(f_i - \overline{f}\right)^2 = \sum_{i=1}^{N_V} f_i^2 - \frac{1}{N_V}\left(\sum_{i=1}^{N_V}\hat{f}_i\right)^2 \tag{3.21}$$

where $\overline{f} = \frac{1}{N_V}\left(\sum_{i=1}^{N_V}\hat{f}_i\right)^2$ is the approximate mean of all sample points.

3.3.1.7 Sum of Squares of Regression (SSR)

$$\mathrm{SSR} = \sum_{i=1}^{N_V}\left(\hat{f}_i - \overline{f}\right)^2 \tag{3.22}$$

3.3.1.8 R Square (R2)

$$\mathrm{R}^2 = 1 - \frac{\mathrm{SSE}}{\mathrm{SSY}} = \frac{\mathrm{SSY}}{\mathrm{SSE}} = 1 - \frac{\sum_{i=1}^{N_V}\left(f_i - \hat{f}_i\right)^2}{\sum_{i=1}^{N_V}\left(f_i - \overline{f}\right)^2} \tag{3.23}$$

When the accuracy of the approximation model is too low, R^2 may be negative. The higher the accuracy of the approximate model, the closer R^2 is to 1.

3.3.1.9 Relative Average Absolute Error (RAAE)

$$\mathrm{RAAE} = 1 - \frac{\sum_{i=1}^{N_V} \left| f_i - \hat{f} \right|}{\sum_{i=1}^{N_V} \left| f_i - \bar{f} \right|} \tag{3.24}$$

3.3.1.10 Relative Maximum Absolute Error (RMAE)

$$\mathrm{RAAE} = \frac{\max\left(\left| f_1 - \hat{f}_1 \right|, \left| f_2 - \hat{f}_2 \right|, \ldots, \left| f_{N_V} - \hat{f}_{N_V} \right| \right)}{\frac{1}{N_V} \sum_{i=1}^{N_V} \left(f_i - \hat{f} \right)^2} \tag{3.25}$$

The above error indicators are all positive, except for the multiple correlation coefficient R^2. The smaller the value of the error-index, the higher the approximate accuracy. If the value of RMAE is too high, the approximation effect of the approximate model in a certain area of the design space is not good enough.

3.3.2 Approximate ability evaluation method

Approximation ability refers to the prediction ability of the approximation model based on the existing training sample points to the unknown points. Various error evaluation methods, such as polynomial response surfaces, can be directly used for fitting approximation methods. However, interpolation approximation methods, such as Kriging and interpolation RBFNN, cannot be evaluated. The approximate model accurately passes through all training sample points, and the RMSE and average relative error are all 0. The multiple correlation coefficient R equals 1, which cannot be evaluated using the existing sample points based on the above-mentioned error evaluation method. Next, the commonly used evaluation methods are introduced.

3.3.2.1 Increase the sampling point method

This method adds a separate sample point for verifying the approximation accuracy and then applies the above error evaluation method to analyze the error based on the verification sample point. This method is simple and effective, but if the high-precision analysis model is complex and time-consuming, increasing the verification sample points will undoubtedly lead to a sharp increase in computational costs.

3.3.2.2 Cross-validation method

This method evaluates the model's generalization error by dividing the training data set to obtain training and validation subsets. For example, *k*-fold cross-validation takes each subset as the validation data and the rest as the training data and averages the results. This can obtain more stable performance indicators. The cross-validation method can better describe the generalization ability of the approximation model and does not need to add additional validation sample points. However, this method requires multiple iterations to construct the approximation model, increasing the computational cost of constructing the approximation model. It is unfavorable, especially for approximation methods with considerable computational complexity or parameter optimization in approximation modeling, such as Kriging and RBFNN.

3.3.2.3 Learning curve method

To evaluate the stability and convergence of the model, the learning curve of the model on the training set and the validation set can be drawn. This curve can show the change in the model's performance with the increase in training times.

3.3.2.4 Visualization method

If the dimension of the data set is allowed, which means it is low, a graphical method can display the model's prediction and actual results. The model's approximation ability can be evaluated intuitively.

In addition, in MDO, an important application of approximate models is to replace high-precision models for MDO. As a result, the approximate model's capacity for generalization and the gradient agreement between it and the exact model is crucial. This is especially obvious for highly nonlinear multimodal optimization problems. Therefore, Yao [5] proposed an approximate evaluation method based on local linear interpolation to use the training sample information and improve the approximation accuracy by enhancing the gradient coincidence. For any point in the definition domain of the exact model, this method approximates the gradient distribution trend of the exact model in the local small range by linear interpolation of the training sample points around the point. If the response value of the linear interpolation model and the approximate model are not much different, the distribution in this region is consistent with the linear model, and the accuracy is high.

3.4 LOCAL AND GLOBAL EXPLORATION OF APPROXIMATE MODELS

This section will take Sequential Approximate Optimization (SAO) as an example to illustrate the method criteria and differences between local and global approximation. SAO is an optimization technique that combines global and local approximation methods. It can perform high-precision modeling in the global range and consider the details of the local area. The modeling process is shown in Figure 3.10.

According to the different accuracy requirements of the approximate model, the sequential point addition strategy is also different. When considering global approximation accuracy, it is necessary to focus on sampling in low-accuracy areas to improve the overall approximation accuracy. When assessing local approximation accuracy, it is crucial to concentrate on sampling the potentially optimal model region.

3.4.1 Local exploration method

In the approximate model, the local approximation method mainly constructs the model for a specific region of the parameter space to approximate the original complex system. Since the focus is on the local exploration near the optimal point, it is necessary to focus on the sampling in the optimal region. Given the similar objectives of general optimization problems, the optimization objective is the minimization of the function value. However, when the actual distribution of the exact model is unknown, the approximate model constructed based on a limited number of sample points may be quite different from the exact model, as shown in the following Figure 3.11.

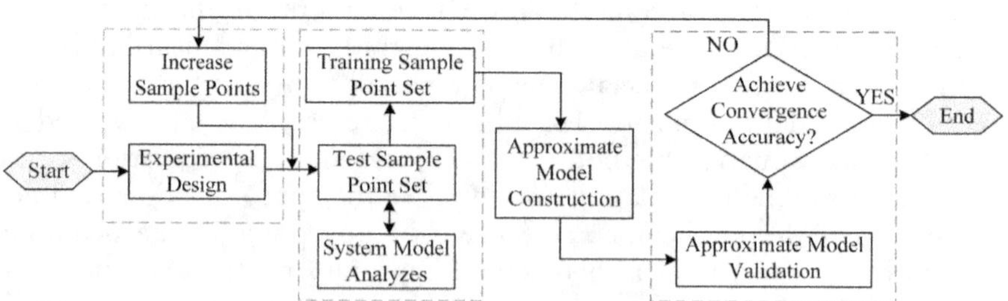

Figure 3.10 SAO flowchart.

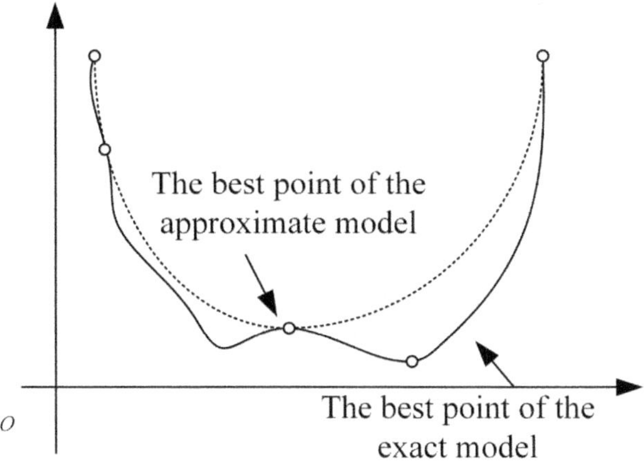

Figure 3.11 The possible difference between the optimal point of the exact model and the approximate model.

This will also cause errors in the search for the best point. Therefore, even if local exploration is to strengthen the optimal region's search point, it is impossible to explore only one region when considering the global optimum. At this time, the point addition strategy needs to add points to the better area of the approximate model, that is, the area where the best point may appear. At the same time, it is necessary to enhance the ability to capture the true global optimal region of the accurate model. There are three main types of sequential point addition strategies for global optimization with enhanced local approximation: Direct search method, indirect search method, and hybrid method [6, 7].

To increase the approximate model's accuracy of approximation in this region, the direct search approach adds the sample point set and searches the optimal point directly within the present optimal region. This can easily lead to the whole model falling into local optimum. The indirect search method mainly searches for areas with less distribution of sample points or areas with low accuracy of approximate models and then adds points. This operation can improve the local approximation accuracy of each region of the approximate model, thereby increasing the probability of obtaining the global optimal point. The hybrid rule combines the two methods to balance the needs of the old donkey to search the potential optimal region directly and improve the accuracy of the global approximation. This can avoid falling into the local optimum in local search, improve the approximation accuracy in the potential optimal region, and avoid the cost of many blind searches. According to the hybrid method principle, various local approximation modeling methods are developed, and different point-addition strategies are introduced.

3.4.1.1 Cramér-Rao lower bound method

The sequential modeling method based on the Kriging model is a commonly used method based on probability statistics, which can directly give the estimated standard deviation of the approximate model. Therefore, the original optimization objective can be replaced by the statistical lower bound $\hat{f}(x) - \tau s(x)$ of the approximate model. The points corresponding to the minimum value of this statistical lower bound are added to the training sample point set to capture the potential optimal region while considering the accuracy characteristics of the model [8]. The coefficient τ is used to control the weight ratio of the direct method and the indirect method. When $\tau = 0$, it is equivalent to a direct search method. When $\tau \to \infty$, it is equivalent to target the area with the largest search error, ignoring the influence of the response value, which is about the same as the indirect search method.

Since the value of the coefficient τ is difficult to determine directly, the potential optimal region can be determined by giving many different values. Then, select the area that needs sequential addition. However, the τ value here is only an estimation of the simulation accuracy of the Kriging model, which may have a large error with the actual situation. If the training samples fail to capture the oscillatory characteristics of the true model, the resultant approximate model constructed using these samples tends to exhibit relatively low oscillations. In this case, the estimated standard deviation will be very small, making it difficult to reflect the actual error characteristics. The RBFNN model's estimated standard deviation model cannot be derived based on the probability method. At this time, its minimum statistical lower bound can be extended, and other error analysis methods can be introduced to estimate the standard deviation model. The resulting value becomes the approximate model's surrogate lower bound.

3.4.1.2 Maximum expected improvement method

The main idea of this method is to find the point that makes the response value of the objective function less than the maximum probability of a predetermined value Y based on the model error of the approximate model. And then add this point to the training sample to update the approximate model. Suppose that the response value $f(\boldsymbol{x})$ of the real model at $\boldsymbol{x}$ is a random number, the expected value is the response value $\hat{f}(\boldsymbol{x})$ of the approximate model, and the standard deviation is $\sigma(\boldsymbol{x})$. Let the minimum response value of the current approximate model be $\hat{f}_{\min}(\boldsymbol{x})$, $Y < \hat{f}_{\min}(\boldsymbol{x})$.

Suppose that the random variable $f(x)$ obeys the normal distribution, then the probability of $f(x) < \hat{f}_{\min}(x)$ is

$$P_I = \Phi\left(\frac{Y - \hat{f}(x)}{\sigma(x)}\right) \tag{3.26}$$

If there are more and more sample points near the current optimal point, the estimated standard deviation $\sigma(x)$ in this region will become smaller and smaller. At the same time, because of $Y < \hat{f}_{\min}(x)$, the probability of increasing sample points near the current optimal point will become smaller and smaller. And turn to the area near the larger $\sigma(x)$ to add points. This method is susceptible to the predetermined value Y. When Y is too small, it will add points too densely in the local area near the current optimal point until the error becomes small. When Y is too large, it will be widely added in the whole region with a lack of pertinence, and the convergence efficiency will be reduced. Here, we mainly face the local approximation requirements, so we should try more Y values in the loop addition point so that it is targeted in the local area and not too dense. This can improve the local approximation accuracy while ensuring the convergence efficiency.

3.4.1.3 Maximum expected improvement function method

This method aims to estimate the potential decrease in the minimum response value of the approximate model when introducing a new sample point. It selects the point that maximizes this expected reduction for addition to the training sample set. This method is mainly for the Kriging model. Define the response value $f(x)$ of the real model at x be a random number, and the expected value and standard deviation correspond to the predicted value is $\hat{f}(x)$ and the standard deviation is $\sigma(x)$ of the Kriging model. Suppose that the minimum response value of the current approximate model is $\hat{f}_{\min}(x)$, then the response value reduction $d(x)$ at the current point x is defined as

$$d(x) = \max\left(\hat{f}_{\min}(x) - f(x), 0\right) \tag{3.27}$$

The probability density function of $f(x)$ is

$$h(x) = \frac{1}{\sqrt{2\pi}\sigma(x)} e^{\frac{-\left[f(x)-\hat{f}(x)\right]^2}{2\sigma^2(x)}} \tag{3.28}$$

Thus, the expected value of $d(x)$ is

$$\begin{aligned}\text{EIF} &= E\left[h(x)\right] = E\left[\max\left(\hat{f}_{\min}(x) - f(x), 0\right)\right] \\ &= \int_{-\infty}^{\hat{f}_{\min}(x)} \left[\hat{f}_{\min}(x) - f(x)\right] \frac{1}{\sqrt{2\pi}\sigma(x)} e^{\frac{-\left[f(x)-\hat{f}(x)\right]^2}{2\sigma^2(x)}} \mathrm{d}f(x) \\ &= \left[\hat{f}_{\min}(x) - f(x)\right] \Phi\left(\frac{Y - \hat{f}(x)}{\sigma(x)}\right) + \sigma(x)\phi\left(\frac{Y - \hat{f}(x)}{\sigma(x)}\right)\end{aligned} \tag{3.29}$$

where Φ represents the cumulative probability distribution function of the standard normal distribution, and ϕ represents the probability density function of the standard normal distribution.

Compared with the previous method, the advantage of this method is that it does not need to determine the d value in advance. However, this method makes it easy to gather points near the current optimal point, and it is easy to converge to the local optimal point in advance. When the global optimal point is determined, this method can better optimize. However, it is necessary to effectively identify the local optimal point in global optimization.

3.4.2 Global exploration method

3.4.2.1 *Global approximation accuracy of the parameter space*

The global approximation approach in the approximation model primarily builds the model for the global approximation accuracy of the parameter space to approximate the original complex linear system.

3.4.2.2 *Maximum prediction error criterion*

This criterion's main goal is to reduce the greatest prediction error value when estimating the model's parameters. The approximation model can provide the expected value of the square of the model's prediction error for the Kriging model and the Bayesian modeling approach based on the Gaussian process. Taking the Kriging function as an example, people define the mean square predicted error as the expected value of the square of the predicted error, as follows:

$$\hat{\Phi}(x) = E\left[\left(\hat{y}(x) - y(x)\right)^2\right] \tag{3.30}$$

This formula can obtain the estimated value of the expected prediction accuracy of the Kriging model at any point in the design space. To a certain extent, it can reflect the size distribution trend of the approximate model prediction accuracy. According to this formula, the distribution trend of the prediction accuracy of the Kriging model in the whole design space can be grasped to a certain extent, which can be used as a priori knowledge for the next cycle to construct the approximate model.

The advantage of this criterion is that it can better highlight the robustness of the model and has better resilience to outliers. Because it focuses on the maximum error, it can better reduce these errors when the model receives the influence of outliers. This method focuses on the maximum error and may ignore the accumulation of small errors. By expanding the sample point set to include the spots with the largest estimation error, the information of the approximate model near this point area can be increased, and the accuracy of the approximate model can be effectively improved.

3.4.2.3 Maximum entropy criterion

This criterion is a statistical reasoning method that is widely used in statistical physics and information theory. In physics, the maximum entropy principle expresses the second law of thermodynamics. Approximate modeling can help researchers infer probability distribution with a given part of information. To quantify the information gathered by an experimental design, Currin et al. incorporated entropy into a computer-simulated experimental design. According to the Bayesian modeling method of the Gaussian random process, it is assumed that the system analysis output based on calculation and simulation is a stationary random process, and then the amount of information obtained by one experiment design is the largest. This is equivalent to the maximum determinant of the prior covariance matrix of the sample point set A in this experiment [9, 10]

$$\begin{gathered}\max\ \det\left[\operatorname{cov}\left(P_D,P_D\right)\right]\\ \operatorname{cov}\left(P_D,P_D\right)=\sigma^2\Gamma\left[R\left(\boldsymbol{x}_i,\boldsymbol{x}_j\right)\right],\boldsymbol{x}_i,\boldsymbol{x}_j\in P_D\end{gathered}\tag{3.31}$$

where σ^2 is the prior variance of the sample point, Γ is the correlation matrix, and R is the prior covariance function that can be selected.

Assuming that the prior covariance function between the sample points is a Gaussian function, it can be seen that

$$R\left(\boldsymbol{x}_i,\boldsymbol{x}_j\right)=e^{-\sum_{k=1}^{n}\theta_k\left|x_i^k-x_j^k\right|^2}\tag{3.32}$$

where n is the dimension of the sample point vector, θ_k is the unknown correlation parameter, and n θ_k constitute the correlation parameter vector $\boldsymbol{\theta}$.

$\boldsymbol{\theta}$ can be obtained by the maximum likelihood estimation method, and the results are obtained by optimizing the following formula

$$\max \quad -\frac{1}{2}\left[N\ln\left(\hat{\sigma}^2\right)+\ln|\Gamma|\right] \tag{3.33}$$

where $\hat{\sigma}^2$ is the estimated value of σ^2, $\hat{\sigma}$ and Γ are functions of $\boldsymbol{\theta}$, and N is the number of sample points.

For the existing sample point set $P_N = \{x_1, x_2, \ldots, x_N\}$, choose to add a sample point $\boldsymbol{x}_{N+1}$ to obtain the maximum amount of information, which is equal to the maximum entropy. It can be obtained by optimizing the following formula

$$\begin{aligned}
&\max \ \det\left[\operatorname{cov}\left(P_{N+1}, P_{N+1}\right)\right]\operatorname{cov}\left(P_{N+1}, P_{N+1}\right) \\
&= \sigma^2\begin{bmatrix} \operatorname{cov}\left(P_{N+1}, P_{N+1}\right) & r\left(x_{N+1}\right) \\ r\left(x_{N+1}\right)^T & 1 \end{bmatrix} \\
&r(x) = \left[R\left(x, x_1\right), \ldots, R\left(x, x_N\right)\right]
\end{aligned} \tag{3.34}$$

3.4.2.4 *The greatest entropy criterion*

The space with fewer sampling points or distance from the existing sample points is more likely to yield sample points when using the greatest entropy criterion. By doing this, we can finish the superb global exploration, get more complete spatial knowledge, and raise the approximate model's global approximation accuracy.

3.4.2.5 *Maximum gradient criterion*

The gradient is the direction of the fastest change of the function value at a point in the function. A new design point position in sequential approximation modeling can be chosen using the maximum gradient criterion. The distribution of many complex system models in the design space is uneven. Some areas are relatively flat, while some areas are more nonlinear. A few sample points are required for the flat area to produce a high-accuracy model. The highly nonlinear region requires more sample points to obtain information about the region in order to achieve reasonable accuracy. Therefore, the new design point can be selected where the gradient of the objective function or constraint function in predicting the search for the optimal point is the largest. This is because the magnitude of the gradient represents the rate of change of the function in all directions. The maximum gradient usually means that the change of the value of the design variable can obtain the maximum change of the function value. Adding a new design point at the position with the largest gradient is expected to achieve a more remarkable improvement in the objective function in this area. The next model incorporates the maximum gradient point from the previous model as a sample point. This enhances the model's accuracy, particularly in highly nonlinear regions.

3.4.2.6 Partial cross-validation error estimation criterion

Partial cross-validation error estimation is a method for predicting the performance evaluation of machine learning models. This method is a variant of cross-validation error estimation and is usually used in scenarios where the cost of cross-validation of all data is too large.

Cross-validation means that for the sample point set $T=\left\{(\boldsymbol{x}_k, y_k): y_k = f(\boldsymbol{x}_k)\right\}_{k=1}^{N_T}$, remove any one of the sample points $\boldsymbol{x}_i, 1 \le i \le k$, and use the remaining sample points to form a set T_{i^-} to establish an approximate model $\hat{f}_{i^-}(\boldsymbol{x})$. The value of sample point $\boldsymbol{x}_i$ is estimated by this model, and the poor verification error of this point is calculated by the following formula:

$$e_{i^-} = \left|f(\boldsymbol{x}_i) - \hat{f}_{i^-}(\boldsymbol{x})\right|, i \in [1, N_T] \tag{3.35}$$

The cross-validation error can reflect the credibility and regularity of the approximate model near each sample point to a certain extent, which means the greater the cross-validation error, the lower the credibility and the greater the probability of irregularity. An approximate function of the global spatial estimation error can be constructed based on the cross-validation error of all sample points. According to this function, the approximate prediction error distribution of the global space can be estimated. The maximum error estimation point is added as a new point to the training sample set, which can increase the number of samples in the untrusted or irregular area and improve the approximation accuracy. This approach eliminates the need for extra validation points to assess the predictive accuracy of the approximate model. It addresses the constraints of conventional approximate modeling, which necessitates verification points for evaluating model accuracy. Thus, the amount of calculation required to obtain the verification point can be omitted, thereby significantly reducing the computational complexity of the approximate modeling.

Li et al. [11] proposed the maximum cumulative error criterion based on the intersection and verification error estimation. With this idea, the whole space error estimation approximation model is constructed as follows:

$$e(x)=\sum_{i=1}^{N^T} e_{i^-} \bullet \mathrm{D}(x,x_i) \tag{3.35}$$

where $\mathrm{D}(x,x_i)$ represents the degree of influence of sample point x_i on point x, which can also be called the correlation between two points. The D function can be expressed in the form of a Gaussian function, as follows:

$$\mathrm{D}(x,x_i)=e^{-\alpha}\left\|x_i-x\right\|^2 \tag{3.36}$$

where $\left\|x_i-x\right\|$ denotes the Euclidean distance between two points, α is a positive constant coefficient, which indicates the changing trend of correlation between two points with distance. The larger the α, the faster the correlation decreases with the increase in distance. The specific example of the relationship in the figure is as follows.

When $\mathrm{D}(x,x_i)$ is less than a certain value β, the influence of the sample point x_i on point x will become very weak or even close to 0. In order to ensure that the estimation error of all points in the design space is greater than 0, a special choice must be made for α. The steps are shown in Figure 3.12.

Step 1. According to the design space and the maximum number of sample points, the average distance d_0 between sample points is estimated.

Step 2. Let the correlation between two points of the distance be the domain value β, then $\alpha=-\frac{\ln\beta}{d_0^2}$.

$$\begin{aligned}\alpha=-\frac{\ln\beta}{d_0^2}\max\ &\sum_{i=1}^{N_T} e_{i^-} \bullet \mathrm{D}(x,x_i)\\ \text{s.t.}\ &\left\|x_i-x\right\|^2>d_c, i\in[1,N_T]\end{aligned} \tag{3.37}$$

With the increase in sample points, if each point is cross-validated, it takes a lot of calculation time and reduces the efficiency. Therefore, from the perspective of improving the efficiency of cross-validation, this method can be improved to cross-validate only some key sample points. This can improve efficiency and better estimate the global approximation error. According to Eq. 3.36, the larger the cross-validation error is, the greater the contribution of the sample points to the full-space error estimation will be.

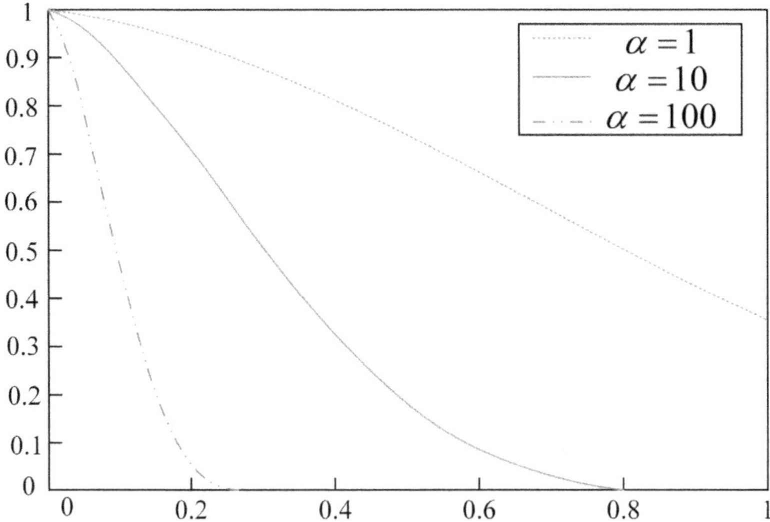

Figure 3.12 The curve of the correlation between two points with the distance.

Simultaneously, a need for additional sample points to gather information becomes more pronounced in regions near these points when the credibility of the approximate model is low. Therefore, these points are the key points to cross-validate the error estimates. The contribution to the misestimation of the whole space is small and negligible for the sample points with small cross-validation errors. Therefore, the error estimation term caused by the sample points can be ignored in Eq. 3.37. Only the key sample points mentioned above are cross-validated to determine the error estimation of the whole space. This is the partial cross-validation method. The key point set selected for partial cross-validation is

$$P_C = \{x_1, x_2, \ldots, x_m\} \subseteq T, 1 \le m \le N_T \tag{3.38}$$

The approximate model of the whole space error estimation is expressed as follows

$$e(\boldsymbol{x}) = \sum_{i=1}^{N^T} e_{i^-} \bullet \mathrm{D}(\boldsymbol{x}, \boldsymbol{x}_i), \boldsymbol{x}_i \in P_C \tag{3.39}$$

The key points of cross-validation can be directly determined by comparing the cross-validation errors of each sample point. According to the key point set P_C, the cross-validation error of each sample point x_i in the point set is e_{i-}, then the average error of all points in the cross-validation key point set is

$$\bar{e} = \frac{1}{m}\sum_{i=1}^{m} e_{i-} \tag{3.40}$$

The ratio of the error of each cross-validation key point to the average error is calculated as follows

$$k_i = \frac{e_{i-}}{\bar{e}} \tag{3.41}$$

If k_i is less than the predetermined value k_c, the sample point x_i can be regarded as a point that weakly impacts the global error estimation. Furthermore, it will be filtered out from the cross-validation key point set. Thus, the key points that need to be cross-validated in the next loop modeling process are determined.

In the next cycle model update, adjustments are made to the approximate model in relation to the previous cycle's approximate model. Some areas may undergo substantial changes. The regions with small changes indicate that the model has higher approximation reliability in these regions, while the regions with large changes have lower approximation reliability. Therefore, it is also necessary to investigate these large change areas and add the sample points in the area and its vicinity to the cross-validation key point set.

Suppose that the approximate model of the jth cycle construction is $\hat{f}_j$, and the approximate model of the $j + 1$th cycle construction is $\hat{f}_{j+1}$. Then, the change of the approximate model constructed by the current cycle relative to the previous cycle approximate model can be calculated as:

$$\Delta\hat{f}_{j+1}(x) = \left|\hat{f}_{j+1}(x) - \hat{f}_j(x)\right| \tag{3.42}$$

The maximum change point $x_{\Delta\max}$ can be obtained by the above optimization. The sample point $x_{\Delta S}$ closest to $x_{\Delta\max}$ is selected from several sample points. If this point does not belong to the cross-validation key point set P_C, this point is added to P_C. When performing the error estimation of cyclic cross-validation each time, the cross-validation key point set determined by the above steps can be used for partial cross-validation. Therefore, the specific implementation steps of partial cross-validation error estimation are as follows.

Step 1. If all sample points in the first cycle are cross-validation points, cross-validation and full-space error estimation are performed. When the

jth cycle is carried out, the model established in this cycle is compared with the model established in the j – 1th cycle. Then, the sample point closest to the maximum change point $\boldsymbol{x}_{\Delta\max}$ of the model is selected and added to the j – 1th cycle to determine the cross-validation key point set. Then, partial cross-validation and full space error estimation are performed from this point set.

Step 2. Obtain the maximum cross-validation error point and add it to the sample point set.

Step 3. Every point in the cross-validation key point set has its cross-validation errors compared, and the points with the largest mistakes are chosen as the key points to create the cross-validation key point set for the following cycle.

Step 4. Go to the next cycle and go back to **Step 1.**

However, it should also be noted that the results of partial cross-validation error estimation may vary due to the different data sets selected. Therefore, the stability and accuracy of the results may be lower than cross-validation using complete data sets. This requires balancing according to the specific situation in practice and selecting the most suitable verification strategy.

3.4.2.7 Maximum sequential cumulative change criterion

Cumulative change of sequential approximate modeling is used to record the cumulative change of the sample points in the whole space with the continuous updating of the model in the sequential approximate modeling. The cumulative change of the point $\boldsymbol{x}$ in the jth cycle modeling is $\Delta_j(\boldsymbol{x})$

$$\Delta_j(\boldsymbol{x}) = \left|\sum_{k=2}^{j}\left[\hat{f}_k(\boldsymbol{x}) - \hat{f}_{k-1}(\boldsymbol{x})\right]\right| = \left|\Delta_{j-1}(\boldsymbol{x}) + \hat{f}_j(\boldsymbol{x}) - \hat{f}_{j-1}(\boldsymbol{x})\right| \tag{3.43}$$

The greater the cumulative change of the region, the lower the credibility of the model in the region or the lower the degree of rules in the region. Therefore, it is proposed to add the cumulative change points of the maximum sequential modeling to the sample point set in each cycle to improve the approximation accuracy of the region. At the same time, because a certain cycle modeling in the cycle history causes a large change in a certain area, the area always has a large cumulative change in the subsequent cycle process, and in fact, the approximate accuracy of the area is already high. To avoid this situation, it is also necessary to comprehensively consider the changes caused by the current cycle model update relative to the previous cycle modeling. Therefore, the cumulative change is adjusted, and the current change information of the approximate model is added based on it, which is recorded as the corrected cumulative change $\Delta_i'(\boldsymbol{x})$, as follows:

$$\Delta_i'(\boldsymbol{x}) = \gamma\Delta_i(\boldsymbol{x}) + (1-\gamma)\Delta\hat{f}_i(\boldsymbol{x}) \tag{3.44}$$

where γ is the weight coefficient, which is used to adjust the proportion of cumulative change and current change in the comprehensive value. This value can be set according to specific modeling tasks or researchers' preferences. By optimizing this formula and setting constraints so that the distance between the optimization point and the existing sample point is not less than the predetermined value d_C, the maximum sequential modeling correction cumulative change point can be obtained. By adding the sample point set to update the model, the approximation accuracy of the model in the irregular region or the low confidence region can be improved to improve the approximation accuracy of the global exploration.

3.4.2.8 Mixed sequential sampling criterion

This hybrid strategy combines a variety of sequential sampling criteria to determine the location of the new design point. This can give full play to the advantages of different criteria to improve the global exploration ability, sampling efficiency, and ability to obtain high-precision model information. Simultaneously, it boosts the applicability and adaptability of the sequential sampling criterion. The partial cross-validation error estimation criterion and the maximum sequential modeling cumulative variation criterion are used to sample the regions with low accuracy and reliability in the approximate model. However, the former only needs to acquire new sample points based on the sample point set information, while the latter requires additional verification points to determine the low-precision region and select the sample points. Moreover, the number of verification points of the latter directly determines the rationality of the selected new points, which increases the difficulty and computational complexity of the determination and construction of the verification point set. The maximum entropy criterion is based on the existing sample points, which are obtained in the area where the sampling points are less and far away from the existing sample points. This has nothing to do with the accuracy of the approximate model itself. It is mainly used to improve the comprehensiveness of the obtained information in the whole space [11]. Therefore, based on the above analysis, the use priority of these three criteria in each cycle modeling can be sorted and selected. The process of determining the mixed sequential sampling criterion is shown in Figure 3.13.

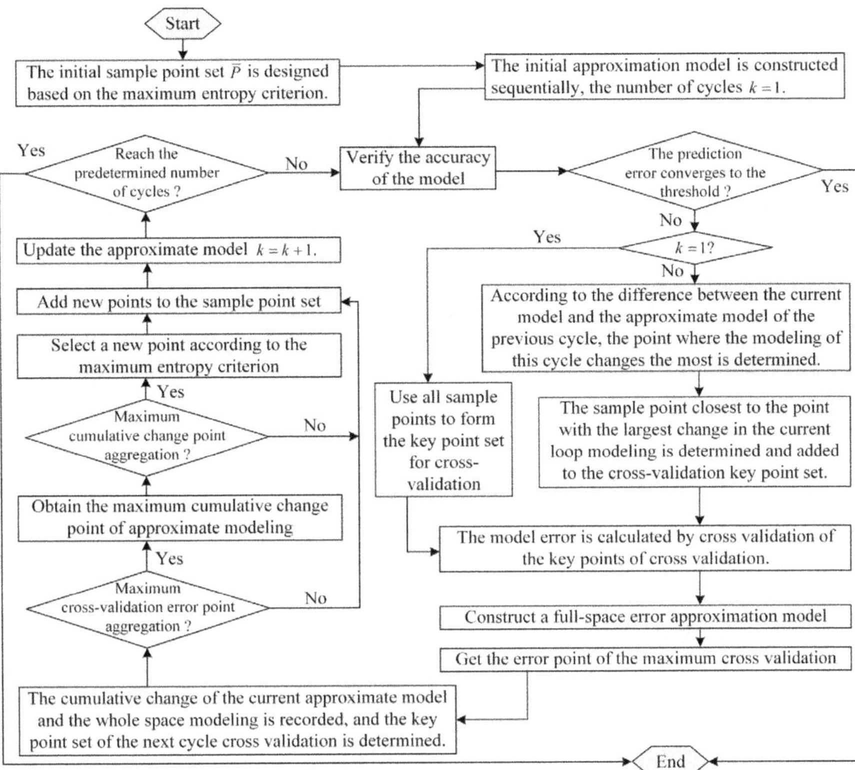

Figure 3.13 Flowchart of mixed sequential sampling criterion.

The basic steps are as follows:

Step 1. Each iteration selects the maximum cross-validation error point according to the partial cross-validation error estimation criterion. If the point does not aggregate with the existing sample points, enter **Step 4**; otherwise, go to **Step 2.**

Step 2. According to the maximum sequential modeling cumulative change criterion, the maximum sequential modeling correction cumulative change point is obtained. If the point does not gather with the existing sample points, enter **Step 4**; otherwise, go to **Step 3.**

Step 3. According to the maximum entropy criterion, to obtain the maximum entropy of the new point, go to **Step 4.**

Step 4. Add new points to the sample point set and update the approximate model.

3.5 MDO METHOD BASED ON APPROXIMATE MODEL

For the optimization system established by the multidisciplinary feasible optimization method, in the optimization process, the analysis of related disciplines should be realized, and the solution of system coupling should be realized iteratively. This will lead to an increase in the time cost of analysis. Using the approximate model to replace the actual multidisciplinary analysis model can avoid the large amount of calculation time consumed by coupling discipline analysis and iteration. This method can effectively shorten the design cycle and improve the design efficiency. The optimization process of the approximate model-based MDO method is shown in Figure 3.14.

Using approximate models instead of complex multidisciplinary coupling simulation programs to deal with MDO problems can significantly reduce system analysis time and effectively carry out multidisciplinary optimization. In numerous engineering applications, the MDO method based on an approximate model has demonstrated a notable enhancement in optimization efficiency when compared to traditional MDO methods, all while maintaining a high level of accuracy. The probability of effectively achieving convergence is also high; the obtained performance indicators can improve the system design. Therefore, the MDO based on the approximate model guarantees accuracy and fast convergence, which is effective and feasible in engineering practice.

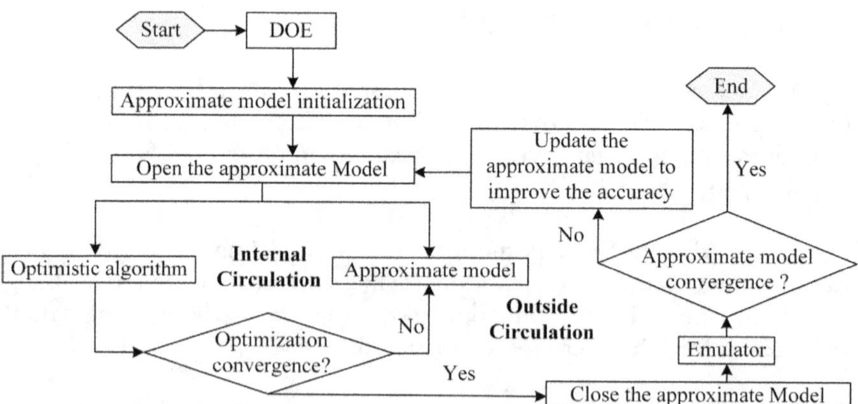

Figure 3.14 Flowchart of multidisciplinary optimization method based on approximate model.

REFERENCES

[1] Wang Z. (2006). Research on the Theory and Application of Aircraft MDO. Beijing, China: National Defense Industry Press.

[2] Chen S., Chng E. S., Alkadhimi K. (1996). Regularized orthogonal least squares algorithm for constructing radial basis function networks. International Journal of Control, 64(5): 829–837.

[3] Orr M. (1998). Optimising the widths of radial basis functions. Proceedings 5th Brazilian Symposium on Neural Networks, 98EX209: 26–29.

[4] Seber G. A. (2009). Multivariate Observations. Hoboken, NJ: John Wiley & Sons.

[5] Yao W. (2007). Uncertainty MDO theory and its application in satellite conceptual design. PhD thesis of National University of Defense Technology.

[6] Jin R., Chen W., Sudjianto A. (2002). On sequential sampling for global metamodeling in engineering design. International Design Engineering Technical Conferences and Computers and Information in Engineering Conference, 36223: 539–548.

[7] Forrester A. I., Keane A. J. (2009). Recent advances in surrogate-based optimization. Progress in Aerospace Sciences, 45(1–3): 50–79.

[8] Cox D. D., John S. (1992). A statistical method for global optimization. Proceedings of 1992 IEEE International Conference on Systems, Man, and Cybernetics, 1241–1246.

[9] Simpson T., Toropov V., Balabanov V., Viana F. (2008). Design and analysis of computer experiments in MDO: A review of how far we have come-or not. 12th AIAA/ISSMO Multidisciplinary Analysis and Optimization Conference, 5802.

[10] Jiang Z., Zhang W., Zhang L. (2007). Sequential maximum entropy method in virtual experiment design. Journal of System Simulation, 19(17): 3876–3879.

[11] Li G., Azarm S. (2006). Maximum Accumulative Error Samplint Strategy for Approximation of Deterministic Engineering Simulations. 11th AIAA/ISSMO Multidisciplinary Analysis and Optimization Conference, 7051.

Chapter 4

Uncertainty analysis method

4.1 AN OVERVIEW OF UNCERTAINTY ANALYSIS

Of all the things that are now recognized, almost everything is full of uncertainty. The consistency of things can be described by deterministic cognition. It can show its basic cognitive characteristics to human beings. However, the differences and uniqueness of things should be understood through uncertainty. In order to grasp the evolution characteristics and development rules of things, in addition to fully understanding their consistency, it is more important to understand their uniqueness. Uncertainty analysis is a technique used to evaluate decision-making or prediction results, considering the possibility and impact of uncertain factors.

In practical engineering applications, the influence of uncertainty factors is also huge. For example, differences in environmental conditions, external loads, metrics, and manual operations can lead to many uncertainties in the production process. According to the sources of these uncertainties, uncertainty is also divided into random and cognitive uncertainty. In product design, to meet the increasingly high and different requirements, the design based on certainty can improve the product's lightweight design. However, the reliability index of the system will show the opposite trend. Therefore, introducing uncertainty analysis into design optimization problems is of great significance. The primary purpose is to make design optimization results more secure and reliable. In addition, with the vigorous development of uncertainty theory, researchers' consideration and analysis of uncertainty factors are becoming more and more comprehensive and reasonable. This can better combine practical problems with theoretical models to achieve the desired results.

In robust optimization design, uncertainty analysis is mainly used to estimate the low-order moment information of the uncertainty distribution of the system output performance. In reliability design, uncertainty analysis is mainly used to estimate the system's reliability to meet the constraints. In some specific cases, obtaining the complete uncertainty distribution

DOI: 10.1201/9781003464792-4

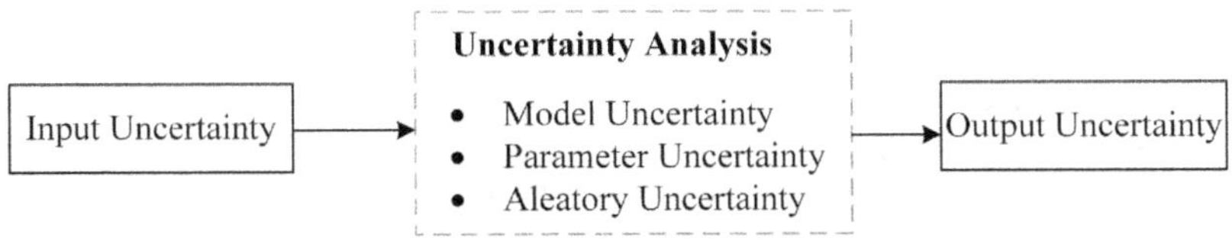

Figure 4.1 Schematic diagram of uncertainty analysis.

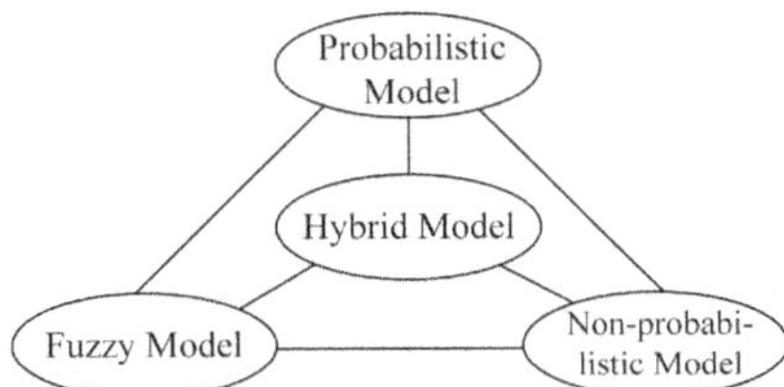

Figure 4.2 The uncertainty triangle.

function of the system performance is also necessary. This involves a focus on uncertainty analysis, that is, uncertainty quantification. The input uncertainty is quantified by uncertainty analysis, and the final output uncertainty is involved in the next analysis, as shown in Figure 4.1.

Although seemingly simple, quantifying uncertainty has always been a key and difficult point in related issues. There are three quantitative models to analyze the uncertainty characteristics of things: The fuzzy, the non-probability, and the probability models. Accordingly, the uncertainty triangle is formed, as shown in Figure 4.2.

Among them, the probability model has a high degree of fitting and can accurately analyze the degree of uncertainty. However, its establishment process is cumbersome and requires a large amount of sample data, which is difficult to carry out in actual engineering operations. When the fuzzy model needs to complete the membership function by virtue of subjective engineering experience, it is difficult to standardize and popularize because of the large individual differences. The non-probability model has a wide range of practical scenarios and is not sensitive to the selection of samples, but the theoretical research is relatively weak. In practical engineering, in most cases, it is a state of coexistence of multiple uncertainties.

This chapter will discuss three main types of uncertainty analysis methods: Stochastic uncertainty analysis, non-probabilistic uncertainty analysis, and mixed uncertainty analysis.

4.2 STOCHASTIC UNCERTAINTY ANALYSIS METHOD

Stochastic uncertainty refers to the inherent variation of the physical system, which is usually caused by the random nature of the input data. When sufficient test data are available, it can be described by a probability distribution.

If the system model is defined as $y = f(\boldsymbol{x})$, where $\boldsymbol{x}$ is a random uncertainty vector, and y is the system response. Let $p(\boldsymbol{x})$ be the joint probability distribution function of the vector $\boldsymbol{x}$ over the domain R. Then the expectation E can be expressed as Eq. 4.1 for any function $\mu(y)$ of y.

$$I = E(\mu(y)) = \int_R \mu(f(\boldsymbol{x}))\, p(\boldsymbol{x}) d\boldsymbol{x} \tag{4.1}$$

When $\mu(y) = y^k$, I is the kth moment of y. When $\mu(y) = y$, I is the expectation of y. When $y \le y_0$, $\mu(y) = 1$; and when $y > y_0$, $\mu(y) = 0$. Then, I is the quantile corresponding to the y_0 quantile of the probability distribution of y.

In practical engineering, the model of the system is often more complex. This will make it difficult to get the expression of $f(\boldsymbol{x})$. At the same time, the integral domain R is also challenging to describe directly. As a result, many new numerical methods for approximate calculation of integrals have been developed, such as the Monte Carlo method, Taylor expansion method, and random expansion method.

4.2.1 Monte Carlo sampling

As a computational method, MCS was first proposed in the mid-1940s by Ulam, S. M. and Von Neumann, J., for the need to develop nuclear weapons. It is a numerical method used to simulate random phenomena. MCS has the characteristics of non-invasive [1, 2]. Therefore, it is widely used in uncertainty-based design, analysis, and optimization. Since MCS is unbiased at any accuracy level, it can be used as a benchmark to evaluate the performance of other methods. Next, the basic idea of the Monte Carlo method is described. In addressing challenges, our primary approach involves constructing a probability model or a stochastic process. Subsequently, we align the model's parameters or numerical characteristics with the solution to the given problem. Then, the digital features are calculated by observing the model or process and sampling tests. Finally, the approximate value of the solution is obtained. The standard error of the estimated value expresses the accuracy of the solution. The basic steps of solving practical problems in the Monte Carlo method are as shown in Figure 4.3.

Step 1. By the characteristics of practical problems a simple and easy-to-implement probability statistical model is constructed, so that the solution is exactly the probability distribution or mathematical expectation of the problem.

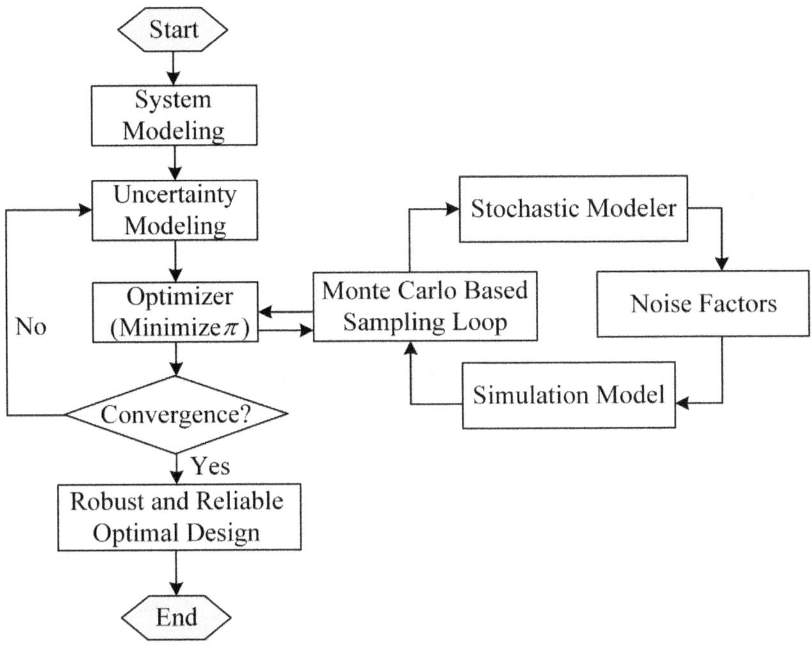

Figure 4.3 General flowchart of uncertainty-based design optimization employing Monte Carlo methods.

Step 2. Give the sampling methods of random variables with different distributions in the model.

Step 3. The simulation results are statistically processed, and the statistical estimation and accuracy estimation of the solution of the problem are given.

In the uncertainty analysis, MCS will first sample in the uncertainty variable value space. And calculate the corresponding system response value of each sample point. Then, the probability distribution characteristics of the system response and other response quantities are analyzed according to the sample information. The specific steps of uncertainty analysis using MCS are as follows. Its general flowchart is shown in Figure 4.3.

Step 1. Using random sampling, Latin hypercube sampling, and other sampling methods, N sample points are randomly generated according to the probability distribution of the uncertainty variable, which is denoted as $\{\boldsymbol{x}_i\}_{1 \le i \le N}$.

Step 2. Calculate the system output response value of each sample point, which is denoted as $\{y_i\}_{1 \le i \le N}$.

Step 3. Calculate the integral value in Eq. 4.1. The result can be expressed as Eq. 4.2.

$$I \approx \tilde{\mu} = \frac{1}{N}\sum_{i=1}^{N}\mu(y_i) \tag{4.2}$$

The variance estimation of $\mu(y)$ can be expressed as Eq. 4.3.

$$\tilde{\sigma}_{\mu}^{2} \approx \frac{1}{N-1}\sum_{i=1}^{N}\left[\mu(y_i)-\tilde{\mu}\right]^2 \tag{4.3}$$

The estimation error of I in Eq. 4.2 can be expressed as Eq. 4.4.

$$e = \frac{\tilde{\sigma}_{\mu}}{\sqrt{N}} \tag{4.4}$$

The Monte Carlo method is based on the universality of things. The larger the sample, the higher the degree and probability that the final result is close to being entirely correct. According to Eq. 4.4, it can also be seen that with the increase of sample size N, the error will also decrease. However, it is precisely because of this that some engineering applications of the MCS are limited. Many system models in modern engineering are relatively complex, such as the system model considering the failure of parts and the finite element analysis of large-scale combinations. Large sample sampling of these problems will lead to a surge in the computational complexity of the MCS to an unbearable range. Therefore, how to grasp the main contradiction in MCS has become a problem worthy of study.

In order to reduce the sample size of MCS to reduce the computational complexity under the premise of ensuring the accuracy required for calculation, the sampling efficiency can be improved by sampling in important areas. Define $h(\boldsymbol{x})$ as the Probability Density Function (PDF) of important region sampling, then Eq. 4.1 can be approximated as Eq. 4.5.

$$I = E(\mu(y)) = \int_R \mu(f(\boldsymbol{x}))\frac{p(\boldsymbol{x})}{h(\boldsymbol{x})}h(\boldsymbol{x})d\boldsymbol{x} \tag{4.5}$$

According to the N sample points generated by $h(\boldsymbol{x})$, the system response values are calculated respectively, and the data pair $\{\boldsymbol{x}_{ai}, y_{ai}\}_{1\le i\le N}$ is obtained. Then Eq. 4.5 can be approximated as Eq. 4.6.

$$I \approx \tilde{\mu}_a = \frac{1}{N}\sum_{i=1}^{N}\mu(y_{ai})\frac{p(\boldsymbol{x}_{ai})}{h(\boldsymbol{x}_{ai})} \tag{4.6}$$

Then the variance and estimation error of $\mu(y)$ can be calculated by Eq. 4.7.

$$\begin{cases} \tilde{\sigma}_{\mu a}^2 \approx \dfrac{1}{N-1}\sum_{i=1}^{N}\left[\mu(y_{ai})\dfrac{p(\boldsymbol{x}_{ai})}{h(\boldsymbol{x}_{ai})} - \tilde{\mu}_a\right]^2 \\ e_a = \dfrac{\tilde{\sigma}_{\mu a}^2}{\sqrt{N}} \end{cases} \tag{4.7}$$

According to Eq. 4.7 analysis, when $h(\boldsymbol{x})$ is appropriate, e_a will be effectively reduced. When N is large enough, e_a even approaches 0. The choice of $h(\boldsymbol{x})$ is described in more detail in reference [3, 4].

4.2.2 Taylor expansion

Taylor's formula is a famous theorem proposed by Brook Taylor, one of the best representatives of Newton's school in the early 18th century. A formula utilizes the data from a function at a certain point to depict an approximate value in its vicinity. When certain conditions are met by the function, the Taylor formula can utilize the derivative values at a certain point to construct a polynomial for approximating the function.

In the uncertainty analysis, the Taylor formula can analyze the low-order moment information of the system response value under the influence of random uncertainty. Define function $y = f(\boldsymbol{x})$, then its Taylor expansion formula at $\boldsymbol{x}_0$ can be expressed as Eq. 4.8.

$$y(\boldsymbol{x}) \approx f(\boldsymbol{x}_0) + \sum_{i=1}^{N}\frac{\partial f(\boldsymbol{x}_0)}{\partial x_i}(x_i - x_{i0}) \tag{4.8}$$

where N is the dimension of the system uncertainty variable vector, x_i and x_{i0} represent the ith component of $\boldsymbol{x}$ and $\boldsymbol{x}_0$, respectively.

The result of the mean and standard deviation of the output $y = f(\boldsymbol{x})$ of the system at point $\boldsymbol{x}_0$ can be calculated by Eq. 4.9.

$$\begin{cases} \mu_y = E(y) \approx f(\boldsymbol{x}_0) + \sum_{i=1}^{N}\dfrac{\partial f(\boldsymbol{x}_0)}{\partial x_i}E(x_i - x_{i0}) \\ \sigma_y = \sqrt{\sum_{i=1}^{N}\left(\dfrac{\partial f(\boldsymbol{x}_0)}{\partial x_i}\right)^2 \sigma_{x_i}^2 + 2\sum_{i=1}^{N}\sum_{j=i+1}^{N}\dfrac{\partial f(\boldsymbol{x}_0)}{\partial x_i}\dfrac{\partial f(\boldsymbol{x}_0)}{\partial x_j}\mathrm{cov}(x_i, x_j)} \end{cases} \tag{4.9}$$

where $\mathrm{cov}(x_i, x_j)$ is the correlation coefficient between x_i and x_j.

If the input variable does not correlate, then its second-order information can be ignored. At this time, the approximation of the mean difference and standard deviation of the system output can be calculated by Eq. 4.10.

$$\begin{cases} \mu_y = f(\boldsymbol{x}_0) \\ \sigma_y = \sqrt{\sum_{i=1}^{N} \left(\dfrac{\partial f(\boldsymbol{x}_0)}{\partial x_i} \right)^2 \sigma_{x_i}^2} \end{cases} \tag{4.10}$$

The practicality of the Taylor formula as a mathematical method is incredibly powerful. This method is simple and intuitive, and it solves the problem directly. In engineering applications, the Taylor formula is still widely used in approximate analysis after considering both complexity and accuracy. However, it also has some limitations. For example, the variance and interval of uncertain variables change too much, or the nonlinearity of the system model is too high. Both of them will seriously affect the accuracy of the analysis. Therefore, the appropriate selection of this method is also necessary.

4.2.3 Random expansion method

The stochastic expansion method is a form of expansion derived from the stochastic process. It can approximate the system response to an expansion of random variables. Based on this, the probability characteristics of the response are analyzed. Polynomial Chaos Expansion (PCE) is a model approximation method using orthogonal polynomials as basic functions. The polynomial chaos method was proposed by Wiener [5] in the study of Gaussian random processes. Ghanem and Spanos [6] applied the spectral expansion of this random variable to the analysis of uncertainty problems in the solid mechanics finite element method. The Generalized Polynomial Chaos (GPC) method generalizes the polynomial chaos method. Xiu and Karniadakis [7] extended the early Hermite polynomial chaos to polynomial chaos suitable for the distribution of general random variables. Next, we will introduce the basic logic of the PCE method.

The model's input parameter is defined as an n-dimensional independent random variable $\boldsymbol{\theta} = (\theta_1, \theta_2, \cdots \theta_n), n \geq 1$. Suppose that each is corresponding to a PDF, and then $\theta_i (i = 1, 2, \cdots, n)$ is $r(\theta_i)$, the joint PDF of a random variable $\boldsymbol{\theta}$ can be expressed as Eq. 4.11.

$$r(\theta) = \prod_{i=1}^{n} r(\theta_i) \tag{4.11}$$

Define $\mathbf{g} = \{g_1, g_2, \cdots, g_N\} (N \geq 1)$ as a response data of the model. N denotes the number of response data, and G denotes an error-free deterministic model, such as a numerical model. Then the relationship between the random variable $\boldsymbol{\theta}$ and the observed value $\mathbf{g}$ can be expressed as Eq. 4.12.

$$G(\boldsymbol{\theta}) = \mathbf{g} \tag{4.12}$$

Taking $N = 1$ as an example, the expansion of GPC is defined. When $N > 1$, each component of G is approximated separately. The pth order approximation of $G(\boldsymbol{\theta})$ is represented by a random polynomial expansion, and the result can be expressed as Eq. 4.13.

$$F_p(\boldsymbol{\theta}) = \sum_{i=1}^{U-1} d_i \boldsymbol{\psi}_i(\boldsymbol{\theta}) \tag{4.13}$$

where d_i is the ith undetermined coefficient, $\boldsymbol{\psi}_i(\boldsymbol{\theta})$ is the ith orthogonal polynomial, and p is the order of the polynomial. In the n-dimensional random variable space, when the highest order of the polynomial $\boldsymbol{\psi}_i(\boldsymbol{\theta})$ is p, the expression of the number of GPC expansion terms can be denoted as Eq. 4.14.

$$U = \frac{(p+n)!}{p!n!} \tag{4.14}$$

The basis function $\boldsymbol{\psi}_i(\boldsymbol{\theta})$ is the vector product of one-dimensional polynomials, which can be expressed as

$$\boldsymbol{\psi}_i(\boldsymbol{\theta}) = \psi_{i_1}(\theta_1) \times \psi_{i_2}(\theta_2), \ldots, \times \psi_{i_n}(\theta_n) \tag{4.15}$$

At the same time, $\boldsymbol{\psi}_i(\boldsymbol{\theta})$ has the following orthogonality

$$\langle \boldsymbol{\psi}_i(\boldsymbol{\theta}) \ \boldsymbol{\psi}_j(\boldsymbol{\theta}) \rangle = \delta_{ij} \langle \boldsymbol{\psi}_i^2(\boldsymbol{\theta}) \rangle \tag{4.16}$$

According to the type of PDF $r(\boldsymbol{\theta})$ of random variables, the GPC method can obtain the optimal approximation accuracy with Askey orthogonal polynomial as the basis function. In the GPC expansion, a variety of orthogonal polynomials can be selected to construct a PCE model according to the actual situation. The optimal orthogonal polynomials corresponding to the common probability distribution types of random variables are shown in Table 4.1.

Table 4.1 The corresponding orthogonal polynomial chaos with different probability distribution

Distribution type of random variable	*Orthogonal polynomial type*	*Interval*
Gaussian distribution	Hermite	$(-\infty,\infty)$
Gamma distribution	Laguerre	$[0,\infty)$
Beta distribution	Jacobi	$[a,b]$
Uniform distribution	Legendre	$[a,b]$
Poisson distribution	Charlie	$\{0,1,2,\ldots\}$
Binomial distribution	Krawtchouk	$\{0,1,\ldots,N\}$
Negative binomial distribution	Meixner	$\{0,1,2,\ldots\}$
Hypergeometric distribution	Hahn	$\{0,1,\ldots,N\}$

The approximation effect of GPC is related to the order p of orthogonal polynomials. Theoretically, the larger the p is, the better the approximation effect is, and the smaller the error between the expansion and the original function is. According to the analysis of classical theory, when $p \to \infty$, if $G(\boldsymbol{\theta})$ is square integrable with respect to $r(\boldsymbol{\theta})$, the approximation error is considered to be convergent, which can be expressed as Eq. 4.17.

$$\left\|G(\boldsymbol{\theta})-F_P(\boldsymbol{\theta})\right\|^2=\int\left(G(\boldsymbol{\theta})-F_P(\boldsymbol{\theta})\right)^2 r(\boldsymbol{\theta})d\boldsymbol{\theta}\to 0 \tag{4.17}$$

The methods for solving PCE coefficients can be divided into two categories [8]: Intrusive and non-intrusive. Galerkin method [6] is a common invasive method. It aims at the mathematical model with random variables and weak solutions. First, the original equation is projected and calculated, and then the equation with coefficients to be solved is constructed. Then, the coefficients are obtained by solving the projection equation to transform the random process into a deterministic equation.

When the deterministic problem is a numerical model, the Galerkin method needs to be invaded into the numerical model program to modify the source program. Therefore, the calculation process of the coefficient solution is relatively complicated. The non-intrusive method only needs to obtain the model output response value, and can directly call the numerical model source program to analyze the random process. This can make the calculation more efficient and more widely used. The commonly used non-invasive methods are the regression and the projection methods.

The regression method is based on the principle of least squares to solve the PCE coefficient. PCE summarizes the values of global random variables as interpolation points of random variables. The combination of interpolation points of different random variables becomes collocation points. Suppose that the value of a random variable becomes the collocation point of a random variable. $\boldsymbol{\theta}^{(k)}(k=1,2,\cdots,Q)$ denotes the kth group of random variables, the corresponding true value is $G\left(\boldsymbol{\theta}^{(k)}\right)$, and the PCE approximate value is $F\left(\boldsymbol{\theta}^{(k)}\right)$. The coefficient to be solved is expressed as vector $d=\left(d_0,d_1,d_2,\ldots,d_U\right)^T$. By minimizing the residual sum between the true value of all collocation points and the approximate value of PCE; the approximate coefficient is obtained. It can be expressed as Eq. 4.18.

$$\begin{aligned} d' &= \arg\min \sum_{k=1}^{Q}\left[G\left(\boldsymbol{\theta}^{(k)}\right)-F\left(\boldsymbol{\theta}^{(k)}\right)\right]^2 \\ &= \arg\min \sum_{k=1}^{Q}\left[G\left(\boldsymbol{\theta}^{(k)}\right)-\sum_{k=1}^{U-1} d_i \boldsymbol{\psi}_i(\boldsymbol{\theta})\right]^2 \end{aligned} \tag{4.18}$$

Define matrix Z as

$$Z=\begin{pmatrix} \boldsymbol{\psi}_0\left(\boldsymbol{\theta}^{(1)}\right) & \boldsymbol{\psi}_1\left(\boldsymbol{\theta}^{(1)}\right) & \cdots & \boldsymbol{\psi}_U\left(\boldsymbol{\theta}^{(1)}\right) \\ \boldsymbol{\psi}_0\left(\boldsymbol{\theta}^{(2)}\right) & \boldsymbol{\psi}_1\left(\boldsymbol{\theta}^{(2)}\right) & \cdots & \boldsymbol{\psi}_U\left(\boldsymbol{\theta}^{(2)}\right) \\ \vdots & \vdots & \ddots & \vdots \\ \boldsymbol{\psi}_0\left(\boldsymbol{\theta}^{(Q)}\right) & \boldsymbol{\psi}_1\left(\boldsymbol{\theta}^{(Q)}\right) & \cdots & \boldsymbol{\psi}_U\left(\boldsymbol{\theta}^{(Q)}\right) \end{pmatrix} \tag{4.19}$$

Then Eq. 4.18 can be calculated by Eq. 4.20 by regression method

$$d'=\left(Z^T Z\right)^{-1} Z^T \mathbf{G} \tag{4.20}$$

where $\mathbf{G}=\left(G\left(\boldsymbol{\theta}^{(1)}\right),G\left(\boldsymbol{\theta}^{(2)}\right),\ldots,G\left(\boldsymbol{\theta}^{(Q)}\right)\right)^T;Q>U$. This requires that the number of collocation points is greater than the number of undetermined coefficients to satisfy $Z^T Z$ reversibility.

To ensure accuracy, Isukapalli et al. [9] proposed the collocation point selection method of two times principle. $Q=2m$ groups of collocation points were randomly selected to balance the influence of each group of collocation points. However, Jiang et al. [10] found that the calculation accuracy of this two times principle is not guaranteed. There is another collocation point selection method. It is to sort the occurrence probability of each collocation point from large to small and then select the collocation points with large occurrence probability to calculate the undetermined coefficient. The projection method solves the PCE coefficient by using the orthogonality of the PCE function. It performs the projection calculation of the original model for each independent basis function and finally obtains the PCE coefficient expression as Eq. 4.21.

$$d_i = \frac{\langle F(\theta)\psi_i(\theta)\rangle}{\langle \psi_i^2(\theta)\rangle} = \frac{1}{\langle \psi_i^2(\theta)\rangle}\int F(\theta)\psi_i(\theta)r(\theta)d\theta \quad (4.21)$$

where $\langle \psi_i^2(\theta)\rangle$ is the inner product of the ith multidimensional orthogonal polynomial. For $i = 0,1,\ldots,U-1$

$$\langle \psi_i^2(\theta)\rangle = \prod_{j=1}^{n}\langle \psi_{j,i}^2(\theta)\rangle \quad (4.22)$$

The inner product of the unit polynomial can be solved analytically, so Eq. 4.22 is easy to solve. According to this result, the amount of calculation of Eq. 4.21 will depend on the difficulty of integral calculation.

The key to the coefficient calculation method mentioned above is to select the appropriate collocation point by some algorithm. Because the number and arrangement of collocation points directly affect the efficiency and accuracy of PCE coefficient calculation. Table 4.2 shows the number of expansion terms corresponding to n-dimensional p-order PCE.

The stochastic collocation method has higher integral calculation efficiency [11]. The commonly used stochastic collocation methods include the product collocation method, stroud-2 collocation method, stroud-3 collocation method, and sparse grid collocation method [12–14]. Shi et al. [15] applied four common collocation methods, and the Smolyak collocation method has the highest calculation accuracy.

Table 4.2 The corresponding expansion terms in n-dimensional PCE of order p

	$n = 1$	$n = 2$	$n = 3$	$n = 4$	$n = 5$	$n = 6$	$n = 7$
$p = 1$	2	3	4	5	6	7	8
$p = 2$	3	6	10	15	21	28	36
$p = 3$	4	10	20	35	56	84	120
$p = 4$	5	15	35	70	126	210	330
$p = 5$	6	21	56	126	252	462	792
$p = 6$	7	28	84	210	462	927	1716

4.3 NON-PROBABILISTIC UNCERTAINTY ANALYSIS METHOD

Uncertainty exists objectively in engineering, but understanding and describing uncertainty are subjective. The stochastic model mentioned in the previous section belongs to the probabilistic analysis method. Still, with the deepening of professional research, many non-probabilistic uncertainty analysis methods are also emerging, such as the interval method, possibility theory, evidence theory, and so on. This section will mainly introduce these methods.

4.3.1 Interval method

Non-probability interval methods are a class of mathematical methods used to deal with uncertainty. They use interval estimation to describe the uncertainty range of variables. These methods are usually used when data is lacking, or the probability distribution of variables cannot be determined.

There are several categories of common non-probabilistic interval methods.

4.3.1.1 Interval estimation

Interval estimation is a method used for estimating the value of parameters or variables. It calculates an interval to describe the uncertainty range of variables. Interval estimation can be based on a confidence interval or prediction interval.

4.3.1.2 Interval analysis

Interval analysis is a method used to analyze the relationship between variables. It describes the uncertainty relationship between variables by calculating the interval of each variable. Interval analysis can be used to evaluate the correlation between variables or predict the value of future variables.

4.3.1.3 Interval optimization

Interval optimization is a method used for maximizing or minimizing the objective function. It describes the uncertainty range by calculating the objective function's maximum or minimum value in each variable's interval. Interval optimization can be used for decision analysis or optimization problems.

4.3.1.4 *Interval fuzzy comprehensive evaluation*

Interval fuzzy comprehensive evaluation is a method for the comprehensive evaluation of multiple variables. It comprehensively evaluates the relationship between variables by calculating the interval fuzzy weight of each variable. Interval fuzzy comprehensive evaluation can be used for decision support systems or risk assessments.

The advantage of the non-probability interval method is that it can deal with uncertainty and ambiguity and provide reliable results. However, their disadvantage is that they usually cannot provide accurate probability estimates because they do not consider the distribution of variables. The emergence of interval variables can be traced back to the third-century BC. Archimedes defined the range of circularity by the area of circumscribed and inscribed regular polygons, and the idea involved was similar to that of interval. Moore [15] introduced this idea into computer science. He limits the irrational number to the interval formed by the two floating-point numbers closest to it. It would ensure that the calculated interval can guarantee the inclusion of accurate results. The floating-point algorithm may cause sequence divergence due to the approximation of irrational numbers. Since the mid-1980s, interval models have been introduced into the engineering field to overcome traditional probability methods' dependence on large sample sizes to describe the fluctuation range of uncertain parameters.

Next, the model of interval analysis in uncertainty analysis is introduced. There is a cognitive uncertainty y in a system, and the system response is set as $z = f(y)$. Suppose $y \in [y^-, y^+]$, then

$$z \in \left[f^-(y), f^+(y)\right] \tag{4.23}$$

where $f^-(y) = \min\limits_{y \in [y^-, y^+]} f(y)$, $f^+(y) = \max\limits_{y \in [y^-, y^+]} f(y)$.

The upper and lower bounds of the uncertainty distribution of the system response can be obtained through this optimization process.

4.3.2 Possibility theory

Possibility theory, a mathematical framework designed for managing uncertainty, shares similarities with probability theory but diverges in its emphasis. While probability theory centers on event frequency and likelihood, possibility theory revolves around assessing the potential and impossibility of events. The possibility theory was originally proposed by the French mathematician Smets and has become an important tool in uncertainty analysis, artificial intelligence, and intelligent control.

Assuming that the system response is $y = f(\boldsymbol{x})$, the input vector $\boldsymbol{x} = \{x_1, x_2, \ldots, x_N\}$ is a fuzzy vector, the membership function of each component x_i in $\boldsymbol{x}$ is μ_{x_i}, and the variables are unrelated. Then, the membership function of the response y can be expressed as Eq. 4.24.

$$\mu_y_{y=y_0} = \sup_{f(x)=y_0} \left\{ \min_{j=1,2,\cdots,n} \left[\mu_{x_i}\left(x_j\right) \right] \right\} \tag{4.24}$$

where sup denotes the upper bound.

According to this formula, to calculate the membership degree of y at the given value y_0, it is necessary to find the value point set of uncertain variables satisfying $f(\boldsymbol{x}) = y_0$ first. Then calculate $\min_{j=1,2,\ldots,n}\left[\mu_{x_i}\left(x_j\right)\right]$ for each point. Finally, the upper bound is obtained by comparing the size of these values to the membership degree at y_0. It is difficult to solve these equations directly, so the method of α^--cut set is often used in engineering applications.

Consider that the α^--cut set of the input fuzzy vector $\boldsymbol{x}$ is

$$X_\alpha(\alpha) = \left\{X_\alpha^1(\alpha), X_\alpha^2(\alpha), \ldots, X_\alpha^N(\alpha)\right\} \tag{4.25}$$

and

$$X_\alpha^j(\alpha) = \left\{x_j \in \Omega_j \middle| \mu_{x_i} \geq \alpha\right\} = \left[x_j^L, x_j^U\right], j = 1,2,\ldots,N \tag{4.26}$$

where Ω_j is the complete set of x_j. Calculate the minimum and maximum values of $y = f(\boldsymbol{x})$ on $X_\alpha(\alpha)$. Then the α^--cut of y is $Y_\alpha(\alpha) = \left[y_{\min}, y_{\max}\right]$. Given a series of values of α, $\mu_y(y)$can be obtained.

$\left[y_{\min}, y_{\max}\right]$ can be determined by sampling method, corner method and optimization method. The sampling method obtains M_i points by discretizing the α^--cut set $X_\alpha^j(\alpha)$. Then, the response values corresponding to $\prod_{i=1}^{N} M_i$ points in N-dimensional space are calculated; $\left[y_{\min}, y_{\max}\right]$ is obtained after comparison.

The sampling method is an algorithm that is relatively easy to implement. However, it requires many calls to the system analysis model to calculate the sample point response value. It is more difficult to apply for complex system analysis models with high computational costs, such as finite element analysis and computational fluid dynamics analysis.

The corner point method first assumes that the extreme value of the response value of the function appears at the corner point of $X_\alpha(\alpha)$. According to this assumption, only the system response values of different corner points need to be compared to confirm. According to this assumption, only the system response values of different corner points must be compared to confirm $\left[y_{\min}, y_{\max}\right]$. Although the corner method seems easy to implement, it can only be used when the function distribution in $X_{\alpha}(\alpha)$ is monotonic.

The optimization method obtains the extreme value $\left[y_{\min}, y_{\max}\right]$ of the function response value by performing two global optimizations. Theoretically, the extreme value obtained by the optimization method is accurate. However, not all optimization methods can achieve the best global optimization effect related to the optimizer. Problems such as local optimization may occur during the optimization process. Nevertheless, the optimization method has higher computational efficiency than the sampling method.

4.3.3 Evidence theory

Evidence theory, also known as Dempster-Shafer evidence theory, was first proposed by Dempster [17]. Later, Shafer [18] further studied and formed an uncertain reasoning theory. It has been widely used in information fusion and uncertain reasoning. This provides a natural and powerful method for expressing and synthesizing uncertain information. Evidence theory can be regarded as a generalization and development based on probability theory. It expresses information such as 'uncertainty' and 'unknown' through trust function and likelihood functions.

When the evidence theory deals with the cognitive uncertainty x, the finite set of possible values is denoted as Ω, the power set is denoted as 2^{Ω}, and the BPA function is denoted as $m: 2^{\Omega} \rightarrow [0,1]$. The uncertainty distribution of system response $f(x)$ can be described by the Cumulative Belief Function (CBF) and Cumulative Plausibility Function (CPF). Define these two functions as

$$CBF(a) = Bel\{f < a\}, CPF = Pl\{f < a\} \tag{4.27}$$

Let the limit state value be a and the failure domain be $D = \{x | f(x) < a\}$. Then the failure credibility $Bel\{f < a\}$ failure likelihood $Pl\{f < a\}$ can be calculated by reliability analysis. CBF and CPF can be constructed by calculating the corresponding credibility and likelihood according to different a.

Based on the uncertainty modeling of evidence theory, $Bel(D)$ is the sum of the BPA of all focal elements entirely located in D in the power set 2^{Ω}, and $Pl(D)$ is the sum of the BPA of all focal elements not empty with D

$$Bel(D) = \sum_{A \in 2^{\Omega}, A \subseteq D} m(A), Pl(D) \sum_{A \in 2^{\Omega}, A \cap D \neq \varnothing} m(A) \tag{4.28}$$

In the process of calculating $Bel(D)$ and $Pl(D)$, it is necessary to determine the distribution range of the response value of the function $f(x)$ in each focal element and compare it with the limit state value a. If the maximum response value is less than a, then the entire focal element is contained in the failure domain. Currently, the calculation of $Bel\ (D)$ and $Pl(D)$ needs to include the focal element. If the value of a is between the maximum response value and the minimum response value, then the focal element part is located in D. Only the calculation of $Pl(D)$ needs to be included. To complete the calculation of $Bel\ (D)$ and $Pl(D)$ once, it is necessary to calculate the minimum and maximum values of the function response for each focal element. When the focal element of the joint distribution of cognitive uncertainty is large, it will sharply increase computing costs. At this time, using the approximate model instead of the exact model for uncertainty analysis can reduce the computational cost. The approximation model includes global and failure domain approximation models. The overall purpose is to reduce the computational cost.

4.4 MIXED UNCERTAINTY ANALYSIS METHOD

In engineering design, it is necessary to consider the random uncertainty in the system and the operating environment and the cognitive uncertainty caused by insufficient cognition. Considering the influence of two kinds of uncertainties, developing a mixed uncertainty analysis method is necessary. Relevant scholars have proposed some meaningful methods in this field. Including the upper and lower limit analysis method of reliability based on probability theory and fuzzy set [19–21], mixed uncertainty analysis method based on FORM for probability analysis and interval analysis based on Unified Uncertainty Analysis (UUA) [22], hybrid method of probability analysis based on PCE and interval analysis based on optimization. This section will mainly introduce two methods. One is A UUA method incorporating Dimension Reduction integral and Effective Global Optimization (DR/EGO-UUA) [23], and the other is A UUA method incorporating FORM and Single-level Optimization (SLO-FORM-UUA).

4.4.1 DR/EGO-UUA

The DR/EGO-UUA method first performs probability analysis on the response function containing random, evidence, fuzzy, and interval variables and applies the dimension reduction integral method to solve the response expectation and variance. Then, the global optimization method is applied to the interval analysis of the first two moments. Finally, the interval algorithm estimates the reliability and plausibility of response expectation and variance.

4.4.1.1 Probabilistic moment analysis based on dimension reduction integral

In accordance with statistical theory, the nth moment of $g(\mathbf{X},\mathbf{Y},\mathbf{Z},\mathbf{P})$ can be expressed as

$$m_{gn} = E\{[g(\mathbf{X},\mathbf{Y},\mathbf{Z},\mathbf{P})]\} = \int_{-\infty}^{\infty}\cdots\int_{-\infty}^{\infty}[g(\mathbf{X},\mathbf{Y},\mathbf{Z},\mathbf{P})]^n f_X(\mathbf{X})d\mathbf{X} \quad (4.29)$$
$$n = 1,2,\ldots$$

where E is the expectation operator and $f_X(\mathbf{X})$ is the joint PDF. However, this formula is a multidimensional integral equation, and it is difficult to obtain an analytical solution. Therefore, when there are only random variables in the uncertainty analysis, the numerical method is often used to approximate the solution.

The Dimensionality Reduction Method (DRM) is one of the most popular and practical methods at present. The main idea of DRM is aimed at approximating high-dimensional integrals by a series of low-dimensional integrals. The univariate dimension reduction integral method is introduced into the UUA method for probability analysis, which can be done as follows

$$m_{gn} = E\left\{\left[\begin{array}{l}\sum_{j=1}^{k_1} g\left(\mu_1,\ldots,\mu_{j-1},X_j,\mu_{j+1},\ldots,\mu_{k_1},\mathbf{Y},\mathbf{Z},\mathbf{P}\right) - \\ (k_1-1)\sum_{j=1}^{k_1} g\left(\mu_1,\ldots,\mu_{k_1},\mathbf{Y},\mathbf{Z},\mathbf{P}\right)\end{array}\right]^n\right\} \quad (4.30)$$

where $g\left(\mu_1,\ldots,\mu_{j-1},X_j,\mu_{j+1},\ldots,\mu_{k_1},\mathbf{Y},\mathbf{Z},\mathbf{P}\right)$ denotes the structural response based on X_j; $\mu_j, j = 1,2,\ldots,k_1$ denotes the mean value of the jth random variable X_j. According to the binomial theorem, Eq. 4.30 can be expanded as

$$\begin{aligned} m_{gn} &= \sum_{i=0}^{n}\binom{n}{i}E\left[\sum_{j=1}^{k_1} g\left(\mu_1,\ldots,\mu_{j-1},X_j,\mu_{j+1},\ldots,\mu_{k_1},\mathbf{Y},\mathbf{Z},\mathbf{P}\right)\right]^i \\ &\quad \left[-(k_1-1)\sum_{j=1}^{k_1} g\left(\mu_1,\ldots,\mu_{k_1},\mathbf{Y},\mathbf{Z},\mathbf{P}\right)\right]^{(n-i)} \\ &= \sum_{i=0}^{n}\binom{n}{i}S_{k_1}^{i}\left[-(k_1-1)\sum_{j=1}^{k_1} g\left(\mu_1,\ldots,\mu_{k_1},\mathbf{Y},\mathbf{Z},\mathbf{P}\right)\right]^{(n-i)} \end{aligned} \quad (4.31)$$

where $\binom{n}{i} = C_n^i = \frac{n!}{i!(n-i)!}$, $S_{k_1}^i$ can be solved according to the recursive formula

$$\begin{aligned}
S_1^i &= E\left[\left(g\left(X_1, \mu_2, \ldots, \mu_{k_1}, \mathbf{Y}, \mathbf{Z}, \mathbf{P}\right)\right)^i\right] \\
S_j^i &= \sum_{i=0}^{n} \binom{n}{i} S_{j-1}^i E\left[\left(g\left(\mu_1, \ldots, \mu_{j-1}, X_j, \mu_{j+1}, \ldots, \mu_{k_1}, \mathbf{Y}, \mathbf{Z}, \mathbf{P}\right)\right)^{(i-k)}\right] \\
S_{k_1}^i &= \sum_{i=0}^{n} \binom{n}{i} S_{k_1-1}^k E\left[\left(g\left(\mu_1, \mu_2, \ldots, \mu_{k_1-1}, X_{k_1}, \mathbf{Y}, \mathbf{Z}, \mathbf{P}\right)\right)^{(i-k)}\right] \\
& i = 1, \ldots, n, j = 2, \ldots, k_1 - 1
\end{aligned} \tag{4.32}$$

The above two formulas only involve one-dimensional integrals.

$$\begin{aligned}
I_{g_i} &= E\left[\left\{g\left(\mu_1, \ldots, \mu_{j-1}, X_j, \mu_{j+1}, \ldots, \mu_{k_1}, \mathbf{Y}, \mathbf{Z}, \mathbf{P}\right)\right\}^i\right] \\
&= \int_{-\infty}^{\infty} \left\{g\left(\mu_1, \ldots, \mu_{j-1}, X_j, \mu_{j+1}, \ldots, \mu_{k_1}, \mathbf{Y}, \mathbf{Z}, \mathbf{P}\right)\right\}^i f_{x_j}\left(X_j\right) dX_j \\
&= \frac{1}{\sqrt{\pi}} \sum_{q=1}^{r} w_q \left\{g\left(\mu_1, \ldots, \mu_{j-1}, X_{jq}, \mu_{j+1}, \ldots, \mu_{k_1}\right)\right\}^i
\end{aligned} \tag{4.33}$$

where $X_{jq} = F_{X_j}^{-1}\left(\Phi\left(\sqrt{2}h_q\right)\right)$, w_q and h_q are the qth Gaussian integral weight and Gaussian integral point, respectively. Substituting Eq. 4.33 and Eq. 4.32 into Eq. 4.31, the statistical moment of the response function $g(\mathbf{X}, \mathbf{Y}, \mathbf{Z}, \mathbf{P})$ can be denoted as an integral function that incorporates the variable $\mathbf{Y}, \mathbf{Z}, \mathbf{P}$. The expectation and variance of the response function can be represented as

$$\begin{aligned}
& E\left(g(\mathbf{X}, \mathbf{Y}, \mathbf{Z}, \mathbf{P})\right) = m_{gn}\big|_{n=1} = S_{k_1}^1 - \left(k_1 - 1\right) g\left(\mu_1, \ldots, \mu_k, \mathbf{Y}, \mathbf{Z}, \mathbf{P}\right) \\
& V\left(g(\mathbf{X}, \mathbf{Y}, \mathbf{Z}, \mathbf{P})\right) = m_{gn}\big|_{n=1} - m_{gn}^2\big|_{n=1} \\
&= \sum_{i=0}^{2} \binom{2}{i} S_{k_1}^i \left[-\left(k_1 - 1\right) g\left(\mu_1, \ldots, \mu_k, \mathbf{Y}, \mathbf{Z}, \mathbf{P}\right)\right]^{2-i} \\
&\quad - \left[S_{k_1}^1 - \left(k_1 - 1\right) g\left(\mu_1, \ldots, \mu_k, \mathbf{Y}, \mathbf{Z}, \mathbf{P}\right)\right]^2 \\
&= S_{k_1}^2 - \left(S_{k_1}^1\right)^2
\end{aligned} \tag{4.34}$$

4.4.1.2 Probabilistic moment analysis based on dimension reduction integral

The evidence analysis of the DR/EGO-UUA method can obtain the first two moments of the response function based on Eq. 4.34.

$$\begin{aligned} E\left(g(\mathbf{X},\mathbf{Y},\mathbf{Z},\mathbf{P})\right) &= \sum_{j=1}^{l}\left[\mathrm{S}_{k_1}^{1}\Big|_{\mathbf{Y}=\mathbf{Y}_{s_j}} - (k_1-1)g\left(\mu_1,\ldots,\mu_{k_1},\mathbf{Y}_{s_j},\mathbf{Z},\mathbf{P}\right)\right] \bullet m_{\mathrm{Y}}(s_j) \\ V\left(g(\mathbf{X},\mathbf{Y},\mathbf{Z},\mathbf{P})\right) &= \sum_{j=1}^{2}\left[\mathrm{S}_{k_1}^{2}\Big|_{\mathbf{Y}=\mathbf{Y}_{s_j}} - \left(\mathrm{S}_{k_1}^{1}\Big|_{\mathbf{Y}=\mathbf{Y}_{s_j}}\right)^2\right] \bullet m_{\mathrm{Y}}(s_j) \end{aligned} \tag{4.35}$$

Then, the fuzzy variables are discretized by α-cut method. Under each membership, the expectation and variance of the response function can be represented as

$$\begin{aligned} E\left(g(\mathbf{X},\mathbf{Y},\mathbf{Z}_\alpha,\mathbf{P})\right) &= \sum_{j=1}^{l}\left[\mathrm{S}_{k_1}^{1}\Big|_{\mathbf{Y}=\mathbf{Y}_{s_j}} - (k_1-1)g\left(\mu_1,\ldots,\mu_{k_1},\mathbf{Y}_{s_j},\mathbf{Z}_\alpha,\mathbf{P}\right)\right] \bullet m_{\mathrm{Y}}(s_j) \\ V\left(g(\mathbf{X},\mathbf{Y},\mathbf{Z}_\alpha,\mathbf{P})\right) &= \sum_{j=1}^{l}\left[\mathrm{S}_{k_1}^{2}\Big|_{\mathbf{Y}=\mathbf{Y}_{s_j},\mathbf{Z}=\mathbf{Z}_\alpha} - \left(\mathrm{S}_{k_1}^{1}\Big|_{\mathbf{Y}=\mathbf{Y}_{s_j},\mathbf{Z}=\mathbf{Z}_\alpha}\right)^2\right] \bullet m_{\mathrm{Y}}(s_j) \end{aligned} \tag{4.36}$$

4.4.1.3 Global optimization analysis

In order to explain the expectation and variance in Eq. 4.36 more conveniently, it can be expressed as

$$\begin{aligned} E\left(g(\mathbf{X},\mathbf{Y},\mathbf{Z}_\alpha,\mathbf{P})\right) &= \sum_{j=1}^{l} H\left(\mathbf{Y}_{s_j},\mathbf{Z}_\alpha,\mathbf{P}\right) \bullet m_{\mathrm{Y}}(s_j) \\ V\left(g(\mathbf{X},\mathbf{Y},\mathbf{Z}_\alpha,\mathbf{P})\right) &= \sum_{j=1}^{l} \tilde{H}\left(\mathbf{Y}_{s_j},\mathbf{Z}_\alpha,\mathbf{P}\right) \bullet m_{\mathrm{Y}}(s_j) \end{aligned} \tag{4.37}$$

where,

$$\begin{aligned} H\left(\mathbf{Y}_{s_j},\mathbf{Z}_\alpha,\mathbf{P}\right) &= \mathrm{S}_{k_1}^{1}\Big|_{\mathbf{Y}=\mathbf{Y}_{s_j},\mathbf{Z}=\mathbf{Z}_\alpha} - (k_1-1)g\left(\mu_1,\ldots,\mu_{k_1},\mathbf{Y}_{s_j},\mathbf{Z}_\alpha,\mathbf{P}\right), j=1,2,\ldots,l \\ \tilde{H}\left(\mathbf{Y}_{s_j},\mathbf{Z}_\alpha,\mathbf{P}\right) &= \mathrm{S}_{k_1}^{2}\Big|_{\mathbf{Y}=\mathbf{Y}_{s_j},\mathbf{Z}=\mathbf{Z}_\alpha} - \left(\mathrm{S}_{k_1}^{1}\Big|_{\mathbf{Y}=\mathbf{Y}_{s_j},\mathbf{Z}=\mathbf{Z}_\alpha}\right)^2, j=1,2,\ldots,l \end{aligned} \tag{4.38}$$

In order to calculate Eq. 4.37, the change interval of $H\left(\mathbf{Y}_{s_j},\mathbf{Z}_\alpha,\mathbf{P}\right)$ and $\tilde{H}\left(\mathbf{Y}_{s_j},\mathbf{Z}_\alpha,\mathbf{P}\right)$ in Eq. 4.38 should be calculated first. Then, the interval analysis algorithm solves the response's expectation and variance. In order to address the issue of local optimization, this method introduced the EGO method proposed by Jones [24]. Through this method, the upper and lower limits can be effectively captured no matter how high the uncertainty of the interval and the nonlinearity of the response function.

$\tilde{H}\left(\mathbf{Y}_{s_j},\mathbf{Z}_\alpha,\mathbf{P}\right)$ can be approximated as a Kriging surrogate model, as Eq. 4.39.

$$\hat{H}(\mathbf{U})=\hat{H}\left(\mathbf{Y}_{s_j},\mathbf{Z}_\alpha,\mathbf{P}\right)=\mathbf{h}(\mathbf{U})\beta+z(\mathbf{U}) \tag{4.39}$$

where U represents all interval variables $\mathbf{Y}_{s_j},\mathbf{Z}_\alpha,\mathbf{P}$, $\mathbf{h}(\mathbf{U})$ is a polynomial with respect to U, it represents the basis function of the regression model, providing a global approximation of the response in the design space. β is a vector composed of trend coefficients. $z(\mathbf{U})$ is a stationary Gaussian random process with a mean of 0, indicating the degree of deviation from the trend. The expectation and variance of the surrogate model $H\left(\mathbf{Y}_{s_j},\mathbf{Z}_\alpha,\mathbf{P}\right)$ can be denoted as

$$\begin{aligned}&\mu_{\hat{H}}(\mathbf{U})=\mathbf{h}^T(\mathbf{U})\beta+\mathbf{r}^T(\mathbf{U})\mathbf{R}^{-1}\left(\mathbf{H}-\mathbf{h}(\mathbf{U})\beta\right)\\&\sigma_{\hat{H}}^2(\mathbf{U})=\sigma_z^2-\sigma_z^2\left[\mathbf{h}^T(\mathbf{U})^T\quad \mathbf{r}(\mathbf{U})^T\right]\begin{bmatrix}0 & \mathbf{F}^T\\ \mathbf{F} & \mathbf{R}\end{bmatrix}^{-1}\begin{bmatrix}\mathbf{h}(\mathbf{U})\\ \mathbf{r}(\mathbf{U})\end{bmatrix}\end{aligned} \tag{4.40}$$

where $\mathbf{r}(\mathbf{U})$ is the covariance function between position U and N training sample points; R is the covariance function between the training sample points, which is a $N\times N$matrix. H is the response output vector corresponding to the training sample point set; F is a $N\times 1$ matrix, and the row vector is $\mathbf{h}^T\left(\mathbf{U}_i\right)$. Parameter σ_z represents the process standard deviation as determined through maximum likelihood estimation.

The Expected Improvement Function (EIF) is defined as

$$\begin{aligned}EI\left(\hat{H}(\mathbf{U})\right)=&\left(\hat{H}_{\min}(\mathbf{U})-\mu_{\hat{H}}(\mathbf{U})\right)\Phi\left[\frac{\left(\hat{H}_{\min}(\mathbf{U})-\mu_{\hat{H}}(\mathbf{U})\right)}{\sigma_{\hat{H}}(\mathbf{U})}\right]\\&+\sigma_{\hat{H}}(\mathbf{U})\phi\left[\frac{\left(\hat{H}_{\min}(\mathbf{U})-\mu_{\hat{H}}(\mathbf{U})\right)}{\sigma_{\hat{H}}(\mathbf{U})}\right]\end{aligned} \tag{4.40}$$

where Φ and ϕ are the PDF and cumulative distribution function of the standard normal, respectively. Eq. 4.39 gives the required conditions. The EIF is used to select a position, so that the point at that position can maximize the value of the expectation-raising function. Thus, this point is added to the sample space as a new sample point to update the model and optimize the extreme value.

It can be concluded that the boundary calculation process of $H\left(\mathbf{Y}_{s_j},\mathbf{Z}_\alpha,\mathbf{P}\right)$ is

Step 1. Collect a small number of $H\left(\mathbf{Y}_{s_j},\mathbf{Z}_\alpha,\mathbf{P}\right)$ by sampling technique.

Step 2. Construct the surrogate model $\hat{H}\left(\mathbf{Y}_{s_j},\mathbf{Z}_\alpha,\mathbf{P}\right)$ based on the sample data.

Step 3. Search for new samples by using the EIF plus point criterion. If the EIF convergence condition is satisfied, stop.

Step 4. Calculate the value of $H\left(\mathbf{Y}_{s_j},\mathbf{Z}_\alpha,\mathbf{P}\right)$ at the new sample point, add the new sample to the original sample space, update the model, and then go back to **Step 3.**

Through these steps, the upper and lower bounds of $H\left(\mathbf{Y}_{s_j},\mathbf{Z}_\alpha,\mathbf{P}\right)$ can be approximately solved. The variation range of $\hat{H}\left(\mathbf{Y}_{s_j},\mathbf{Z}_\alpha,\mathbf{P}\right)$ can also be obtained.

4.4.1.4 Reliability and plausibility of statistical moments

According to Eq. 4.37, the reliability and plausibility of $E\left(g\left(\mathbf{X},\mathbf{Y},\mathbf{Z}_\alpha,\mathbf{P}\right)\right)$ are calculated by interval operation. The results are as follows:

$$\begin{aligned} Bel\left\{E\left(g\left(\mathbf{X},\mathbf{Y},\mathbf{Z}_\alpha,\mathbf{P}\right)\right)\right\} &= \underline{E}\left(g\left(\mathbf{X},\mathbf{Y},\mathbf{Z}_\alpha,\mathbf{P}\right)\right) \\ &= \sum_{j=1}^{l} \underline{H}\left(\mathbf{Y}_{s_j},\mathbf{Z}_\alpha,\mathbf{P}\right) \bullet m_Y\left(s_j\right) \end{aligned} \tag{4.41}$$

$$\begin{aligned} Pl\left\{E\left(g\left(\mathbf{X},\mathbf{Y},\mathbf{Z}_\alpha,\mathbf{P}\right)\right)\right\} &= \bar{E}\left(g\left(\mathbf{X},\mathbf{Y},\mathbf{Z}_\alpha,\mathbf{P}\right)\right) \\ &= \sum_{j=1}^{l} \bar{H}\left(\mathbf{Y}_{s_j},\mathbf{Z}_\alpha,\mathbf{P}\right) \bullet m_Y\left(s_j\right) \end{aligned} \tag{4.42}$$

where $\underline{H}\left(\mathbf{Y}_{s_j},\mathbf{Z}_\alpha,\mathbf{P}\right)$ and $\bar{H}\left(\mathbf{Y}_{s_j},\mathbf{Z}_\alpha,\mathbf{P}\right)$ represent the lower and upper bounds, respectively.

4.4.2 SLO-FORM-UUA

Studies have shown that the method based on optimizing the search for the most probable point is more robust and effective. However, if this method is used to solve the outer maximum possible point, the problem will become a two-layer nested optimization problem, which is more difficult to solve. So, Yao [25] proposed SLO-FORM-UUA. For the optimization problem

$$\begin{cases} find\ \boldsymbol{u} \\ \min\ \|\boldsymbol{u}\| \\ s.t.\ G(\boldsymbol{u},z) = a, z = \underset{z \in c_k}{\arg\min} G(\boldsymbol{u},z) \end{cases} \tag{4.43}$$

Through theoretical analysis, it is known that when $Pl_k(D) \leq 0.5$, the double-level optimization problem can be simplified to a single-level optimization problem as follows:

$$\begin{cases} find\ \boldsymbol{u} \\ \min\ \|\boldsymbol{u}\| \\ s.t.\ G(\boldsymbol{u},z) = a, z \in c_k \end{cases} \tag{4.44}$$

By optimizing $\boldsymbol{u}$ and z at the same time, this form removes the constraints of the inner layer optimization in Eq. 4.43. This can greatly simplify the optimization problem. For the case of $Pl_k(D) > 0.5$, $Pl_k(D)$ can be calculated by Eq. 4.45.

$$\begin{aligned} Pl_k(D) &= \Pr\left\{\boldsymbol{x} \middle| f_{\min}(\boldsymbol{x}, z < \boldsymbol{a}), z \in c_k\right\} \\ &\approx \begin{cases} \Phi\left(\beta_{k\_Pl}\right) = \Phi\left(-\left\|\boldsymbol{u}^*_{k\_Pl}\right\|\right) \\ 1 - \mathrm{Bel}_k\left(\bar{D}\right) \end{cases} \end{aligned} \tag{4.45}$$

Since the optimization problem in Eq. 4.44 is a single-layer equality-constrained optimization problem, it can be solved based on existing mature algorithms, such as the gradient method, Lagrange method, sequential quadratic programming, penalty function method, and so on. Comparing Eq. 4.44 with the maximum possible point search problem in FORM under the condition of a single random variable, the difference is that the cognitive uncertainty variable z is added to the optimization variable. Therefore, the search methods dedicated to the maximum possible point, such as HLRF and its derived methods [26, 27], can be slightly extended to solve this equation.

For the optimization problem

$$\begin{cases} find\ \boldsymbol{u} \\ \min\ \|\boldsymbol{u}\| \\ s.t.\ G(\boldsymbol{u},z) = a, z = \arg\max_{z\in c_k} G(\boldsymbol{u},z) \end{cases} \tag{4.46}$$

The first-order KKT necessary condition of the optimal solution of the inner layer optimization is used as a constraint to act on the outer layer optimization, which is to ensure that the outer layer optimization summary search $\boldsymbol{u}$ and z always satisfies the necessary condition of the optimal solution of the inner layer optimization. Thus, the original double-level optimization problem is equivalently transformed into a single-level optimization problem. This idea has also been applied in RBDO. In order to transform the optimization-reliability analysis nested loop problem into a single-layer optimization problem, Agarwal et al. [28] proposed the first-order KKT necessary condition as the outer optimization constraint condition, where the first-order KKT necessary condition is the optimal solution of the optimization problem of the inner maximum possible point.

When $1 \le i \le N_z$, there is

$$\begin{cases} \dfrac{\partial G}{\partial z_i} = 0, \dfrac{\partial^2 G}{\partial z_i^2} < 0, z_{i\_k}^1 < z_i < z_{i\_k}^u \\ \operatorname{sgn}\left(\dfrac{\partial G}{\partial z_i}\right) = 1, z_i = z_{i\_k}^u \\ \operatorname{sgn}\left(\dfrac{\partial G}{\partial z_i}\right) = -1, z_i = z_{i\_k}^1 \end{cases} \tag{4.47}$$

Replace the inner loop in Eq. 4.47 with Eq. 4.46 as a constraint. Then the optimization problem can be re-expressed as

$$\begin{cases} find\ \boldsymbol{u} \\ \min\ \|\boldsymbol{u}\| \\ s.t.\ G(\boldsymbol{u},z) = a, z \in c_k \\ for\ 1 \le i \le N_z, \begin{cases} \dfrac{\partial G}{\partial z_i} = 0, \dfrac{\partial^2 G}{\partial z_i^2} < 0, z_{i\_k}^1 < z_i < z_{i\_k}^u \\ \operatorname{sgn}\left(\dfrac{\partial G}{\partial z_i}\right) = 1, z_i = z_{i\_k}^u \\ \operatorname{sgn}\left(\dfrac{\partial G}{\partial z_i}\right) = -1, z_i = z_{i\_k}^1 \end{cases} \end{cases} \tag{4.48}$$

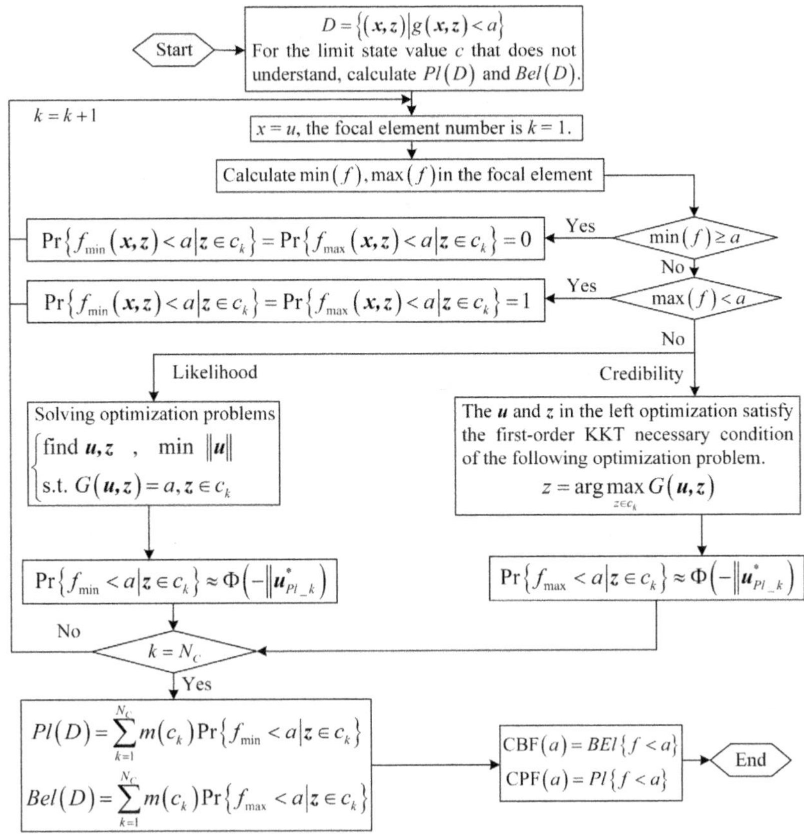

Figure 4.4 The flowchart of SLO-FORM-UUA.

where $z^1_{i\_k}$ and $z^u_{i\_k}$ are the lower and upper limits of the epistemic uncertainty variable z_i corresponding to the focal element c_k.

The specific flowchart of the SLO-FORM-UUA method is shown in Figure 4.4.

The SLO-FORM-UUA method has the following advantages:

I. Mixed uncertainty analysis only involves single-level optimization problems, which can be solved directly by existing optimization algorithms
II. Compared with the double-layer nested optimization, the single-layer optimization is more efficient
III. Single-layer optimization is easier to program than double-layer nested optimization

REFERENCES

[1] Helton J. C., Johnson J. D., Sallaberry C. J., Storlie C.B. (2006). Survey of sampling-based methods for uncertainty and sensitivity analysis. Reliability Engineering & System Safety, 91(10–11): 1175–1209.
[2] Landau D. P., Binder K. (2014). A Guide to Monte Carlo Simulations in Statistical Physics. Cambridge: Cambridge University Press.
[3] George L. A., Alfredo H. S. A., Wilson H. T. (1992). Optimal importance-sampling density estimator. Journal of Engineering Mechanics, 118(6): 1146–1163.
[4] Au S. K., Beck J. L. (2003). Important sampling in high dimensions. Structural Safety, 25(2): 139–163.
[5] Wiener N. (1938). The homogeneous chaos. American Journal of Mathematics, 60(4): 897–936.
[6] Ghanem R. G., Spanos P. D. (1991). Stochastic Finite Elements: A Spectral Approach. New York: Springer.
[7] Xiu D., Karniadakis G. E. (2002). The Wiener-Askey polynomial chaos for stochastic differential equations. SIAM Journal on Scientific Computing, 24(2): 619–644.
[8] Crestaux T., Le Maı^tre O., Martinez, J. M. (2009). Polynomial chaos expansion for sensitivity analysis. Reliability Engineering & System Safety, 94(7): 1161–1172.
[9] Isukapalli S. S., Roy A., Georgopoulos P. G. (1998). Stochastic response surface methods (SRSMs) for uncertainty propagation: Application to environmental and biological systems. Risk Analysis, 18(3): 351–363.
[10] Jiang S., Li D., Zhou C. (2012). Optimal probability collocation number analysis of stochastic response surface method. Journal of Computational Mechanics, 03: 345–351.
[11] Babuška I., Nobile F., Tempone R. (2007). A stochastic collocation method for elliptic partial differential equations with random input data. SIAM Journal on Numerical Analysis, 45(3): 1005–1034.
[12] Xiu D. (2009). Fast numerical methods for stochastic computations: A review. Communications in Computational Physics, 5(2–4): 242–272.
[13] Zhang G., Lu D., Ye M., Gunzburger M., Webster, C. (2013). An adaptive sparse-grid high-order stochastic collocation method for Bayesian inference in groundwater reactive transport modeling. Water Resources Research, 49(10): 6871–6892.
[14] Liao Q., Zhang D., Tchelepi H. (2017). A two-stage adaptive stochastic collocation method on nested sparse grids for multiphase flow in randomly heterogeneous porous media. Journal of Computational Physics, 330: 828–845.
[15] Shi L., Cai S., Yang J. (2010). Collocation method for stochastic analysis of three-dimensional groundwater flow. Journal of Hydraulic Engineering, 01: 47–54.
[16] Moore R. E. (1966). Interval Analysis. Englewood Cliffs, NJ: Prentice-Hall.
[17] Dempster A. P. (1967). Upper and lower probabilities induced by a multivalued mapping. Annals of Mathematical Statistics, 38: 325–339.
[18] Shafer G. (1976). A Mathematical Theory of Evidence. Princeton, NJ: Princeton University Press, 19–63.

[19] Du L., Choi K. K. (2008). An inverse analysis method for design optimization with both statistical and fuzzy uncertainties. Structural and Multidisciplinary Optimization, 37(2): 107–119.
[20] Adduri P. R., Penmetsa R. C. (2009). System reliability analysis for mixed uncertain variables. Structural Safety, 31(5): 375–382.
[21] Zhang X., Huang H., Xu H. (2010). MDO with discrete and continuous variables of various uncertainties. Structural and Multidisciplinary Optimization, 42: 605–618.
[22] Du X. (2008). Unified uncertainty analysis by the first order reliability method. Journal of Mechanical Design, 130(9): 91401.
[23] Mao D. (2019). Research on unified uncertainty analysis method under mixed uncertainty. Master thesis of Hunan University.
[24] Jones D. R., Schonlau M., Welch W. J. (1998). Efficient global optimization of expensive black-box functions. Journal of Global Optimization, 13(4): 455–492.
[25] Yao W., Chen X., Luo W., Van Tooren M., Guo J. (2011). Review of uncertainty-based MDO methods for aerospace vehicles. Progress in Aerospace Sciences, 47(6): 450–479.
[26] Hasofer A. M., Lind N. C. (1974). Exact and invariant second-moment code format. Journal of Engineering Mechanics, 100(1): 111–121.
[27] Radkwistz R., Fiessler B. (1978). Structural reliability under combined random load sequences. Computers and Structures, 9(5): 489–494.
[28] Agarwal H., Renaud J., Lee J., et al. (2004). A unilevel method for reliability based design optimization. Proceedings of the 45th AIAA/ASME/ASCE/AHS Structures, Structural Dynamics, and Materials Conference, Palm Springs.

Chapter 5

Deterministic MDO method

5.1 OVERVIEW OF THE MDO METHOD

The optimization method and calculus appeared in the same era, dating back to the middle of the 19th century. Cauchy first used the steepest descent method to solve the unconstrained minimum problem. Despite the early contributions, substantial progress in these methods was limited until the mid-20th century. During this period, the advent of high-speed digital computers facilitated the development of optimization programs, catalyzing additional research into innovative computational methods. In the 1970s, Haftka et al. had a deep understanding of complex system design, and the idea of Multidisciplinary Design Optimization (MDO) began to take shape. Sobieski formally proposed the concept of MDO in 1982 [1]. As a design methodology, the main idea of MDO is to focus on the early conceptual design stage of the system, gather the knowledge theory methods of each subsystem (or discipline), and consider the needs of each subsystem to achieve the overall optimal method [2].

As an emerging technology, MDO's theoretical system has not yet been fully formed, and some definitions and concepts are not entirely unified [3]. Therefore, before expounding on MDO's basic methods, it is necessary to clarify some basic concepts in MDO technology. The most confusing are 'optimization methods' and 'optimization algorithms'.

For MDO technology, the 'optimization method' and 'optimization algorithm' are two essentially different concepts, representing two crucial research directions of MDO technology [4]. The 'optimization method' focuses on expressing MDO problems, including the decomposition/coordination of problems and the transmission of design information. Its purpose is to establish a reasonable optimization system for specific problems, select appropriate optimization strategies to reduce the computational and communication burden during optimization, and solve the computational complexity and organizational complexity problems faced by MDO technology structurally. The 'optimization algorithm' describes the search of the design space, the local convergence, and the global convergence characteristics of

DOI: 10.1201/9781003464792-5

the iteration [5]. When the MDO method is determined, the optimizer of the method can be combined with the 'optimization algorithm' to solve the MDO problem.

Compared with the traditional single-disciplinary design optimization, the MDO method must consider the coupling effect between the disciplines in the system during the optimization process so that it will produce many complex problems [6]. The foremost challenges revolve around computational complexity and organizational intricacy. In MDO, there are a large number of design variables and coupling variables, and the computational complexity of system analysis and optimization algorithms increases superlinearly with the increase of the scale of the problem, so the computational cost is often more significant than the sum of the optimization computational costs of various disciplines. In addition, for the MDO process of complex systems, the organization and management of information exchange between disciplines are critical, and effective methods and strategies must be adopted to organize system optimization [7]. A series of methods and technologies around the two significant difficulties of MDO constitute the development direction and main content of MDO.

5.2 SINGLE-STAGE OPTIMIZATION METHOD

5.2.1 Multidisciplinary feasible method

The Multidisciplinary Feasible (MDF) method is one of the most basic methods for solving the MDO problem. It is also known as the nested analysis and design method, the fully integrated optimization method, or the all-in-one method, which belongs to the single-stage optimization method [8].

Integrating optimization and multidisciplinary analysis is the most fundamental idea of the MDF method. This method ensures multidisciplinary consistency by performing a complete multidisciplinary analysis, uses the design variable d to obtain the output variable $Y_{i.}$ after multidisciplinary analysis, and then uses d and $Y_{i.}$ to evaluate the objective function f, the inequality constraint function g, and the equality constraint function h. The optimization model of the MDF method is

$$
\begin{aligned}
&\min_{\mathrm{DV}} f\left(\mathbf{d}_{\mathrm{s}},\mathbf{d},\mathbf{Y}\left(\mathbf{d}_{\mathrm{s}},\mathbf{d},\mathbf{Y}\right)\right)\\
&\text{s.t. } g_0\left(\mathbf{d}_{\mathrm{s}},\mathbf{d},\mathbf{Y}\left(\mathbf{d}_{\mathrm{s}},\mathbf{d},\mathbf{Y}\right)\right)\le 0,\\
&\quad h_0\left(\mathbf{d}_{\mathrm{s}},\mathbf{d},\mathbf{Y}\left(\mathbf{d}_{\mathrm{s}},\mathbf{d},\mathbf{Y}\right)\right)=0,\\
&\quad g_i\left(\mathbf{d}_{\mathrm{s}},\mathbf{d}_i,\mathbf{Y}_{\bullet i},\mathbf{Y}_{i\bullet}\left(\mathbf{d}_{\mathrm{s}},\mathbf{d}_i,\mathbf{Y}_{\bullet i}\right)\right)\le 0,\\
&\quad h_i\left(\mathbf{d}_{\mathrm{s}},\mathbf{d}_i,\mathbf{Y}_{\bullet i},\mathbf{Y}_{i\bullet}\left(\mathbf{d}_{\mathrm{s}},\mathbf{d}_i,\mathbf{Y}_{\bullet i}\right)\right)=0,\\
&\quad \mathrm{DV}=\left\{\mathbf{d}_{\mathrm{s}},\mathbf{d}\right\},\ i=1,2,\cdots,n
\end{aligned}
\tag{5.1}
$$

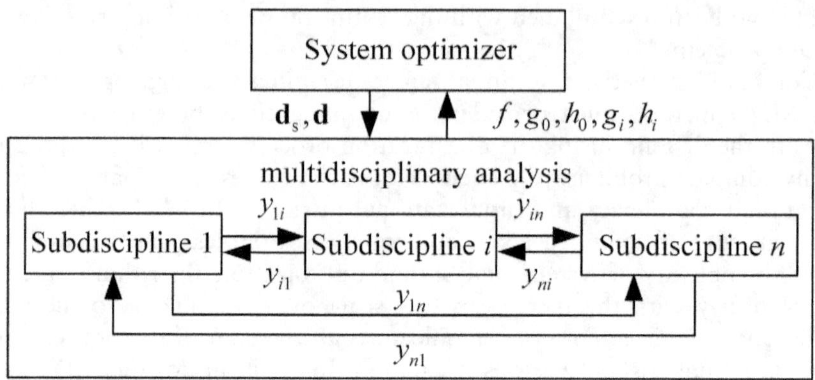

Figure 5.1 The MDF method strategy.

As shown in Figure 5.1, the MDF method integrates the optimizer and nested multidisciplinary analysis. The optimizer outputs a set of design variables $\mathbf{d}_s$ and $\mathbf{d}$ to the multidisciplinary analysis module for each iteration in the optimization process.

The multidisciplinary analysis module outputs the system objective function value, inequality constraint value, and equality constraint value to the optimizer by analyzing each sub-discipline. In general, the multidisciplinary analysis process is carried out by using the fixed-point iteration method. Each sub-discipline is calculated to obtain the value of the coupling variable that satisfies the inter-disciplinary consistency requirements shown in Eq. 5.2.

$$\mathbf{Y} = \mathbf{Y}\left(\mathbf{d}_s, \mathbf{d}, \mathbf{Y}\right) \tag{5.2}$$

A significant advantage of the MDF method is that its optimization model is simple and only contains design variables and system objective functions. Inequality and equality constraints are directly controlled by the optimizer, which is conducive to establishing the optimization model of the engineering system. Another advantage is that in the optimization process of the MDF method, the output design variable values of each iteration meet the inter-disciplinary consistency requirements. In practical engineering, constrained by factors such as time and cost, engineers frequently aim to achieve a product's design scheme that satisfies specific improvement requirements. This pragmatic approach prioritizes practical feasibility over attaining the theoretical optimal value in the mathematical sense. This feature of the MDF method is beneficial in shortening the optimization iteration time and speeding up the design optimization process while obtaining the product design scheme that meets the requirements of performance improvement.

The MDF method's disadvantage is that its computational cost is too high. In each optimization iteration of the MDF method, multidisciplinary analysis is required. The multidisciplinary analysis process requires multiple sub-disciplinary analyses until the coupling variable values that meet the consistency requirements of the coupling disciplines are obtained. The multidisciplinary analysis process significantly increases the optimization time of the MDF method and reduces the computational efficiency.

5.2.2 Individual discipline feasible

The Individual Discipline Feasible (IDF) method is one of the basic methods in the early development stage of MDO, which belongs to the single-stage optimization algorithm [9]. As its name indicates, the IDF method offers a means to evade multidisciplinary analysis in each MDO cycle. While maintaining the feasibility of a single discipline, the IDF method realizes the drive of the optimizer to multidisciplinary feasibility and optimization approximation by controlling the coupling variables.

The IDF method's most fundamental idea is to treat coupling variables as optimization design variables, equivalent to single-disciplinary analysis design variables. As each discipline is analyzed in isolation, variations in coupling variables between disciplines arise. The IDF method eliminates this difference by enhancing the control of consistency constraints so that the coupling variables of each discipline are compatible with each other. The IDF method optimization model is expressed as follows:

$$
\begin{aligned}
&\min_{\mathrm{DV}} f\left(\boldsymbol{d}_s, \boldsymbol{d}, \mathbf{Y}\left(\boldsymbol{d}_s, \boldsymbol{d}, \widehat{\mathbf{Y}}\right)\right) \\
&\text{s.t. } g_0\left(\boldsymbol{d}_s, \boldsymbol{d}, \mathbf{Y}\left(\boldsymbol{d}_s, \boldsymbol{d}, \widehat{\mathbf{Y}}\right)\right) \le 0, \\
&\quad h_0\left(\boldsymbol{d}_s, \boldsymbol{d}, \mathbf{Y}\left(\boldsymbol{d}_s, \boldsymbol{d}, \widehat{\mathbf{Y}}\right)\right) = 0, \\
&\quad g_i\left(\boldsymbol{d}_s, \boldsymbol{d}_i, \widehat{\mathbf{Y}}_{i}, \mathbf{Y}_{i\cdot}\left(\boldsymbol{d}_s, \boldsymbol{d}_i, \widehat{\mathbf{Y}}_{\cdot i}\right)\right) \le 0, \\
&\quad h_i\left(\boldsymbol{d}_s, \boldsymbol{d}_i, \widehat{\mathbf{Y}}_{\cdot i}, \mathbf{Y}_{i\cdot}\left(\boldsymbol{d}_s, \boldsymbol{d}_i, \widehat{\mathbf{Y}}_{i}\right)\right) = 0, \\
&\quad \hat{g}_i = \widehat{\mathbf{Y}}_{i\cdot} - \mathbf{Y}_{i\cdot}\left(\boldsymbol{d}_s, \boldsymbol{d}_i, \widehat{\mathbf{Y}}_{\cdot i}\right) = 0, \\
&\quad \mathrm{DV} = \left\{\boldsymbol{d}_s, \boldsymbol{d}, \widehat{\mathbf{Y}}\right\}, i = 1, 2, \ldots, n
\end{aligned}
\tag{5.3}
$$

where $\hat{\mathbf{Y}}$ is the auxiliary coupling variables, $\hat{g}_i$ is the auxiliary equality constraint of the ith discipline.

As shown in Figure 5.2, the IDF method does not perform multidisciplinary analysis but introduces auxiliary coupling variables and establishes auxiliary equality constraints to meet the consistency requirements of coupling disciplines.

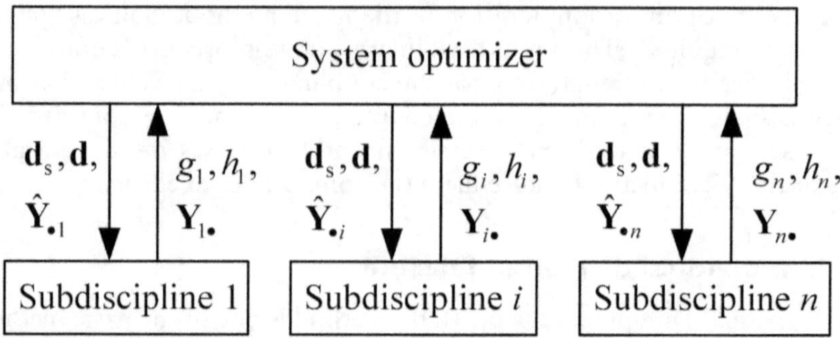

Figure 5.2 The IDF method strategy.

In the optimization process, for each iteration, the optimizer outputs a set of design variables $\mathbf{d}_s$, $\mathbf{d}$, and $\hat{\mathrm{Y}}$ values for each sub-discipline, which performs sub-discipline analysis in parallel and independently. After sub-discipline analysis, each sub-discipline then provides the optimizer with the optimization constraints in the discipline and the output coupling variable values. In each optimization iteration process, the auxiliary coupling variable value provided by the optimizer needs to be equal to the value of the coupling variable obtained after the discipline analysis.

In the IDF method, the necessity for parallel and independent sub-discipline analysis streamlines the process, eliminating the need to analyze and calculate coupling relationships between sub-disciplines. This contributes to an enhanced optimization efficiency within the IDF method. The distributed sub-discipline analysis strategy is conducive to using various professional discipline analysis tools, aligning with modern engineering systems' organizational structure.

Nonetheless, when the system's coupling variables become excessively numerous, the optimization problem increases, thereby impinging on the optimization efficiency. By re-dividing sub-disciplines and reducing information transmission between sub-disciplines, the above problems can be alleviated. In addition, if the optimizer uses a gradient-based optimization method when the sub-discipline model is complex and the analysis cost is high, it will cause the gradient information of the objective function and the constraint function to be difficult to obtain, resulting in a reduction in the optimization efficiency of the IDF method.

5.2.3 Consistency optimization method

The All-At-Once (AAO) method was proposed as early as the mid-1980s. At that time, the application object was a non-hierarchical system. The basic idea was to combine all variables and optimize them by a processor called the main control machine, while multiple controlled machines processed the analysis and calculation of constraints and objectives of various disciplines simultaneously. This strategy ideally solves the influence of coupling factors. Still, it significantly weakens the authority of domain experts to participate, especially when the number of variables is large, and it is difficult to deal with the system layer. Later, several scholars put forward a series of improvement measures to enhance the participation ability of experts in the field.

The AAO method, also known as Simultaneous Analysis and Design, is a single-stage optimization method. It simultaneously optimizes all system variables (design variables, coupling variables, and state variables of various disciplines) in the system. It performs disciplinary calculations at each iteration step until the optimization is completed. Disciplines and systems are feasible. The strength of the AAO method lies in the fact that it does not require the results of the multidisciplinary analysis in each optimization iteration process to be feasible, as long as the multidisciplinary optimization problem is feasible at the optimal point to avoid wasting most of the operation time on repeated the multidisciplinary analysis to determine a feasible solution. Moreover, the AAO method does not require complete multidisciplinary analysis in the optimization process and does not need to ensure consistency between disciplines through simulation analysis iteration. Instead, it ensures the global feasibility of the design solution through the optimizer.

The optimization model of the AAO method is

$$\begin{aligned}
&\min_{\mathrm{DV}} f\left(d_s,d,Y,\hat{Y}\right)=f_0\left(d_s,d,Y\right)+f_i\left(d_s,d_i,\hat{Y}_{\bullet i}\right)\\
&\text{s.t. } g_0\left(d_s,d,Y\right)\le 0,\\
&\quad h_0\left(d_s,d,Y\right)=0,\\
&\quad g_i\left(d_s,d_i,\hat{Y}_{\bullet i}\right)\le 0,\\
&\quad h_i\left(d_s,d_i,\hat{Y}_{\bullet i}\right)=0,\\
&\quad \hat{g}_i=\hat{Y}_{i\bullet}-Y_{i\bullet}=0,\\
&\quad w_i=\hat{Y}_{i\bullet}\left(d_s,d_i,\hat{Y}_{\bullet i}\right)-Y_{i\bullet}\left(d_s,d_i,Y_{\bullet i}\right)=0,\\
&\quad \mathrm{DV}=\left\{d_s,d,\hat{Y},Y\right\},i=1,2,\ldots,n
\end{aligned}\tag{5.4}$$

where w_i is the residual analysis of the i discipline, $f_0(\bullet)$ is a part of the system objective function, which is a function of local variables and coupling variables of two or more disciplines, f_i is a part of the system objective function, which is a function of local variables and auxiliary input coupling variables of the $i$discipline.

As shown in Figure 5.3, the AAO method has neither systematic nor subdisciplinary analysis. In the optimization process, for each iteration, the optimizer outputs a set of values of the design variable $\mathbf{d}_s,\mathbf{d}_i,\hat{\mathbf{Y}}_{\bullet i},\mathbf{Y}_{\bullet i}$ to the

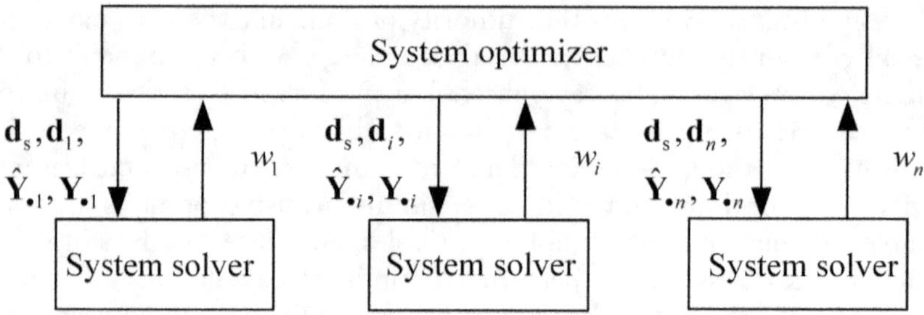

Figure 5.3 The AAO method strategy.

solver, and the solver outputs the residual analysis of each sub-discipline to the optimizer to meet the consistency of the coupling information between each sub-discipline.

The AAO method is suitable for MDO problems with high system analysis or sub-discipline analysis costs. Multiple system or sub-discipline analyses are usually carried out in the optimization process. Upon attaining the ultimate outcome, the optimization process concludes, rendering the analysis process devoid of further impact. The AAO method abandons complex systems or sub-disciplinary analyses and reduces the computational burden through the system solver.

However, the use of coupling variables and auxiliary variables as additional design variables, the scale of the AAO method is increased, which is not conducive to the convergence of the optimization process. Therefore, the application of the AAO method is limited. In addition, the linear relationship in the objective function cannot accurately describe all the object targets in practical engineering.

5.2.4 Multidisciplinary optimization method based on independent subspaces

MDO of Independent Subspaces (MDOIS) is a single-stage method for solving multidisciplinary problems, which requires the decomposition of disciplines and systematic analysis. In multidisciplinary optimization, the system can be disassembled into separate subsystems. Each sub-discipline has a relatively independent optimization space and subject analysis, which can make the complex system simple. MDOIS can decompose the system into this simple operation [10].

The MDOIS method applies to the case where the system does not contain $f_0(\bullet)$, g_0, and h_0, which means the system can be strictly divided into sub-disciplines according to the discipline boundary. Each sub-discipline contains only local variables and input coupling variables without shared variables, and the objective function of each sub-discipline is a function of local variables and input coupling variables.

Certain conditions must be satisfied to break down the intricate MDO problem into more manageable disciplinary calculations. Firstly, each sub-discipline should exclusively involve local design variables. Secondly, every sub-discipline must possess its independent constraint function and objective function. However, modern multidisciplinary optimization problems are complex systems, and many problems cannot meet the above conditions. As a single-stage MDO method, the MDOIS optimization model is

$$
\begin{aligned}
&\min_{\mathrm{DV}} f_i = f_i\left(d_i, \mathbf{Y}_{i\cdot}\left(d_i\right)\right) \\
&\text{s.t. } g_i\left(d_i, \mathbf{Y}_{i\cdot}\left(d_i\right)\right) \le 0, \\
&\quad\;\; h_i\left(d_i, \mathbf{Y}_{i\cdot}\left(d_i\right)\right) = 0, \\
&\quad\;\; \mathrm{DV} = \left\{d_i\right\}, i = 1, 2, \ldots, n
\end{aligned} \tag{5.5}
$$

As shown in Figure 5.4, each sub-discipline of the MDOIS method contains a separate optimizer. The input coupling variable $\mathbf{Y}_{\bullet i}$ of each sub-discipline is obtained by the multidisciplinary analysis of the system and is input from the system analyzer to each sub-discipline as an optimization parameter. The optimizer in the sub-discipline only needs to use the local variables of the sub-discipline as the design variables to optimize their respective objective functions, so the problem is relatively simple.

Like the IDF method, the MDOIS method has a parallel and independent sub-disciplinary optimization process, which is conducive to using analytical tools for various professional disciplines. At the same time, the distributed optimization strategy conforms to the organizational structure of modern engineering systems. In addition, if the global sensitivity of the system is

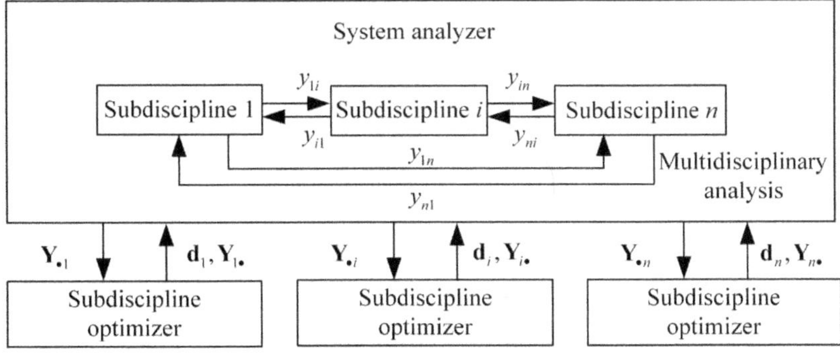

Figure 5.4 The MDOIS method strategy.

easy to obtain, the MDOIS method can further reflect good applicability. As shown in Eq. 5.6, in each optimization iteration, the input coupling variable $\mathbf{Y}_{\bullet i}$ of the sub-discipline can be corrected according to the change of local variables without multidisciplinary analysis, thus reducing the computational burden of system analysis.

$$\mathbf{Y}_{\bullet i}^{(k+1)} = \mathbf{Y}_{\bullet i}^{(k)} + \frac{\partial \mathbf{Y}_{\bullet i}}{\partial \mathbf{d}_i} \Delta \mathbf{d}_i \tag{5.6}$$

where the superscript k is the kth optimization iteration, $\Delta \mathbf{d}_i$ is the change value of the local variable.

Suppose the global sensitivity of the system is not easy to obtain. In that case, the MDOIS method will still need multidisciplinary analysis like the MDF method, which will inevitably increase the computational burden. At the same time, the scope of application of the MDOIS method is limited due to the lack of consideration of shared variables in the optimization problem.

5.3 MULTI-LEVEL OPTIMIZATION METHOD

5.3.1 Concurrent subspace optimization

The Concurrent Subspace Optimization (CSSO) method is a non-hierarchical two-level MDO method. The CSSO method consists of a system-level optimizer and multiple Subsystem Optimizers (SSO) [11]. The CSSO method assigns system design variables to each subspace. Experts in different disciplines use their optimization algorithms to optimize each subspace in parallel. The optimization variables within each discipline remain non-overlapping, allowing for the transmission of global variables within the system through coupling functions. The optimization of each subspace (subsystem) needs to satisfy the constraints of the current subsystem and can also include the constraints of other subsystems.

The basic idea of the CSSO method is to decompose the design variables and adopt the approximate model technology so that multidisciplinary feasibility can be guaranteed at each step of the design optimization process. Its basis is the linear programming method based on sensitivity information. The strategy adopted is to use the linear programming method to find a better feasible solution in a small neighborhood (small enough to be approximated by linearization) based on a feasible solution and repeat this process continuously to find a better solution finally. The CSSO optimization process is equivalent to splitting one step in the overall optimization process into several consecutive small steps. The optimization objectives of the system and the subspace are the same, but the design variables and constraints are different.

Unlike the CO method, the CSSO method does not significantly differ in selecting optimization objective functions for system-level and subsystem-level optimization problems. The optimization problem at the system level can also be regarded as another subject problem, and each subject problem has its optimizer.

In subspace optimization, the Global Sensitivity Equation (GSE) is employed to estimate the objective function and constraint function, and the correlation between the subspace objective function and the system objective function is established.

The constraints of subspace include discipline constraints and discipline consistency constraints. The CSSO method makes the MDA in each cycle feasible, and all design variables are processed simultaneously at the system level. The optimization process is alternately performed between subspace optimization and system-level optimization.

In the CSSO method, each subspace independently optimizes a set of disjoint design variables. In the optimization process of each subspace (subsystem), the objective function of the subspace is analyzed using the analysis method of this discipline. In contrast, other objective functions and constraints are calculated using an approximation based on GSE. Each subspace only optimizes a part of the design variables of the whole system, and each subspace's design variables do not overlap.

The design optimization results of each subspace are combined to form a new design scheme of the CSSO method, which is used as the next initial value of the iterative process.

The system layer optimization model of CSSO is as follows:

$$\begin{aligned}
&\min_{\mathrm{DV}} f = f\left(d_s, d, \tilde{Y}\left(d_s, d, \tilde{Y}\right)\right) \\
&\text{s.t. } g_0\left(d_s, d, \tilde{Y}\left(d_s, d, \tilde{Y}\right)\right) \le 0, \\
&\quad h_0\left(d_s, d, \tilde{Y}\left(d_s, d, \tilde{Y}\right)\right) = 0, \\
&\quad g_i\left(d_s, d_i, \tilde{Y}_{i\bullet}\left(d_s, d_i, \tilde{Y}_{\bullet i}\right)\right) \le 0, \\
&\quad h_i\left(d_s, d_i, \tilde{Y}_{i\bullet}\left(d_s, d_i, \tilde{Y}_{\bullet i}\right)\right) = 0, \\
&\quad \mathrm{DV} = \left\{d_s, d\right\}, i = 1, 2, \ldots, n
\end{aligned} \tag{5.7}$$

where $\tilde{\mathrm{Y}}$ is the approximate coupling variable value obtained by response surface method. The optimization model of the i discipline in its sub-discipline level is as follows:

$$
\begin{aligned}
&\min_{\mathrm{DV}} f = f\left(d_s, d, Y_{i\bullet}\left(d_i, \tilde{Y}_{\bullet i}\right), \tilde{Y}_{\bullet i}\right) \\
&\text{s.t. } g_0\left(d_s, d, \tilde{Y}\left(d_s, d, \tilde{Y}\right)\right) \le 0, \\
&\quad h_0\left(d_s, d, \tilde{Y}\left(d_s, d, \tilde{Y}\right)\right) = 0, \\
&\quad g_i\left(d_s, d_i, Y_{i\bullet}\left(d_s, d_i, \tilde{Y}_{\bullet i}\right)\right) \le 0, \\
&\quad h_i\left(d_s, d_i, Y_{i\bullet}\left(d_s, d_i, \tilde{Y}_{\bullet i}\right)\right) = 0, \\
&\quad g_j\left(d_s, \tilde{Y}_{\bullet i}\left(d_s, \tilde{Y}\right)\right) \le 0, \\
&\quad h_j\left(d_s, \tilde{Y}_{\bullet i}\left(d_s, \tilde{Y}\right)\right) = 0, \\
&\quad \mathrm{DV} = \left\{d_s, d\right\}, i, j = 1, 2, \ldots, n, i \ne j
\end{aligned}
\tag{5.8}
$$

where $g_j(\bullet)$ and $h_j(\bullet)$ are inequality constraints and equality constraints in all sub-disciplines except the i discipline.

As shown in Figure 5.5, in the CSSO method, each sub-discipline's optimization objective function is the system-level optimization objective function, and the sub-discipline's design variable set is a subset of the design variable set of the system-level optimization problem. In the optimization problem of sub-disciplines, the input coupling variables of this discipline exist as design parameters, and their values are obtained by the response surface method. The output coupling variables of this discipline are obtained by the disciplined analysis of the corresponding sub-disciplines.

In the optimization process of the CSSO method, due to the introduction of response surface analysis technology, the solution process is fast. The optimization problems of each sub-discipline can be solved in parallel and independently, reducing the number of system analyses and conforming to modern engineering systems' organizational structure.

The CSSO method's convergence has not been mathematically proved, and the optimization iteration process may have oscillating fluctuations, limiting the method's broad application.

5.3.2 Bi-level integrated system synthesis

Bi-Level Integrated System Synthesis (BLISS) is an MDO method for engineering systems based on decomposition technology proposed by Sohieszczanski-Sobieski, Agte, and Sandusky in 1998. It is considered an MDO method with strong development potential because it deals with multiple variables and allows for easy human intervention in the optimization process.

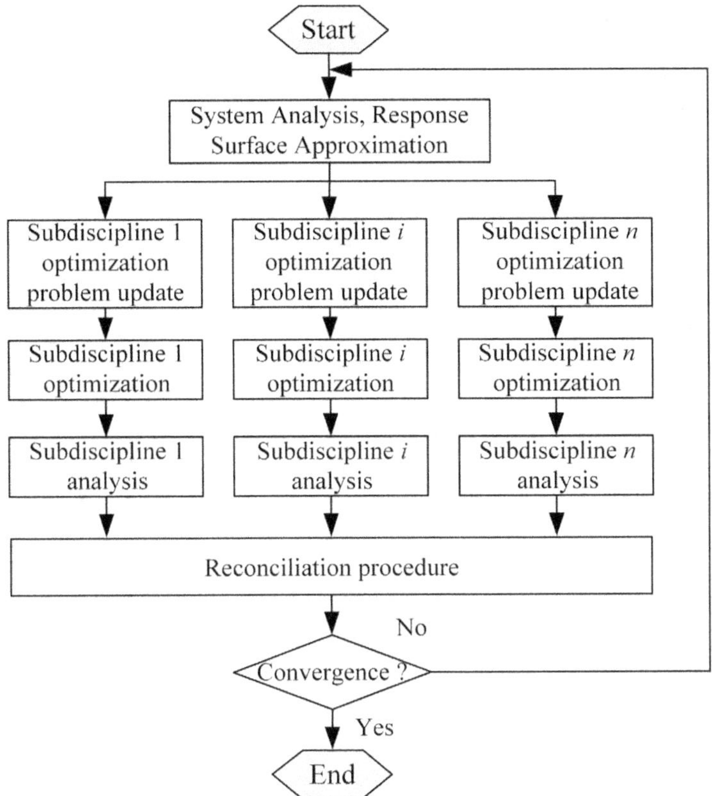

Figure 5.5 The CSSO method strategy.

The BLISS method uses GSE for sensitivity analysis. The optimal autonomy within the subsystem minimizes the system objectives under local constraints, while the coordination problem involves only a relatively small number of common design variables of each module. The derivative of the system objective to the global design variable controls the solution of the coordination problem. These derivatives can be calculated in two ways, resulting in two versions of BLISS.

The BLISS method divides the design variables of multidisciplinary problems into two layers: system-level design variables and modular design variables. The two variables exist in the system and discipline-level optimization processes. The system-level optimization process optimizes a small number of global design variables, and parallel discipline optimization optimizes the local design variables of the discipline.

In the BLISS optimization process, the optimal sensitivity analysis data links the discipline optimization results with the system optimization. Much like the CSSO method, the optimization process commences with a thorough system analysis to guarantee multidisciplinary feasibility. Employing gradient guidance enhances system design, and an iterative optimization cycle ensues between the discipline design space and the system design space.

The basic strategy of the BLISS method is to separate the system-level optimization from the potential optimization of many subsystems so that the system-level design variables are significantly reduced. At the beginning of the BLISS operation, it is necessary to assign initial values to the system design variables and then improve the design variables through loops to achieve optimization. Each cycle consists of two steps: The first step involves freezing the system layer variables and independently optimizing the local design variables within the subsystem layer, conducted in parallel and autonomously. Subsequently, in the second step, building upon the first, the optimization focuses on the system layer variables to achieve further refinement.

The BLISS method belongs to the MDO multi-level method. It divides the MDO problem into a system-level optimization problem and a sub-discipline-level optimization problem. The system-layer optimization model is as follows:

$$
\begin{aligned}
&\min_{\mathrm{DV}} f = f\left(\boldsymbol{d}_s, \boldsymbol{d}\right) + \left(\frac{\partial f}{\partial \boldsymbol{d}_s}\right)\Delta \boldsymbol{d}_s \\
&\text{s.t.}\quad g_0\left(\boldsymbol{d}_s, \boldsymbol{d}\right) + \left(\frac{\partial g_0}{\partial \boldsymbol{d}_s}\right)\Delta \boldsymbol{d}_s \le 0, \\
&\qquad h_0\left(\boldsymbol{d}_s, \boldsymbol{d}\right) + \left(\frac{\partial h_0}{\partial \boldsymbol{d}_s}\right)\Delta \boldsymbol{d}_s = 0, \\
&\qquad g_i\left(\boldsymbol{d}_s, \boldsymbol{d}_i\right) + \left(\frac{\partial g_i}{\partial \boldsymbol{d}_s}\right)\Delta \boldsymbol{d}_s \le 0, \\
&\qquad g_i\left(\boldsymbol{d}_s, \boldsymbol{d}_i\right) + \left(\frac{\partial g_i}{\partial \boldsymbol{d}_s}\right)\Delta \boldsymbol{d}_s \le 0, \\
&\qquad h_i\left(\boldsymbol{d}_s, \boldsymbol{d}_i\right) + \left(\frac{\partial h_i}{\partial \boldsymbol{d}_s}\right)\Delta \boldsymbol{d}_s = 0, \\
&\qquad \Delta \mathrm{d}_s^L \le \Delta \boldsymbol{d}_s \le \Delta \boldsymbol{d}_s^U, \\
&\mathrm{DV} = \left\{\Delta \boldsymbol{d}_s\right\}, i = 1, 2, \ldots, n
\end{aligned}
\tag{5.9}
$$

where $\Delta \mathbf{d}_s$ is the change value of the shared variable, $\Delta \mathbf{d}_s^L$ and $\Delta \mathbf{d}_s^U$ represent the lower and upper bounds of the change value of the shared variable, respectively. The optimization model of the ith discipline in its sub-disciplinary layer is

$$\begin{aligned}
&\min_{\mathrm{DV}} f = f(\boldsymbol{d}_s, \boldsymbol{d}) + \left(\frac{\partial f}{\partial \boldsymbol{d}_i}\right)\Delta \boldsymbol{d}_i \\
&\text{s.t.} \quad g_0(\boldsymbol{d}_s, \boldsymbol{d}) + \left(\frac{\partial g_0}{\partial \boldsymbol{d}_s}\right)\Delta \boldsymbol{d}_i \le 0, \\
&\qquad h_0(\boldsymbol{d}_s, \boldsymbol{d}) + \left(\frac{\partial h_0}{\partial \boldsymbol{d}_i}\right)\Delta \boldsymbol{d}_i = 0, \\
&\qquad g_i(\boldsymbol{d}_s, \boldsymbol{d}_i) + \left(\frac{\partial g_i}{\partial \boldsymbol{d}_i}\right)\Delta \boldsymbol{d}_i \le 0, \\
&\qquad h_i(\boldsymbol{d}_s, \boldsymbol{d}_i) + \left(\frac{\partial h_i}{\partial \boldsymbol{d}_i}\right)\Delta \boldsymbol{d}_i = 0, \\
&\qquad \Delta \mathbf{d}_i^L \le \Delta \boldsymbol{d}_i \le \Delta \boldsymbol{d}_i^U, \\
&\qquad \mathrm{DV} = \{\Delta \boldsymbol{d}_i\}, i = 1, 2, \ldots, n
\end{aligned} \tag{5.10}$$

where $\Delta \mathbf{d}_i$ is the change value of the local variable of the sub-discipline, $\Delta \mathbf{d}_i^L$ and $\Delta \mathbf{d}_i^U$ represent the lower and upper bounds of the change value of the local variable, respectively.

As shown in Figure 5.6, the BLISS method uses the gradient guidance method to solve the optimization problems in the system and sub-discipline layers. Finally, the system objective function is solved. At the beginning of the optimization, the system analysis is used to obtain a design scheme that meets the consistency requirements between sub-disciplines. In the system-level optimization problem, the global sensitivity equation or response surface analysis is used to obtain the optimal value of the system objective function by changing the shared variables within the allowable range. The optimization problems of each sub-discipline in the sub-discipline layer are solved independently and in parallel. The coupling relationship between disciplines uses coupling variables to process the derivatives of design variables.

The BLISS method does not directly consider the consistency requirement of the coupling variables. Still, it uses the system objective function and the constraint function as the partial derivative information of the design variables, respectively. In the optimization process, the optimal value of the system objective function is obtained by solving the change value of the design variables. The consistency of coupling information between sub-disciplines can be guaranteed by systematic analysis at the beginning of

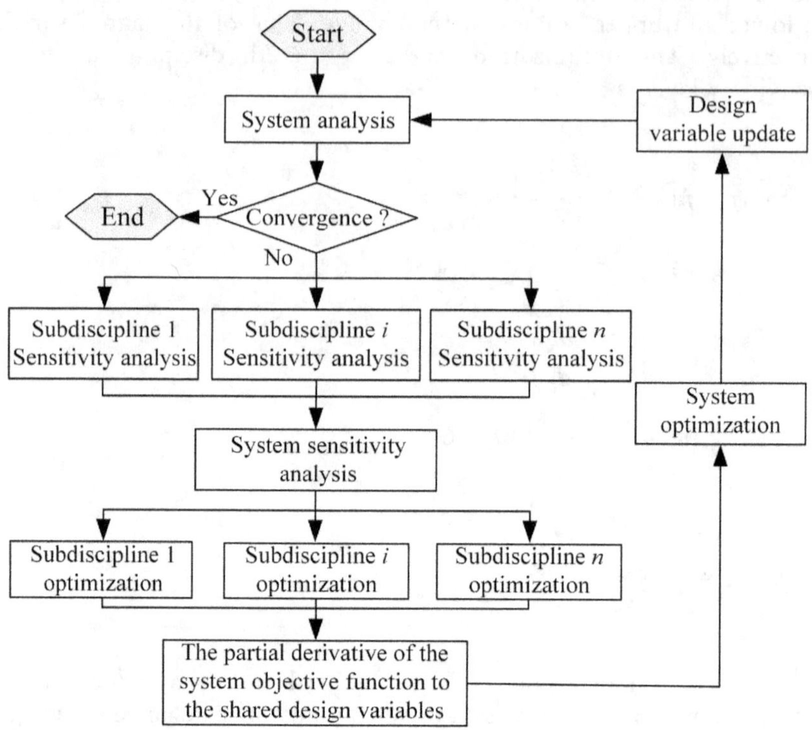

Figure 5.6 The BLISS optimization strategy.

the optimization process. Therefore, the optimization model of the BLISS method is more straightforward than other methods.

However, the calculation cost of the BLISS method is increased by the sensitivity analysis process of the sub-discipline and the partial derivative calculation process of the system objective function to the design variables. This problem can be solved by response surface analysis and establishing a surrogate model.

5.3.3 Collaborative optimization method

The Collaborative Optimization (CO) method is a two-level optimization strategy proposed by Braun and Kroo based on the AAO method. In the CO method, the analysis and design optimization should be carried out among the subsystems. Braun proposed two forms of CO methods in his doctoral dissertation, and CO_2 is widely studied and used.

The basic idea of the CO method is to construct a system layer to coordinate the inconsistency of the results of each sub-task. When each sub-task (discipline) is optimized, the influence of other disciplines can be temporarily ignored, and only the constraints of this discipline need to be satisfied. The goal of discipline-level optimization is to minimize the difference between the optimization results of the discipline and the target values provided to the discipline by system-level optimization. The inconsistency of the optimization results at each discipline level is coordinated by system-level optimization. Through multiple iterations between system-level optimization and discipline-level optimization, a system optimal design scheme that meets the requirements of inter-disciplinary consistency is finally converged.

The top layer of CO is the system-level optimizer, which optimizes the multidisciplinary variables (system-level design variable Z) to meet the consistency J^* of interdisciplinary constraints and minimize the system objective F. Each subsystem optimizer in the subspace design variable X_i; the calculation result Y_j of subset and subspace analysis; the minimum mean square error is used as the optimization objective of the subsystem. While satisfying the subspace constraint g_j, the system-level design variable Z is obtained. The system-level design variable Z is treated as a fixed value in the subspace optimization process. In practical applications, the interdisciplinary consistency constraint J_j; usually, inequality is used. J_i is defined as follows:

$$J_i = \left|X_j - Z_j^s\right|^2 + \left|Y_j - Z_j^c\right|^2 \tag{5.11}$$

where Z_j^s is the system design variable; Z_j^c is the system coupling variable.

The system layer optimization model of CO method is

$$\begin{aligned}
&\min_{\mathrm{DV}} f = f\left(\hat{\boldsymbol{d}}_s, \hat{\boldsymbol{d}}, \hat{\boldsymbol{Y}}\right) \\
&\text{s.t.} \quad g_0\left(\hat{\boldsymbol{d}}_s, \hat{\boldsymbol{d}}, \hat{\boldsymbol{Y}}\right) \le 0, \\
&\qquad\; h_0\left(\hat{\boldsymbol{d}}_s, \hat{\boldsymbol{d}}, \hat{\boldsymbol{Y}}\right) = 0, \\
&J_i = \left\|\boldsymbol{d}_{s,i} - \hat{\boldsymbol{d}}_s\right\|_2^2 + \left\|\boldsymbol{d}_i - \hat{\boldsymbol{d}}_i\right\|_2^2 + \left\|\boldsymbol{Y}_{\bullet i} - \hat{\boldsymbol{Y}}_{\bullet i}\right\|_2^2 \\
&\qquad + \left\|\boldsymbol{Y}_{i\bullet}\left(\boldsymbol{d}_{s,i}, \boldsymbol{d}_i, \boldsymbol{Y}_{\bullet i}\right) - \hat{\boldsymbol{Y}}_{i\bullet}\right\|_2^2 = 0, \\
&\mathrm{DV} = \left\{\hat{\boldsymbol{d}}_s, \hat{\boldsymbol{d}}, \hat{\boldsymbol{Y}}\right\}, i = 1, 2, \ldots, n
\end{aligned} \tag{5.12}$$

where $\hat{\mathbf{d}}_s, \hat{\mathbf{d}}, \hat{\mathbf{Y}}$ are the auxiliary design variable, J_i is the compatibility constraint of the ith discipline, $\mathbf{d}_{s,i}$ is the shared variable of the ith discipline. The optimization model of the ith discipline in its sub-discipline layer is

$$\begin{aligned}
&\min_{\mathrm{DV}} J_i = J_i\left(\boldsymbol{d}_{s,i}, \boldsymbol{d}_i, \boldsymbol{Y}_{\bullet i}\right) \\
&\text{s.t.} \quad g_i\left(\boldsymbol{d}_{s,i}, \boldsymbol{d}_i, \boldsymbol{Y}_{\bullet i}\right) \le 0, \\
&\qquad\; h_i\left(\boldsymbol{d}_{s,i}, \boldsymbol{d}_i, \boldsymbol{Y}_{\bullet i}\right) = 0, \\
&\qquad \mathrm{DV} = \left\{\boldsymbol{d}_{s,i}, \boldsymbol{d}_i, \boldsymbol{Y}_{\bullet i}\right\}, i = 1, 2, \ldots, n
\end{aligned} \tag{5.13}$$

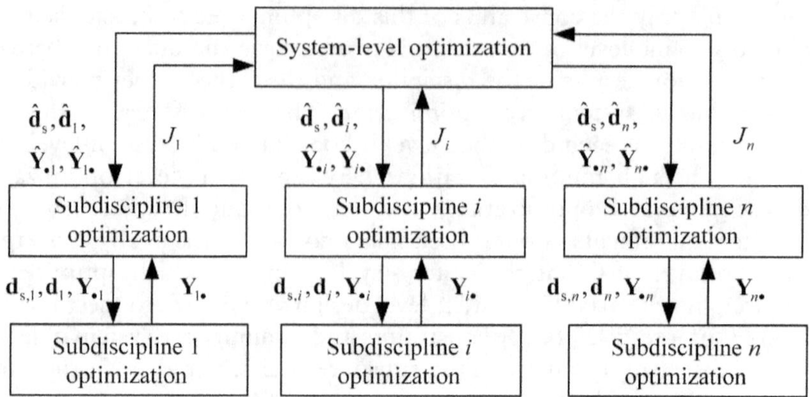

Figure 5.7 The CO method strategy.

As shown in Figure 5.7, the CO method does not perform systematic analysis, and the sub-discipline analysis can be performed in parallel and independently. In the sub-disciplinary optimization problem, under the premise of satisfying the discipline constraints of the discipline, the disparity between the discipline design variables and the design variables within the system-level optimization problem is minimized by solving the sub-disciplinary objective function. In the system-level optimization problem, the differences in design variables between sub-disciplines are coordinated through compatibility constraints while solving the system's objective function. Finally, a design scheme that meets the consistency requirements between sub-disciplines is obtained.

The CO method is simple in structure, and its distributed design optimization system is easy to be applied in modern engineering. Each sub-discipline maintains its optimizer so that discipline designers can better participate in the overall design process of the system. The system-level optimization problem coordinates the difference of design variables between sub-disciplines through compatibility constraints, which avoids the direct transmission of coupling information between sub-disciplines.

Yet, as the number of design variables within the system increases, the task becomes more challenging for the system optimizer to explore the global optimal solution for the system target. Simultaneously, meeting the compatibility constraint condition becomes more difficult. In addition, the sub-disciplinary optimization objective considered by the optimizer of each sub-disciplinary is the compatibility constraint, which is not the subject objective in the practical sense. When dealing with problems with multiple sub-disciplinary objectives, different sub-disciplinary objectives need to be integrated into a system-level objective function.

REFERENCES

[1] Wei L. (2020). Research on multidisciplinary robust design optimization method considering parameter and model uncertainty. PhD thesis of Huazhong University of Science and Technology (in Chinese), Wuhan, China.
[2] Sobieszczanski-Sobieski J. (1995). Multidisciplinary design optimization: An emerging new engineering discipline. Advances in Structural Optimization, 25: 483–496.
[3] Corrado G., Ntourmas G., Sferza M., et al. (2022). Recent progress, challenges and outlook for multidisciplinary structural optimization of aircraft and aerial vehicles. Progress in Aerospace Sciences, 135: 100861.
[4] Wang C., Fan H., Qiang X. (2023). A review of uncertainty-based multidisciplinary design optimization methods based on intelligent strategies. Symmetry, 15(10): 1875.
[5] Bharti V., Biswas B., Shukla K. K. (2023). MDO: A novel murmuration-flight based dispersive optimization algorithm and its application to image security. Journal of Ambient Intelligence and Humanized Computing, 14 (5): 4809–4826.
[6] Yang M., Wang Y., Liang Y., Wang C. (2022). A new approach to system design optimization of underwater gliders. IEEE/ASME Transactions on Mechatronics, 27(5): 3494–3505.
[7] Wu Y., Zhou L., Zheng P., Sun Y., Zhang K. (2022). A digital twin-based multidisciplinary collaborative design approach for complex engineering product development. Advanced Engineering Informatics, 52: 101635.
[8] Li W., Gao L., Garg A., Xiao M. (2022). Multidisciplinary robust design optimization considering parameter and metamodeling uncertainties. Engineering with Computers, 38: 191–208.
[9] Yang M., Wang Y., Liang Y., Wang C. (2022). A new approach to system design optimization of underwater gliders. IEEE/ASME Transactions on Mechatronics, 27(5): 3494–3505.
[10] Wang X. (2018). Research on modeling and solving methods of MDO considering uncertainty factors. Master thesis of University of Electronic Science and Technology of China (in Chinese), Chengdu, China.
[11] Xie C., Yao W. (2023). Concurrent subspace complex coordination optimization method for structural layout synthetical optimization problem. Journal of Aerospace Engineering, 36(5): 04023037.

Chapter 6

UBMDO method

6.1 UBMDO METHOD BASED ON PROBABILITY THEORY

6.1.1 UBMDO method based on first-order saddle point approximation

The Saddle Point Approximation (SPA) method is a more commonly used tool for asymptotic analysis, originating from studying complex functions. This method's most prominent feature is that the probability distribution's approximation effect is still accurate under minor sample points. Daniels [1] introduced this method into the study of statistical inference and gave the density function approximation formula of the mean value of multidimensional independent identical random variables, thus improving the calculation accuracy. Barndorff and Cox [2] used this method to propose an SPA method for the maximum likelihood estimation of the density function. Since then, the SPA method has received extensive attention from the academic community [3–4]. In particular, Du and Sudjianto [5] introduced the SPA method into the reliability analysis technique and proposed the SPA. In this method, under the condition of the existence of the cumulative generating function of random variables, the Cumulant Generating Function (CGF) or the Moment Generating Function (MGF) of random variables is used to directly approximate the probability distribution characteristics of the linear limit state function in the non-normal distribution space, which avoids the conversion process of random variables from non-normal to normal space, thereby improving the reliability evaluation accuracy.

6.1.1.1 Reliability analysis of saddle point approximation and first-order saddle point approximation

The SPA method can accurately approximate the cumulative distribution function of random variables [6–15]. It was initially applied to the Second-Order Reliability Method (SORM) [16]. Although the introduction of SPA makes SORM more accurate, it still cannot avoid the problem of low

DOI: 10.1201/9781003464792-6

computational efficiency of SORM-based Reliability-based Multidisciplinary Design Optimization (RBMDO). Therefore, this paper will use the First-order Saddle Point Approximation (FOSPA) method to study RBMDO.

Suppose that the Probability Density Function (PDF) of the random variable X_R is $f(x_R)$, then the MGF of X_R is [17]

$$M_{X_R}(t) = \int_{-\infty}^{\infty} e^{tx_R} f(x_R) dx_R \tag{6.1}$$

The CGF of the random variable X_R is [18]

$$K_{X_R}(t) = \ln\left[M_{X_R}(t)\right] = \ln\left[\int_{-\infty}^{\infty} e^{tx_R} f(x_R) dx_R\right] \tag{6.2}$$

The common distribution of CGF is shown in Table 6.1.

Here, we will use two important properties of CGF:

Property (1): If the random variables $\mathbf{X}_R = (X_{R1}, X_{R2}, \cdots, X_{Rn})$ are independent of each other and their CGF are $K_{X_{Ri}}(t)(i = 1 \sim n)$, respectively, then the CGF of the limit state function $G = \sum_{i=1}^{n} X_{Ri}$ is

Table 6.1 Common distribution of CGF

Distribution type	*PDF*	*CGF*
Normal	$f(x) = (1/\sqrt{2\pi}\sigma)\cdot \exp\left[(x-\mu)^2/2\sigma^2\right]$	$K(t) = \mu t + \frac{1}{2}\sigma^2 t^2$
Gamma	$f(x) = \beta^{\alpha}/\Gamma(\alpha) x^{\alpha-1} e^{-\beta x}$	$K(t) = \alpha\left[\ln\beta - n(\beta - t)\right]$
Extreme value	$f(x) = (1/\sigma)\exp\left[(x-\mu)/\sigma\right]\cdot \exp\left\{-\exp\left[(x-\mu)/\sigma\right]\right\}$	$K(t) = \mu t + \ln\Gamma(1-\sigma t)$
Uniform	$f(x) = 1/(b-a)$	$K(t) = \ln(e^{bt} - e^{at}) - \ln(b-a) - \ln(t)$
Exponential	$f(x) = \alpha\exp(-\alpha x)$	$K(t) = -\ln(1 - t/\alpha)$
χ^2	$f(x) = \left[1/\Gamma(n/2)2^{n/2}\right]\cdot x^{n/2-1} e\left(-\frac{1}{2}x\right)$	$K(t) = -\frac{1}{2}n\ln(1-2t)$

$$K_G(t)=\sum_{i=1}^{n}K_{X_{Ri}}(t) \tag{6.3}$$

Property (2): If the CGF of the random variable X_R is $K_{X_R}(t)$, then the CGF of the limit state function $G=aX_R+b$ (a and b are constants) is

$$K_G(t)=K_{X_R}(at)+bt \tag{6.4}$$

If the random variable X_R obeys the χ^2 distribution, and its CGF is $K_{X_R}(t)=-(1/2)n\ln(1-2t)$, then the CGF of the random variable $G=aX_R+b$ is $K_G(t)=-(1/2)n\ln(1-2at)+bt$.

Considering the general case, that is, for the limit state function $G=G(X_R)$, if the CGF of the random variable X_R is $K_{X_R}(t)$, the PDF and the Cumulative Density Function (CDF) of the limit state function G can be obtained by using the SPA method. The PDF of the limit state function G is [17]

$$f_G(X_R)=\left[2\pi K_G''(t^*)\right]^{-\frac{1}{2}}e^{\left[K_G(t^*)\right]} \tag{6.5}$$

where K_G'' is the second derivative of CGF; t^* is a saddle point, which can be solved by Eq. 6.6 [15]

$$K_G'(t)=0 \tag{6.6}$$

where K_G' is the first derivative of CGF.

In References [9, 19], two approximate methods for solving the CDF of the limit state function G are given, respectively

$$F_G(0)=\Pr\left[G(X_R)\le 0\right]\approx\Phi(w)+\phi(w)\left(\frac{1}{w}-\frac{1}{v}\right) \tag{6.7}$$

or

$$F_G(0)=\Pr\left[G(X_R)\le 0\right]\approx\Phi\left[w+\frac{1}{w}\log\frac{v}{w}\right] \tag{6.8}$$

where $\Phi(\bullet)$ and $\phi(\bullet)$ are CDF and PDF of standard normal distribution, respectively; $w = \text{sign}(t)\left\{2\left[-K_G(t)\right]\right\}^{\frac{1}{2}}\Big|_{t^*}$; $v = t\left\{\frac{d^2\left[K_G(t)\right]}{dt^2}\right\}^{\frac{1}{2}}\Bigg|_{t^*}$; $\text{sign}(t) = \begin{cases} 1, & t > 0 \\ 0, & t = 0 \\ -1, & t < 0 \end{cases}$.

The above two properties of CGF explain how to obtain the limit state function CGF through the design variable CGF when the limit state function and the design variable are linear and how to obtain the PDF and CDF of the limit state function through saddle point analysis. The following will explain how to approximate the limit state function linearly in the FOSPA and calculate the reliability of the corresponding approximate limit state function under the specified conditions.

First, the limit state function's Most Likelihood Point (MLP) is searched. MLP is $\mathrm{X}_{\mathrm{R}}^* = \left(\mathrm{X}_{\mathrm{s}}^*, \mathrm{X}^*, \mathrm{P}^*\right)$, which is the point with the maximum probability density on the limit state function $G_i\left(\mathbf{d}_{\mathrm{s}}, \mathbf{d}_i, \mathbf{X}_{\mathrm{s}}, \mathbf{X}_i, \mathbf{P}, \mathbf{Y}_{\bullet i}\right)$. The model for determining MLP can be expressed as

$$\begin{aligned} &\max_{\mathrm{X}_{\mathrm{R}}} \prod f_{\mathrm{X}_{\mathrm{s}}}\left(\mathrm{X}_{\mathrm{s}}\right) f_{\mathrm{X}}(\mathrm{X}) f_{\mathrm{P}}(\mathrm{P}) \\ &\text{s.t. } G_i\left(\mathbf{d}_{\mathrm{s}}, \mathbf{d}_i, \mathbf{X}_{\mathrm{R}}, \mathbf{Y}\right) = 0, \\ &\qquad \mathbf{Y}_{i\bullet} = \mathbf{Y}_{i\bullet}\left(\mathbf{d}_{\mathrm{s}}, \mathbf{d}_i, \mathbf{X}_{\mathrm{R}}, \mathbf{Y}_{\bullet i}\right) \end{aligned} \tag{6.9}$$

Linearizing the limit state function at the MLP can minimize the loss of reliability analysis accuracy caused by the linearized limit state function [20]. The first-order Taylor approximation of the limit state function $G_i\left(\mathbf{d}_{\mathrm{s}}, \mathbf{d}_i, \mathbf{X}_{\mathrm{s}}, \mathbf{X}_i, \mathbf{P}, \mathbf{Y}_{\bullet i}\right)$ at the MLP is expanded

$$\begin{aligned} G_i\left(\mathbf{d}_{\mathrm{s}}, \mathbf{d}_i, \mathbf{X}_{\mathrm{R}}, \mathbf{Y}_{\bullet i}\right) &\approx \widehat{G}_i\left(\mathbf{d}_{\mathrm{s}}, \mathbf{d}_i, \mathbf{X}_{\mathrm{R}}, \mathbf{Y}_{\bullet i}\right) \\ &= G_i\left(\mathbf{d}_{\mathrm{s}}, \mathbf{d}_i, \mathbf{X}_{\mathrm{R}}^*, \mathbf{Y}_{\bullet i}\right) + \nabla G_i\left(\mathbf{X}_{\mathrm{R}}^*\right)\left(\mathbf{X}_{\mathrm{R}} - \mathbf{X}_{\mathrm{R}}^*\right)^T \end{aligned} \tag{6.10}$$

$\nabla G_i\left(\mathbf{X}_{\mathrm{R}}^*\right)$ is the gradient of $G_i\left(\mathbf{d}_{\mathrm{s}}, \mathbf{d}_i, \mathbf{X}_{\mathrm{R}}, \mathbf{Y}_{\bullet i}\right)$ at $\mathbf{X}_{\mathrm{R}}^*$, as shown in Eq. 6.11.

$$\nabla G_i\left(\mathbf{X}_{\mathrm{R}}^*\right) = \left(\frac{\partial G_i}{\partial \mathrm{X}_{\mathrm{s}}}, \frac{\partial G_i}{\partial \mathrm{X}_i}, \frac{\partial G_i}{\partial \mathrm{P}}\right)\Bigg|_{\mathbf{X}_{\mathrm{R}}^*} \tag{6.11}$$

After obtaining the first-order linear approximation of the limit state function, the CGF of the approximate limit state function $\widehat{G}_i$ is

$$K_{\widehat{G}}(t) = K_{X_R}\left(\nabla G_i\left(\mathbf{X}_R^*\right)t\right) + \left(G_i\left(\mathbf{d}_s, \mathbf{d}_i, \mathbf{X}_R^*, \mathbf{Y}_{\bullet i}\right) - \nabla G_i\left(\mathbf{X}_R^*\right)\mathbf{X}_R^*\right)t \quad (6.12)$$

The saddle point t^* in the above formula can be determined by the following formula

$$d\left[K_{\widehat{G}}(t)\right]/dt = 0 \quad (6.13)$$

The approximate reliability evaluations of the limit state function satisfying the performance requirements obtained by Eqs. 6.7 and 6.8 are as follows:

$$R_i \approx \Pr[G_i\left(\mathbf{d}_s, \mathbf{d}_i, \mathbf{X}_R, \mathbf{Y}_{\bullet i}\right) \le 0] = \Phi(w) + \phi(w)\left(\frac{1}{w} - \frac{1}{v}\right) \quad (6.14)$$

or

$$R_i \approx \Pr[G_i\left(\mathbf{d}_s, \mathbf{d}_i, \mathbf{X}_R, \mathbf{Y}_{\bullet i}\right) \le 0] = \Phi\left[w + \frac{1}{w}\log\frac{v}{w}\right] \quad (6.15)$$

where $w = \mathrm{sign}(t)\left\{2\left[-K_{\widehat{G}}(t)\right]\right\}^{\frac{1}{2}}\Big|_{t^*}$; $v = t\left\{\dfrac{d^2\left[K_{\widehat{G}}(t)\right]}{dt^2}\right\}^{\frac{1}{2}}\Bigg|_{t^*}$.

6.1.1.2 Case study: simple math example

The mathematical example in Figure 6.1 contains two subdisciplines: Discipline 1 and 2. The mathematical model of the optimization problem is as follows:

$$\begin{aligned}
&\min_{\mathrm{DV}} f = f_1 + f_2 \\
&\text{s.t.} \ -1 \le x_1 \le 1, \\
&\qquad -1 \le x_2 \le 1, \\
&\qquad 0 \le x_3 \le 3, \\
&\qquad 0 \le y_{12} \le 1, \\
&\qquad 0 \le y_{21} \le 2, \\
&\mathrm{DV} = \left\{x_1, x_2, x_3\right\}
\end{aligned} \quad (6.16)$$

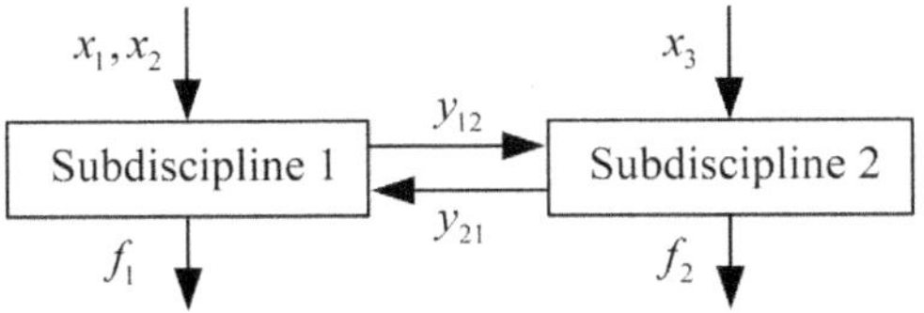

Figure 6.1 Mathematical example MDO problem.

Table 6.2 Uncertainty information of design variables in mathematical examples

Variable	*Mean value*	*Standard deviation*	*Distribution type 1*	*Distribution type 2*
x_1	x_1^M	$0.001x_1^M$	Normal distribution	Extreme value distribution
x_2	x_2^M	$0.001x_2^M$	Normal distribution	Extreme value distribution
x_3	x_3^M	$0.001x_3^M$	Normal distribution	Extreme value distribution

where $f_1 = (y_{12} - 1)^2 + x_1^2 + x_2^2$, $f_2 = (y_{21} - 2)^2 + x_3^2$, $y_{12} = x_1 - x_2 + 2y_{21}$, and $y_{21} = x_3 - y_{12}$.

Here, the input variables are assumed to have random uncertainties belonging to normal and extreme value distributions. Figure 6.2 shows a flowchart of the UBMDO method based on first-order SPA. The uncertainty information of random variables is shown in Table 6.2.

The mathematical model of the optimization problem is

$$\begin{aligned} &\min_{\mathrm{DV}} f = f_1 + f_2 \\ &\text{s.t. } \Pr\left[G_1 = x_1 x_2^2 + y_{12} - 1 \le 0\right] \ge 0.99, \\ &\quad\quad \Pr\left[G_2 = x_3^2 + y_{12}^2 + y_{21} - 5 \le 0\right] \ge 0.99, \\ &\quad\quad -1 \le x_1^M \le 1, \\ &\quad\quad -1 \le x_2^M \le 1, \\ &\quad\quad 0 \le x_3^M \le 3, \\ &\quad\quad 0 \le y_{12} \le 1, \\ &\quad\quad 0 \le y_{21} \le 2, \\ &\mathrm{DV} = \left\{x_1^M, x_2^M, x_3^M\right\} \end{aligned} \tag{6.17}$$

where $f_1 = (y_{12} - 1)^2 + (x_1^M)^2 + (x_2^M)^2$, $f_2 = (y_{21} - 2)^2 + (x_3^M)^2$, $y_{12}^M = x_1^M - x_2^M + 2y_{21}^M$, and $y_{21}^M = x_3^M - y_{12}^M$. The deterministic MDO model in kth ring is

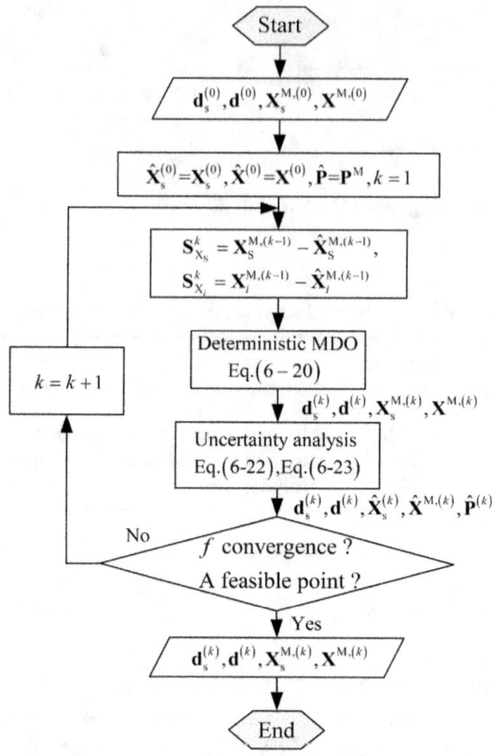

Figure 6.2 Flowchart of UBMDO method based on first-order saddle point approximation.

$$\begin{aligned}
&\min_{\mathrm{DV}}\ f = \left(y_{12}^{\mathrm{M},(k)} - 1\right)^2 + \left(x_1^{\mathrm{M},(k)}\right)^2 + \left(x_2^{\mathrm{M},(k)}\right)^2 + \left(y_{21}^{\mathrm{M},(k)} - 2\right)^2 + \left(x_3^{\mathrm{M},(k)}\right)^2 \\
&\text{s.t. } G_{\Pi 1} = \left(x_1^{\mathrm{M},(k)} - \left(x_1^{\mathrm{M},(k-1)} - \hat{x}_1^{(k-1)}\right)\right)\left(x_2^{\mathrm{M},(k)} - \left(x_2^{\mathrm{M},(k-1)} - \hat{x}_2^{(k-1)}\right)\right)^2 + y_{12}^{\mathrm{M},(k)} - 1 \le 0, \\
&\qquad G_{\Pi 2} = \left(x_3^{\mathrm{M},(k)} - \left(x_3^{\mathrm{M},(k-1)} - \hat{x}_3^{(k-1)}\right)\right)^2 + \left(y_{12}^{\mathrm{M},(k)}\right)^2 + y_{21}^{\mathrm{M},(k)} - 5 \le 0, \\
&\qquad -1 \le x_1^{\mathrm{M}} \le 1, \\
&\qquad -1 \le x_2^{\mathrm{M}} \le 1, \\
&\qquad 0 \le x_3^{\mathrm{M}} \le 3, \\
&\qquad 0 \le y_{12}^{\mathrm{M}} \le 1, \\
&\qquad 0 \le y_{21}^{\mathrm{M}} \le 2, \\
&\mathrm{DV} = \left\{x_1^{\mathrm{M}}, x_2^{\mathrm{M}}, x_3^{\mathrm{M}}\right\}
\end{aligned} \tag{6.18}$$

where $y_{12}^{M,(k)} = x_1^{M,(k)} - x_2^{M,(k)} + 2y_{21}^{M,(k)}$, $y_{21}^{M,(k)} = x_3^{M,(k)} - y_{12}^{M,(k)}$.

The mathematical model of MLP in the kth ring is

$$\begin{aligned}
&\max_{\mathrm{DV}} \prod_{i=1}^{3} f_{x_i}(x_i) \\
&\text{s.t. } G_1 = x_1^{(k)}\left(x_2^{(k)}\right)^2 + y_{12}^{(k)} - 1 = 0, \\
&\quad G_2 = \left(x_3^{M,(k)}\right)^2 + \left(y_{12}^{M,(k)}\right)^2 + y_{21}^{M,(k)} - 5 = 0, \\
&\quad y_{12}^{M,(k)} = x_1^{M,(k)} - x_2^{M,(k)} + 2y_{21}^{M,(k)}, \\
&\quad y_{21}^{M,(k)} = x_3^{M,(k)} - y_{12}^{M,(k)}, \\
&\quad -1 \le x_1^{M,(k)} \le 1, \\
&\quad -1 \le x_2^{M,(k)} \le 1, \\
&\quad 0 \le x_3^{M,(k)} \le 3, \\
&\quad 0 \le y_{12}^{M,(k)} \le 1, \\
&\quad 0 \le y_{21}^{M,(k)} \le 2, \\
&\quad \mathrm{DV} = \left\{x_1^{M,(k)}, x_2^{M,(k)}, x_3^{M,(k)}\right\}
\end{aligned} \tag{6.19}$$

The mathematical model of the expected MLP solution in the kth ring is

$$\begin{aligned}
&\max_{\mathrm{DV}} \prod_{i=1}^{3} f_{x_i}(x_i) \\
&\text{s.t. } G_{\Pi 1} = (x_1^{M,(k)} - (x_1^{M,(k-1)} - \hat{x}_1^{(k-1)}))(x_2^{M,(k)} - (x_2^{M,(k-1)} - \hat{x}_2^{(k-1)}))^2 + y_{12}^{M,(k)} - 1 = 0, \\
&\quad G_{\Pi 2} = \left(x_3^{M,(k)} - \left(x_3^{M,(k-1)} - \hat{x}_3^{(k-1)}\right)\right)^2 + \left(y_{12}^{M,(k)}\right)^2 + y_{21}^{M,(k)} - 5 = 0, \\
&\quad y_{12}^{M,(k)} = x_1^{M,(k)} - x_2^{M,(k)} + 2y_{21}^{M,(k)}, \\
&\quad y_{21}^{M,(k)} = x_3^{M,(k)} - y_{12}^{M,(k)}, \\
&\quad -1 \le x_1^{M,(k)} \le 1, \\
&\quad -1 \le x_2^{M,(k)} \le 1,
\end{aligned}$$

Table 6.3 Optimization results of mathematical examples under different distribution types

	Normal distribution			*Extreme value distribution*		
	FOSPA	*FORM*	*MCS*	*FOSPA*	*FORM*	*MCS*
x_1^M	−0.2775	−0.2775	−0.2773	−0.2777	−0.2774	−0.2773
x_2^M	0.3142	0.3142	0.3144	0.3173	0.3151	0.3179
x_3^M	1.0021	1.0021	1.0028	1.0057	1.0036	1.0061
y_{12}^M	0.4708	0.4708	0.4713	0.4721	0.4716	0.4723
y_{21}^M	0.5312	0.5312	0.5315	0.5336	0.5320	0.5338
f	3.6171	3.6171	3.6174	3.6183	3.6176	3.6185
k	3	3	–	3	3	–

$$
\begin{aligned}
&0 \le x_3^{M,(k)} \le 3, \\
&0 \le x_{12}^{M,(k)} \le 1, \\
&0 \le x_{21}^{M,(k)} \le 1, \\
\mathrm{DV} = &\left\{ x_1^{M,(k)}, x_2^{M,(k)}, x_3^{M,(k)} \right\}
\end{aligned}
\tag{6.20}
$$

Here, Eq. 6.18 is solved by the IPO method. The optimization results of random variables obeying normal distribution and extreme value distribution are given in Table 6.3.

In the table, n_{sys}, n_1, and n_2 are the number of iterations of system optimization, sub-discipline 1 optimization, and sub-discipline 2 optimization, respectively. It can be seen that when the random variables obey the normal distribution, the same results can be obtained by using First Order Saddle Point Approximation (FOSPA) and First Order Reliability Method (FORM) methods. This is because FORM is a special case of FOSPA in the normal distribution of random variables, and FOSPA is a more general case of FORM. When the random variables obey the extreme value distribution, the results obtained by the RBMDO with FOSPA are closer to the standard values than the results obtained by the RBMDO with FORM. This is because, in FOSPA, the transformation of random variables from non-normal space to normal space is canceled. This avoids the defect of increasing the degree of nonlinearity to the original optimization problem and makes the reliability analysis in the RBMDO problem more accurate.

6.1.2 UBMDO method based on subset simulation

6.1.2.1 The basic idea of the subset simulation method

The subset simulation method converts the reliability assessment problem of small probability failure events into a reliability assessment problem of an initial failure event and a range of intermediate failure events with a high probability of occurrence. The product of the occurrence probability of the initial failure event and the conditional occurrence probability of each intermediate failure event can express the final failure probability.

Define the intermediate failure domains F_j, $j = 1 \sim m$, and $F_1 \supset F_2 \supset \cdots \supset F_m = F$, therefore.

$$F_k = \bigcap_{j=1}^{k} F_j \tag{6.21}$$

The reasonable definition of each intermediate failure domain realizes the conditional failure probability. Therefore, according to the needs of practical problems, each intermediate failure domain is defined as

$$F_j = \left\{ G(\mathbf{X}_{\mathrm{R}}) \le G_j \right\},\ j = 1 \sim m \tag{6.22}$$

where $G_m = 0 < \cdots < G_2 < G_1$. By evaluating the probability of occurrence of each intermediate failure event, the final failure probability can be expressed as

$$\begin{aligned} p_f &= P(F_m) = P\left(\bigcap_{j=1}^{m} F_j\right) = P\left(F_m \middle| \bigcap_{j=1}^{m-1} F_j\right) P\left(\bigcap_{j=1}^{m-1} F_j\right) \\ &= P(F_m | F_{m-1}) P\left(\bigcap_{j=1}^{m-1} F_j\right) \\ &= \cdots = P(F_1) \prod_{j=1}^{m-1} P(F_{j+1} | F_j) \end{aligned} \tag{6.23}$$

Through the subset simulation method, although the failure probability p_f is very small, the relatively large values of $P(F_{j+1}|F_j)$ and $P(F_1)$ can be obtained by defining a reasonable intermediate failure domain, and finally the probability evaluation with satisfactory accuracy can be obtained by using a small number of sample points. For example, if the failure probability is

10^{-3}, according to the error calculation formula $\xi = \left(\frac{1-p_f}{Np_f}\right)^{\frac{1}{2}}$, when evaluating the accuracy $\xi = 0.01$, the number of sample points required to use Monte Carlo Simulation (MCS) is approximately 10^7. If the original failure domain is divided into three intermediate failure domains $P(F_j) = 10^{-1}$, $j = 1 \sim 3$, the number of sample points required to use the subset simulation method is approximately $j = 1 \sim 3$ under the same evaluation accuracy. The selection of the intermediate failure region has no significant influence on the accuracy of the subset simulation method, so it can be artificially defined in advance or adaptively selected according to the actual situation [21].

In the calculation of $P(F_1)$, the MCS method can be used directly. Its reliability can be approximately expressed as

$$P(F_1) \approx \frac{1}{N_1}\sum_{i=1}^{N_1} I_G(x_{Ri}) \tag{6.24}$$

where x_{Ri}, $i = 1 \sim N_1$ is an independent and identically distributed sample point with PDF of $f(x_R)$.

When calculating $P(F_{j+1}|F_j)$, it is necessary to obtain the sample points that obey PDF as $f(\bullet|F_j), j = 1 \sim m-1$. Among them,

$$f(x_R|F_j) = \frac{f(x_R)I_G(x_R)}{P(F_j)} \tag{6.25}$$

Through the Markov Chain Monte Carlo (MCMC) strategy, the sample points whose distribution approximates to $f(x_R|F_j)$ can be obtained. After obtaining the sample points, each intermediate failure event can be obtained by Eq. 6.26

$$P(F_{j+1}|F_j) \approx \frac{1}{N_j}\sum_{i=1}^{N_j} I_G(x_{Ri}) \tag{6.26}$$

The specific process of the subset simulation method is shown in Figure 6.3. In the next section, the MCMC strategy will be described.

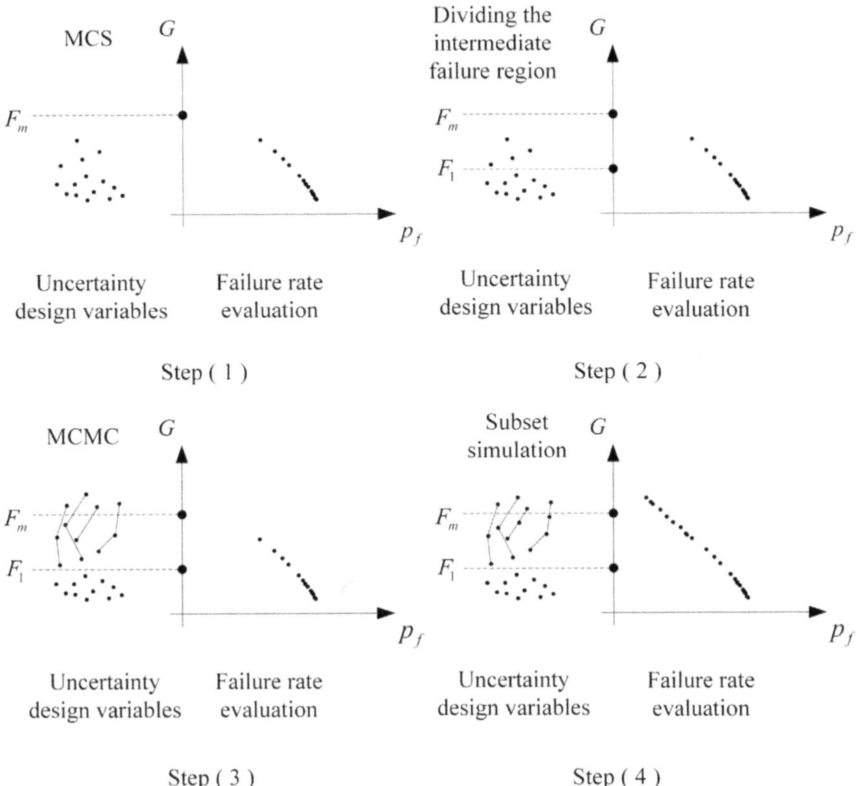

Figure 6.3 Subset simulation analysis method schematic diagram.

6.1.2.2 Basic principles of MCMC strategy

The MCMC strategy is a method that can generate sample points that approximately meet the distribution requirements. The basic method is to use the intermediate state point of the Markov chain with ergodicity as the sample point and approximately obtain a required distribution through the limit stable distribution property of the Markov chain. The key of the MCMC strategy is to generate a transition from the current state point to the next state point by transferring the PDF. This paper will use the Modified Metropolis Algorithm (MMA) sampling method in [22] to construct the corresponding transition PDF.

Let the sample point in the current state j in the Markov chain be $\mathbf{X}^{(j)} = \left[x_1^{(j)}, x_2^{(j)}, \cdots, x_n^{(j)}\right] \in F$, which obeys the distribution of PDF as $f\left(\bullet \middle| F_j\right)$,

$j = 1 \sim m-1$. Through the joint transfer PDF $p\left(\mathbf{X}^{(j+1)} \middle| \mathbf{X}^{(j)}\right)$, $i = 1 \sim n$, the specific steps to obtain the sample point $\mathbf{X}^{(j+1)} = \left[x_1^{(j+1)}, x_2^{(j+1)}, \cdots, x_n^{(j+1)}\right] \in F$ in the next state $j+1$ are as follows:

Step 1. Generate candidate sample point $\tilde{\mathbf{X}}^{(j+1)} = \left[\tilde{x}_1^{(j+1)}, \tilde{x}_2^{(j+1)}, \cdots, \tilde{x}_n^{(j+1)}\right]$.

① The one-dimensional PDF $p^*\left(x_i^{(j+1)} \middle| x_i^{(j)}\right)$ is selected as the proposal PDF. For all sample points in the intermediate state j, with $x_i^{(j)}$ as the center, the random walk is performed to the state point $\hat{x}_i^{(j+1)'}$ that conforms to the proposed PDF and satisfies the symmetry condition $p^*\left(\hat{x}_i^{(j+1)} \middle| x_i^{(j)}\right) = p^*\left(x_i^{(j)} \middle| \hat{x}_i^{(j+1)}\right)$.

② The acceptance rate of candidate sample points is calculated, respectively.

$$r_i = \min\left\{1, \frac{f\left(\hat{x}_i^{(j+1)}\right)}{f\left(x_i^{(j)}\right)}\right\} \tag{6.27}$$

③ The acceptance rate r_i accepts $\hat{x}_i^{(j+1)}$ as the candidate sample point $\tilde{x}_i^{(j+1)} = \hat{x}_i^{(j+1)}$; the acceptance rate $1 - r_i$ accepts $x_i^{(j)}$ as the candidate sample point $\tilde{x}_i^{(j+1)} = x_i^{(j)}$.

Step 2. Generate the sample point $\mathbf{X}^{(j+1)} = \left[x_1^{(j+1)}, x_2^{(j+1)}, \cdots, x_n^{(j+1)}\right]$ of the next state $j+1$.

Check the position of point $\tilde{x}_i^{(j+1)}$. If $\tilde{x}_i^{(j+1)} \in F_i$, then accept the candidate sample point $\tilde{x}_i^{(j+1)}$ as the sample point of the next state $j+1$, $x_i^{(j+1)} = \tilde{x}_i^{(j+1)}$; otherwise, the sample point of the current state j is accepted as the sample point of the next state $j+1$, $x_i^{(j+1)} = x_i^{(j)}$.

Reference [21] gives a more general case and cancels the symmetry condition of the proposed PDF. The acceptance rate calculation method of candidate sample points is modified to

$$r_i = \min\left\{1, \frac{f\left(\hat{x}_i^{(j+1)}\right) p^*\left(x_i^{(j)} \middle| \hat{x}_i^{(j+1)}\right)}{f\left(x_i^{(j)}\right) p^*\left(\hat{x}_i^{(j+1)} \middle| x_i^{(j)}\right)}\right\} \tag{6.28}$$

Step 1 can be seen as a random walk process of the candidate sample points around the current sample points. **Step 2** ensures that the next generation of sample points can be located in the failure domain with a high probability. The processing method of **Step 2** is similar to that of directly using the MCS method to count the number of sample points falling into the failure domain. However, the MMA method has a much higher utilization rate of sample points than the MCS method. The reason is that the next generation of sample points is generated near the current sample point. Since the current sample points are all in the failure domain, the probability that the next generation of sample points will also fall into the failure domain is high. Therefore, the MCMC strategy is more efficient in exploring the failure domain. However, the higher analysis efficiency is obtained at the same high correlation cost between the two generations of sample points. The higher the correlation is, the lower the evaluation accuracy is [23]. The process of generating sample points from **Step 1** to **Step 2** is shown in Figure 6.4.

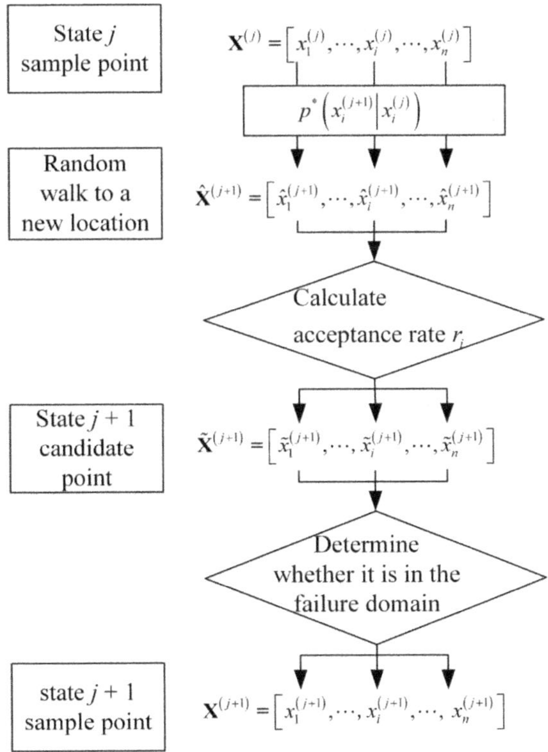

Figure 6.4 Modified Metropolis-Hastings sampling method diagram.

The correlation between the sample points of two consecutive states generated by the MMA sampling method is inevitably introduced [21]:

1. Through the way of random walk, there must be a correlation between the new sample points and the previous generation of sample points
2. If the new sample point and the previous generation of the sample point are the same, there is a correlation

Engineering experience shows that [21] the distribution type of the proposed PDF has little effect on the sampling method, and the normal distribution or uniform distribution centered on the current state point can be selected as the distribution type of the proposed PDF. The proposed PDF's variance dramatically influences the sampling method's efficiency. The variance is large, and the correlation between the two generations of sample points is low, but the acceptance rate of candidate sample points is not high. The variance is small, and the acceptance rate of candidate sample points is relatively high, but the correlation between the two generations of sample points is high. Reference [23] studied the classification of random variables and gave the recommended PDF selection method for random variables. At the same time, considering the symmetry requirements, this paper selects the one-dimensional uniform distribution as the recommended PDF.

The following will show that the newly generated state $j+1$ sample point $\mathbf{X}^{(j+1)} = \left[x_1^{(j+1)}, x_2^{(j+1)}, \cdots, x_n^{(j+1)}\right]$ still obeys the distribution of PDF as $f\left(\bullet \middle| F_j\right)$.

In **Step 1**, the sample point $\mathbf{X}^{(j)} = \left[x_1^{(j)}, x_2^{(j)}, \cdots, x_n^{(j)}\right]$ of the current state j obeys independent identical distribution. Therefore, the joint transition PDF of each sample point in the failure domain F_j from state j to state $j+1$ is

$$p\left(\mathbf{X}^{(j+1)} \middle| \mathbf{X}^{(j)}\right) = \prod_{i=1}^{n} p\left(x_i^{(j+1)} \middle| x_i^{(j)}\right) \tag{6.29}$$

where $p\left(x_i^{(j+1)} \middle| x_i^{(j)}\right)$ is the transition PDF of the sample point x_i from state j to state $j+1$

$$p\left(x_i^{(j+1)} \middle| x_i^{(j)}\right) = p^*\left(x_i^{(j+1)} \middle| x_i^{(j)}\right) \min\left\{1, \frac{f\left(x_i^{(j+1)}\right)}{f\left(x_i^{(j)}\right)}\right\} \tag{6.30}$$

According to the symmetry condition of the proposed PDF and $\min\{1,\ a/b\}\times b=\min\{1,\ b/a\}\times a$ when a and b are positive, the following relationship can be obtained

$$p\left(x_i^{(j+1)}\middle|x_i^{(j)}\right)f\left(x_i^{(j)}\right)=p\left(x_i^{(j)}\middle|x_i^{(j+1)}\right)f\left(x_i^{(j+1)}\right) \tag{6.31}$$

Considering that the sample points are all in the failure domain F_j, from the Eq. 6.29 and the Eq. 6.31, there are

$$p\left(\mathbf{X}^{(j+1)}\middle|\mathbf{X}^{(j)}\right)f\left(\mathbf{X}^{(j)}\middle|F_j\right)=p\left(\mathbf{X}^{(j)}\middle|\mathbf{X}^{(j+1)}\right)f\left(\mathbf{X}^{(j+1)}\middle|F_j\right) \tag{6.32}$$

If the sample point of the current state j obeys the distribution of the PDF $f\left(\bullet\middle|F_j\right)$, the PDF $f\left(\mathbf{X}^{(j+1)}\right)$ of the sample point of the state $j+1$ is

$$\begin{aligned}f\left(\mathbf{X}^{(j+1)}\right)&=\int p\left(\mathbf{X}^{(j+1)}\middle|\mathbf{X}^{(j)}\right)f\left(\mathbf{X}^{(j)}\middle|F_j\right)d\mathbf{X}^{(j)}\\&=\int p\left(\mathbf{X}^{(j)}\middle|\mathbf{X}^{(j+1)}\right)f\left(\mathbf{X}^{(j+1)}\middle|F_j\right)d\mathbf{X}^{(j)}\\&=f\left(\mathbf{X}^{(j+1)}\middle|F_j\right)\int p\left(\mathbf{X}^{(j)}\middle|\mathbf{X}^{(j+1)}\right)d\mathbf{X}^{(j)}\\&=f\left(\mathbf{X}^{(j+1)}\middle|F_j\right)\end{aligned} \tag{6.33}$$

That is, the newly generated state $j+1$ sample points still obey the distribution of PDF as $f\left(\bullet\middle|F_j\right)$.

6.1.2.3 *Case study: gear reducer design*

As one of the classic examples, the gear reducer design problem is widely used to verify the effectiveness of different optimization methods [24]. As shown in Figure 6.5, the optimization problem contains seven design variables. The physical meaning of each design variable and the design interval are given in Table 6.4. The MDO problem of the gear reducer includes the gear discipline and the shaft discipline, and its multidisciplinary relationship is shown in Figure 6.6.

The goal of system optimization is to reduce the volume of the gear reducer. The mathematical model of the optimization problem is

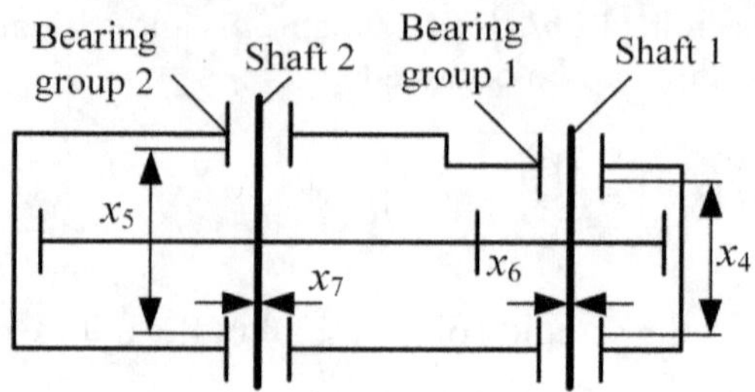

Figure 6.5 Structure diagram of gear reducer.

Table 6.4 Gear reducer design variables

Variable	*Physical meaning (unit)*	*Design lower bound*	*Design upper bound*	*Variable*	*Physical meaning (unit)*	*Design lower bound*	*Design upper bound*
x_1	Coefficient of face width	2.6	3.6	x_5	Axis 2 length (cm)	7.3	8.3
x_2	Gear modulus (cm)	0.3	1.0	x_6	Axis 2 length (cm)	2.9	3.9
x_3	Number of pinion teeth (pieces)	17	28	x_7	Axis 2 length (cm)	5	5.5
x_4	Axis 1 length (cm)	7.3	8.3				

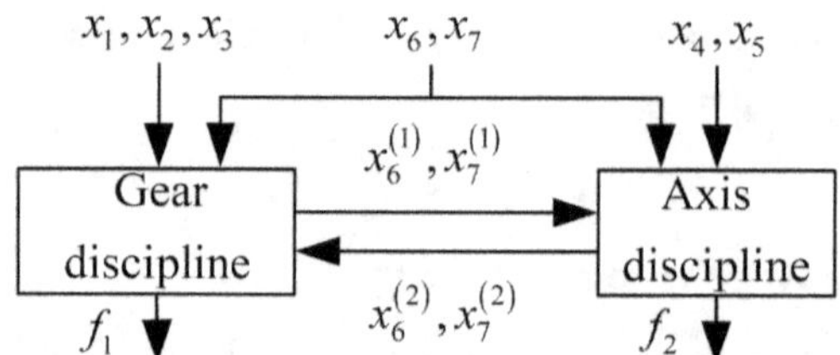

Figure 6.6 The MDO problem of gear reducer.

$$\begin{aligned}
\min\quad & f = 0.7854x_1x_2^2\left(3.333x_3^2 + 14.933x_3 - 43.0934\right) \\
& \quad - 1.508x_1\left(x_6^2 + x_7^2\right) + 7.477\left(x_6^3 + x_7^3\right) + 0.7854\left(x_4x_6^2 + x_5x_7^2\right) \\
\text{s.t.}\quad & g_1 = 27/\left(x_1x_2^2x_3\right) - 1 \le 0, \\
& g_2 = 397.5/(x_1x_2^2x_3^2) - 1 \le 0,
\end{aligned}$$

$$g_3 = 1.93x_4^3/(x_2x_3x_6^4) - 1 \le 0,$$
$$g_4 = 1.93x_5^3/(x_2x_3x_7^4) - 1 \le 0,$$
$$g_5 = A_1/B_1 - 1100 \le 0,$$
$$g_6 = A_2/B_2 - 850 \le 0,$$
$$g_7 = x_2x_3 - 40 \le 0,$$
$$g_8 = x_1/x_2 - 12 \le 0,$$
$$g_9 = -x_1/x_2 + 5 \le 0,$$
$$g_{10} = (1.5x_6 + 1.9)/x_4 - 1 \le 0,$$
$$g_{11} = (1.1x_7 + 1.9)/x_5 - 1 \le 0, \tag{6.34}$$

where $A_1 = \left[\left(\frac{745x_4}{x_2x_3}\right)^2 + 16.9\times10^6\right]^{0.5}$, $A_2 = \left[\left(\frac{745x_5}{x_2x_3}\right)^2 + 157.5\times10^6\right]^{0.5}$, $B_1 = 0.1x_6^3$, and $B_2 = 0.1x_7^3$.

Here, it is assumed that the input variables have random uncertainties that obey the normal distribution. In the reliability optimization model, the reliability of the uncertainty constraint needs to be greater than or equal to 0.9999. Under the Sequential Optimization and Reliability Assessment (SORA) strategy, the deterministic design optimization adopts the Identification-based Optimization (IBO) method. At the same time, the MCS and Subset Simulation-based Reliability Analysis (SSRA) numerical simulation reliability methods are used to evaluate the reliability. After the loop calculation, the optimization process converges. Table 6.5 shows the optimization results of MCS-MDO-SORA and SSRA-MDO-SORA. Similar to mathematical examples, the two methods can obtain similar calculation results. In order to avoid directly evaluating the reliability of the minimal failure probability, the latter divides the failure domain. Therefore, the utilization rate of sample points is improved, which in turn improves the efficiency of RBMDO.

Table 6.5 Design results of gear reducers using numerical simulation reliability analysis

	x_1	x_2	x_3	x_4^M	x_5^M	x_6^M	x_7^M	f	N	k
MCS-MDO-SORA	3.906	0.7711	18	7.415	7.832	3.510	5.450	3565	6×10^5	6
SSRA-MDO-SORA	3.907	0.712	18	7.413	7.835	3.511	5.453	3573	1.8×10^4	6

6.2 NON-PROBABILISTIC UBMDO METHOD

6.2.1 UBMDO method based on interval theory

6.2.1.1 Interval number and its basic algorithm

As a non-probabilistic convex set model, the interval number model plays an essential role in uncertainty optimization. The general definition of interval number is shown in Eq. 6.35

$$X^I = \left[X^L, X^R\right] = \left\{x \middle| X^L \leqslant x \leqslant X^R, x : x \in R\right\} \tag{6.35}$$

where X^I denotes the interval range, X^L denotes the upper bound of the interval, and X^R denotes the lower bound of the interval. If the two endpoints of the interval are equal, that is, $X^L = X^R$, the interval is called a point interval. The interval number geometric model is shown in Figure 6.7.

For any two intervals $X^I = \left[X^L, X^R\right]$ and $Y^I = \left[Y^L, Y^R\right]$, the basic four-operation rule can be expressed as Eq. 6.36 [25]

$$\begin{cases} X^I + Y^I = \left[X^L, X^R\right] + \left[Y^L, Y^R\right] = \left[X^L + Y^L, X^R + Y^R\right] \\ X^I - Y^I = \left[X^L, X^R\right] - \left[Y^L, Y^R\right] = \left[X^L - Y^R, X^R - Y^L\right] \\ X^I \bullet Y^I = \left[X^L, X^R\right] \bullet \left[Y^L, Y^R\right] \\ \qquad = \begin{bmatrix} \min\left(X^L Y^L, X^L Y^R, X^R Y^L, X^R Y^R\right), \\ \max\left(X^L Y^L, X^L Y^R, X^R Y^L, X^R Y^R\right) \end{bmatrix} \\ X^I / Y^I = \left[X^L, X^R\right] / \left[Y^L, Y^R\right] = \left[X^L, X^R\right] \bullet \left[1/Y^R, 1/Y^L\right] \\ \qquad = \begin{bmatrix} \min\left(X^L/Y^L, X^L/Y^R, X^R/Y^L, X^R/Y^R\right), \\ \max\left(X^L/Y^L, X^L/Y^R, X^R/Y^L, X^R/Y^R\right) \end{bmatrix}, \\ \qquad \text{where}\left[Y^L, Y^R\right] \neq 0 \end{cases} \tag{6.36}$$

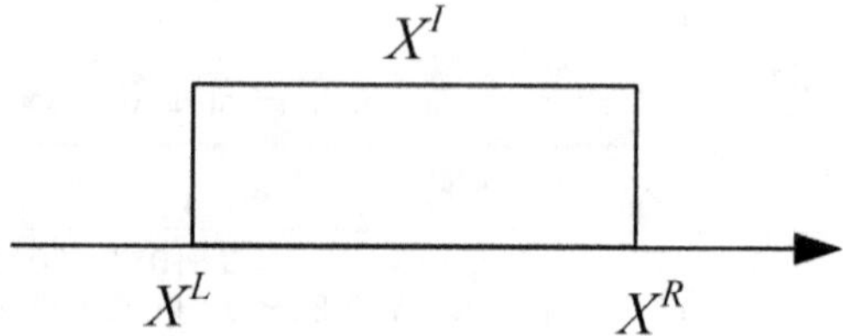

Figure 6.7 Interval number geometric model.

There is a certain difference between the four operations of interval number and the four operations of real number, and the interval expansion problem will occur in the process of interval number operation. In particular, the more the number of times the interval number appears in the function, the greater the error will occur in the application of practical engineering problems, as shown in Eq. 6.37 [26]

$$\begin{cases} f(X_1, X_2, X_3) = \dfrac{X_1 + X_2}{X_1 - X_2} X_3 \\ X_1 = X_1^I = [1,3] \\ X_2 = X_2^I = [4,6] \\ X_3 = X_3^I = [7,10] \end{cases} \tag{6.37}$$

The true value range of the original function $f(X_1, X_2, X_3)$ is [−70, −8.4]. Suppose the four arithmetic rules of interval number are used to calculate directly. In that case, the calculated interval function value is [−90, -6], which is quite different from the actual function value, and the error is large. At present, interval uncertainty optimization is an optimization method based on interval uncertainty analysis. The interval uncertainty analysis is an uncertainty analysis method based on the basic properties of interval numbers. How to establish a more accurate method to overcome the error in the interval calculation process is of great significance to the optimization model and uncertainty analysis model. According to research at home and abroad, the methods currently used to deal with interval calculation errors include the combination method, truncation method, direct interval expansion method, improved interval expansion method, perturbation method, sub-interval method, etc. Most relevant methods so far only apply to some nonlinear engineering problems. In the future, for complex high-dimensional nonlinear engineering problems, it is still necessary to further explore how to adapt to the interval model analysis method of practical engineering.

6.2.1.2 Research on analysis method of interval uncertainty optimization model

6.2.1.2.1 Construction of interval uncertainty optimization model

In the deterministic optimization problem, the specific value of the optimal solution of the optimization model can be directly obtained by the optimization algorithm. However, there are many unavoidable uncertainties in engineering problems, which leads to the fact that the optimal solution obtained by deterministic optimization cannot meet the actual situation in many cases, so it is particularly crucial to study uncertainty optimization. Before studying the uncertainty optimization model, the deterministic optimization model is given, as shown in Eq. 6.38

$$\begin{cases} \text{Find DV} \\ \min\limits_{\text{DV}} \ f(X) \\ \text{s.t.} \quad h(X) = 0 \\ \qquad\ \ g(X) \geq 0 \\ \text{DV} = \{X\} \end{cases} \tag{6.38}$$

where $f(\bullet)$ represents the objective function, $h(\bullet)$ represents the equality constraint, $g(\bullet)$ represents the inequality constraint, DV represents the design variable space, X represents the design variable. In this paper, the default $f(\bullet)$, $h(\bullet)$, and $g(\bullet)$ are real-valued continuous functions that can obtain second-order continuous partial derivatives.

The uncertainty optimization method considers the uncertainty factors based on the traditional deterministic optimization method to model, analyze, and optimize the solution. The nested loop optimization strategy is usually used to nest the uncertainty analysis process into the optimization solution, and the optimization process is shown in Figure 6.8.

The optimization model considering uncertainty factors can be expressed as Eq. 6.39

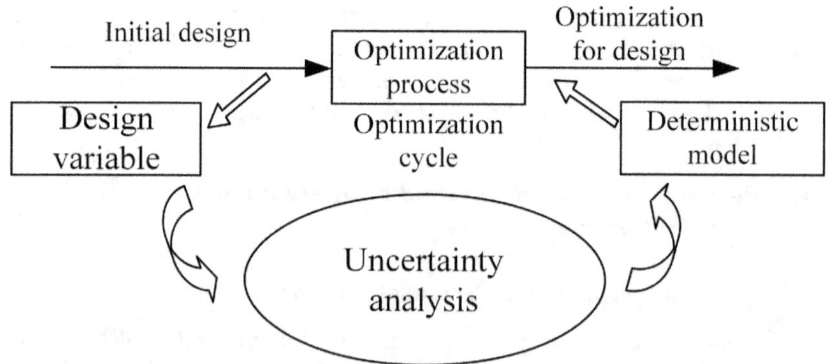

Figure 6.8 Optimization process considering uncertainty.

$$\begin{cases} \min\limits_{\text{DV}} & f(X, X_U) \\ \text{s.t.} & h(X, X_U) = 0 \\ & g(X, X_U) \ge 0 \\ \text{DV} = \{X, X_U\} \end{cases} \tag{6.39}$$

Except for the design variables already included in Eq. 6.38, X_U represents the uncertain design variables of the optimization problem.

The non-probabilistic uncertainty method's interval uncertainty analysis and optimization method are essential. This chapter mainly studies the interval uncertainty analysis method and optimization solution strategy. The optimization model considering interval uncertainty can be expressed as Eq. 6.40

$$\begin{cases} \min\limits_{\text{DV}} & f(X, X_U^I) \\ \text{s.t.} & h(X, X_U^I) = a^I, \\ & g(X, X_U^I) \ge b^I \\ & X_U^I \in [X_U^L, X_U^R],\ X \in R^n \\ \text{DV} = \{X, X_U^I\} \end{cases} \tag{6.40}$$

where X_U^I represents the interval variable, X_U^L and X_U^R represent the upper and lower bounds of the interval variable, a^I and b^I represent the possibility interval of the constraint.

6.2.1.2.2 Analysis method of interval uncertainty optimization model

For the uncertainty optimization model, this paper uses the interval possibility degree method and the interval order method for analysis. In the study of interval possibility degree transformation, Japanese scholars Nakahara et al. [27] proposed an interval possibility degree formula, Zhang et al. [28] proposed an interval possibility degree method based on three interval positional relations, Jiang et al. [25] based on the three positional relations proposed by Zhang, further supplemented and improved the three positional relations, and proposed an interval possibility degree method based on six positional relations. The possibility degree calculation method based on probability theory for positional relations is generally more objective and accurate. The size relationship of the interval is expressed as shown in Figure 6.9.

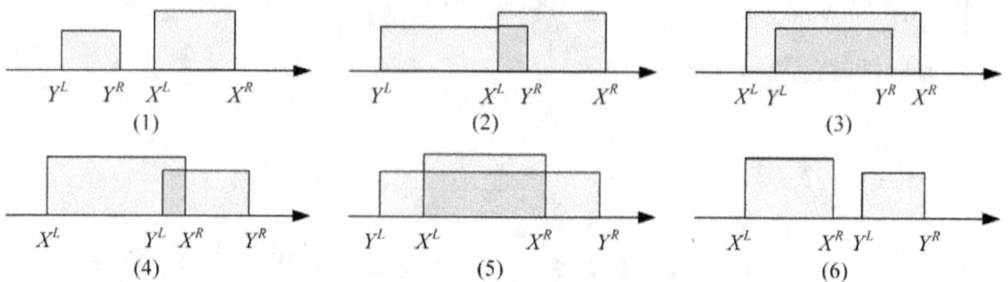

Figure 6.9 Interval positional relationship diagram.

According to this positional relationship, Eq. 6.41 shows the interval possibility calculation method established by Jiang [9].

$$P\left(X^I \le Y^I\right)=\begin{cases} 0, & X^L \ge Y^R \\ \dfrac{Y^R-X^L}{2\left(X^R-X^L\right)} \bullet \dfrac{Y^R-X^L}{Y^R-Y^L}, & Y^L \le X^L < Y^R \le X^R \\ \dfrac{Y^L-X^L}{X^R-X^L}+\dfrac{Y^R-Y^L}{2\left(X^R-X^L\right)}, & X^L < Y^L < Y^R \le X^R \\ \dfrac{Y^L-X^L}{X^R-X^L}+\dfrac{X^R-Y^L}{X^R-X^L} \bullet \dfrac{Y^R-X^R}{Y^R-Y^L} & \\ +\dfrac{X^R-Y^L}{2\left(X^R-X^L\right)} \bullet \dfrac{X^R-Y^L}{Y^R-Y^L}, & X^L < Y^L \le X^R < Y^R \\ \dfrac{Y^R-X^R}{Y^R-Y^L}+\dfrac{X^R-X^L}{2\left(Y^R-Y^L\right)}, & Y^L \le X^L < X^R < Y^R \\ 1, & X^R < Y^L \end{cases} \tag{6.41}$$

The above probability-based possibility calculation method has excellent advantages in practical engineering, but in practical engineering, there are not only cases where $X^I \le Y^I$ probability is required, but also cases where $X^I \ge Y^I$ probability is required. Therefore, in order to meet the optimization model in practical engineering, this chapter further supplements the possibility solution model for the above interval position relationship, as follows:

$$P(X^I \geq Y^I) = \begin{cases} 1, & X^L \geq Y^R \\ \dfrac{X^R - Y^R}{X^R - X^L} + \dfrac{Y^R - X^L}{X^R - X^L} \bullet \dfrac{X^L - Y^L}{Y^R - Y^L} \\ + \dfrac{Y^R - X^L}{2(X^R - X^L)} \bullet \dfrac{Y^R - X^L}{Y^R - Y^L}, & Y^L \leq X^L < Y^R \leq X^R \\ \dfrac{X^R - Y^R}{X^R - X^L} + \dfrac{Y^R - Y^L}{2(X^R - X^L)}, & X^L < Y^L < Y^R \leq X^R \\ \dfrac{X^R - Y^L}{2(Y^R - Y^L)} \bullet \dfrac{X^R - Y^L}{X^R - X^L}, & X^L < Y^L \leq X^R < Y^R \\ \dfrac{X^L - Y^L}{Y^R - Y^L} + \dfrac{X^R - X^L}{2(Y^R - Y^L)}, & Y^L \leq X^L < X^R < Y^R \\ 0, & X^R < Y^L \end{cases} \tag{6.42}$$

After transforming the constraint conditions by the interval possibility method, the optimization model can be expressed as shown in the Eq. 6.43.

$$\begin{cases} \min\limits_{\mathrm{DV}} \ f(X, X_U^I) \\ \text{s.t.} \quad P\left(h\ (X, X_U^I) = a^I\right) \geq \lambda \\ \qquad\ P\left(g\ (X, X_U^I) \geq b^I\right) \geq \theta \\ \qquad\ X_U^I \in \left[X_U^L, X_U^R\right], X \in R^n \\ \mathrm{DV} = \{X, X_U^I\} \end{cases} \tag{6.43}$$

where λ and θ are the interval possibility level given according to the actual demand.

Using the interval order relation transformation method to figure out the interval uncertainty optimization model in practical engineering problems is of great significance. The interval order relation is mainly used to judge the advantages and disadvantages of the two intervals and express the preference of the decision for the interval. In interval uncertainty optimization, there are interval variables in the objective function. Therefore, after the uncertainty analysis, the objective function value is generally an interval, and the objective function interval needs to be compared and analyzed in the optimization process. Ma [29] summarized the maximization problem of interval order relation, and Jiang [25] further supplemented and improved the interval order relation based on interval midpoint and interval radius. In this paper, aiming at the minimization optimization problem, the interval order relationship is biased toward the upper bound of the interval, and the midpoint of the interval is further improved. The comprehensive summary is as follows:

1. The interval order relationship that is biased toward the maximum and minimum values of the interval

$$\begin{cases} A^I \leq_{LR} B^I \text{ and only if } A^L \geq B^L \text{ and } A^R \geq B^R \\ A^I <_{LR} B^I \text{ and only if } A^I \leq_{LR} B^I \text{ and } A^I \neq B^I \end{cases} \quad (6.44)$$

2. Interval order relations biased toward interval expectation and radius

$$\begin{cases} A^I \leq_{cw} B^I \text{ and only if } A^c \geq B^c \text{ and } A^w \geq B^w \\ A^I <_{cw} B^I \text{ and only if } A^I \leq_{cw} B^I \text{ and } A^I \neq B^I \end{cases} \quad (6.45)$$

3. The interval order relation, which is biased toward the interval expectation and the interval minimum

$$\begin{cases} A^I \leq_{Lc} B^I \text{ and only if } A^L \geq B^L \text{ and } A^c \geq B^c \\ A^I <_{Lc} B^I \text{ and only if } A^I \leq_{Lc} B^I \text{ and } A^I \neq B^I \end{cases} \quad (6.46)$$

4. The interval order relation, which is biased toward the interval expectation and the interval maximum

$$\begin{cases} A^I \leq_{cR} B^I \text{ and only if } A^c \geq B^c \text{ and } A^R \geq B^R \\ A^I <_{cR} B^I \text{ and only if } A^I \leq_{cR} B^I \text{ and } A^I \neq B^I \end{cases} \quad (6.47)$$

5. The interval order relation, which is biased toward the interval minimum value

$$\begin{cases} A^I \leq_{L} B^I \text{ and only if } A^L \geq B^L \\ A^I <_{L} B^I \text{ and only if } A^I \leq_{L} B^I \text{ and } A^I \neq B^I \end{cases} \quad (6.48)$$

6. The interval order relation biased toward the interval maximum value

$$\begin{cases} A^I \leq_{R} B^I \text{ and only if } A^R \geq B^R \\ A^I <_{R} B^I \text{ and only if } A^I \leq_{R} B^I \text{ and } A^I \neq B^I \end{cases} \quad (6.49)$$

7. Interval order relations biased toward interval expectations

$$\begin{cases} A^I \leq_c B^I \text{ and only if } A^c \geq B^c \\ A^I <_c B^I \text{ and only if } A^I \leq_c B^I \text{ and } A^I \neq B^I \end{cases} \tag{6.50}$$

8. Interval order relation biased toward interval radius

$$\begin{cases} A^I \leq_w B^I \text{ and only if } A^w \geq B^w \\ A^I <_w B^I \text{ and only if } A^I \leq_w B^I \text{ and } A^I \neq B^I \end{cases} \tag{6.51}$$

In this paper, we mainly focus on the upper and lower bounds of the interval to analyze the uncertainty of the objective function. We can calculate the objective function of the interval uncertainty optimization model as follows:

$$f\left(X, X_U^I\right) = f^I(X) = \left[f^L(X), f^R(X)\right] \tag{6.52}$$

By transforming the objective function, the transformed optimization model can be expressed as follows:

$$\begin{cases} \min\limits_{\text{DV}} \ \left(f^L(X),\ f^R(X)\right) \\ \text{s.t.} \quad P\left(h\left(X, X_U^I\right) = a^I\right) \geq \lambda \\ \qquad\ \ P\left(g\left(X, X_U^I\right) \geq b^I\right) \geq \theta \\ \text{DV} = \left\{X, X_U^I\right\}, \quad X_U^I \in \left[X_U^L, X_U^R\right], \ X \in R^n \end{cases} \tag{6.53}$$

where

$$\begin{cases} f^L(X) = \min f\left(X, X_U^I\right) \\ f^R(X) = \max f\left(X, X_U^I\right) \end{cases} \tag{6.54}$$

Further processing the constraints in the optimization model, the optimization model can be expressed as follows:

$$\begin{cases} \min\limits_{\text{DV}} \ \left(f^L(X),\ f^R(X)\right) \\ \text{s.t.} \quad P\left(h^I(X) = a^I\right) \geq \lambda \\ \qquad\ \ P\left(g^I(X) \geq b^I\right) \geq \theta \\ \text{DV} = \left\{X, X_U^I\right\}, \quad X \in R^n \end{cases} \tag{6.55}$$

The specific calculation process of $h^I(X)$ and $g^I(X)$ is shown in Eq. 6.56.

$$\begin{cases} h^I(X) = \left(\min\limits_{X_U^I} h(X, X_U^I), \max\limits_{X_U^I} h(X, X_U^I) \right) \\ g^I(X) = \left(\min\limits_{X_U^I} g(X, X_U^I), \max\limits_{X_U^I} g(X, X_U^I) \right) \end{cases} \tag{6.56}$$

The uncertainty optimization model is transformed into a deterministic multi-objective optimization model based on the upper and lower bounds of the interval through uncertainty analysis and model transformation. This model can more intuitively quantify the uncertainty of engineering problems and is more convenient for calculation.

6.2.1.2.3 Mathematical examples

The mathematical example optimization model is shown in Eq. 6.57

$$\begin{cases} \min \ f(X, X_U) = 300 - (X_1 - 3)^2 X_{U1} + X_2^3 X_{U2}^2 + (X_3^2 + 5) X_{U3} \\ \text{s.t.} \ \ h_1(X, X_U) = X_1 X_{U1}^2 + 3X_2 X_{U2} - X_3^2 X_{U3} = [45, 55] \\ \qquad g_1(X, X_U) = X_1^2 X_{U1} - X_2 X_{U2}^2 + X_3^2 X_{U3}^2 \le [115, 130] \\ \qquad g_2(X, X_U) = X_1 X_{U1} + X_2^2 X_{U2}^2 - X_3 X_{U3} \le [30, 40] \\ \qquad 4 \le X_1 \le 15, 2 \le X_2 \le 8, -2 \le X_3 \le 10 \\ \qquad X_{U1} \in [0.9, 1.1], X_{U2} \in [0.8, 1.2], X_{U3} \in [0.85, 1.15] \end{cases} \tag{6.57}$$

The uncertainty variables $X_{U1} \in [0.9, 1.1]$, $X_{U2} \in [0.8, 1.2]$, and $X_{U3} \in [0.85, 1.15]$ are set, respectively, and the uncertainty level is defined as the ratio of the interval radius to the interval midpoint. The uncertainty levels are 10%, 20%, and 15%, respectively. Next, the interval possibility method and the interval order relation method are used to deal with the interval uncertainty variables in the optimization model to solve the above uncertainty model.

First, the above model is transformed, and the model of Eq. 6.57 can be transformed into Eq. 6.58 by interval possibility degree method and interval order relation method

$$\begin{cases} \min\limits_{X} \begin{pmatrix} \underset{\min X_U}{f^L(X)} = 300-(X_1-3)^2 X_{U1} + X_2^3 X_{U2}^2 + (X_3^2+5) X_{U3} \\ \underset{\max X_U}{f^R(X)} = 300-(X_1-3)^2 X_{U1} + X_2^3 X_{U2}^2 + (X_3^2+5) X_{U3} \end{pmatrix} \\ \text{s.t.} \quad P\left(h_1(X,X_U) = X_1 X_{U1}^2 + 3X_2 X_{U2} - X_3^2 X_{U3} = [45,55]\right) \geq \lambda \\ \qquad P\left(g_1(X,X_U) = X_1^2 X_{U1} - X_2 X_{U2}^2 + X_3^2 X_{U3}^2 \leq [115,130]\right) \geq \theta_1 \\ \qquad P\left(g_2(X,X_U) = X_1 X_{U1} + X_2^2 X_{U2}^2 - X_3 X_{U3} \leq [30,40]\right) \geq \theta_2 \\ \qquad 4 \leq X_1 \leq 15, 2 \leq X_2 \leq 8, -2 \leq X_3 \leq 10 \\ \qquad X_{U1} \in [0.9,1.1], X_{U2} \in [0.8,1.2], X_{U3} \in [0.85,1.15] \end{cases} \tag{6.58}$$

The multi-objective optimization problem is transformed by the evaluation function method, and the optimization model under the constraint condition is processed by the penalty function method. The optimization model of Eq. 6.58 can be transformed into Eq. 6.59

$$\begin{cases} \min\limits_{\mathrm{DV}} \ f = f_M(X) + \sigma P(X) \\ \mathrm{DV} = \{X\},\ X \in R^n \end{cases} \tag{6.59}$$

where

$$\begin{cases} f_M(X) = \min\left(\mu \left| f^L(X) - f_L^* \right| + (1-\mu) \left| f^R(X) - f_R^* \right| \right) \\ P(X) = \begin{pmatrix} \phi_1\left(P\left(g_1(X,X_U) \geq [115,130]\right) - \theta_1\right) \\ +\phi_2\left(P\left(g_2(X,X_U) \geq [30,40]\right) - \theta_2\right) \\ +\varphi\left(P\left(h_1(X,X_U) = [45,50]\right) - \lambda\right) \end{pmatrix} \end{cases} \tag{6.60}$$

Before optimization, the relevant optimization parameters are first set, and the specific parameters are set as follows: The penalty factor σ is set to 10^7, the population size of the improved PSO algorithm is set to 20, and the inertia weight ω is usually set to $[0.8,1.2]$. In order to better search for the global optimal solution, ω in this section and below it is set to 1.1, and the acceleration constants c_1 and c_2 are generally equal, that is, $c_1 = c_2 = 1.5$ is taken, and the number of iterations of the inner and outer optimization algorithms is set to 50 and 100, respectively.

After the parameter setting is completed, the change of the only constraint possibility of the importance factor of the objective function and the change of the importance factor of the only objective function of the constraint possibility will be discussed separately, and the optimization results will be obtained by analyzing and solving.

First, the importance factor of the objective function is set to be unique. When the importance factor μ of the objective function is set to be 0.5, that is, in the case of interval midpoint preference, different constraint possibility degrees are set, and the interval uncertainty mathematical optimization model is solved under different constraint possibility degrees. The specific setting values and calculation results are shown in Table 6.6.

It can be seen from the calculation results that when the importance factor μ of the objective function is a fixed value of 0.5, the constraint possibility is different, and there are some differences between the obtained optimal design vector and the objective function interval. It can be seen from the data in the table that the greater the constraint possibility, the greater the restriction on the function, so the data show that the greater the constraint possibility, the greater the minimum value of the function corresponding to the optimal solution. In order to better express the solution process, as shown in Figure 6.10, the convergence curves of the solution process corresponding to different constraint possibilities are given.

The above analysis shows that the importance factor of the objective function is unique, and the constraint possibility is different. Subsequently, the constraint possibility remains unchanged, and the situation of different importance factors of the objective function is analyzed and solved. Given the constraint possibility degree $\lambda = \theta_1 = \theta_2 = 1.0$, in the case of this constraint possibility degree, the importance factor μ of the objective function is solved in four cases: 0, 1/3, 2/3, and 1. The calculation results are shown in Table 6.7.

It can be seen from the analysis of the above calculation results that when the constraint possibility degree is set to a fixed value of 1.0, the optimal design vector and the interval range of the objective function are different by

Table 6.6 Solution results of different constraint possibility

Constraint possibility degree	*Optimal design vector*	*Range of the objective function*	*Adaptation degree (10^9)*
0.0	[11.69,5.92,0.12]	[16.87,26.27]	6.29
0.2	[11.55,5.74,0.13]	[38.24,46.92]	5.64
0.4	[11.41,5.57,0.13]	[58.01,66.15]	5.49
0.6	[11.27,5.38,0.14]	[77.32,84.76]	5.04
0.8	[11.13,5.19,0.14]	[96.71,103.25]	4.49
1.0	[10.98,4.99,0.14]	[114.13,119.85]	3.97

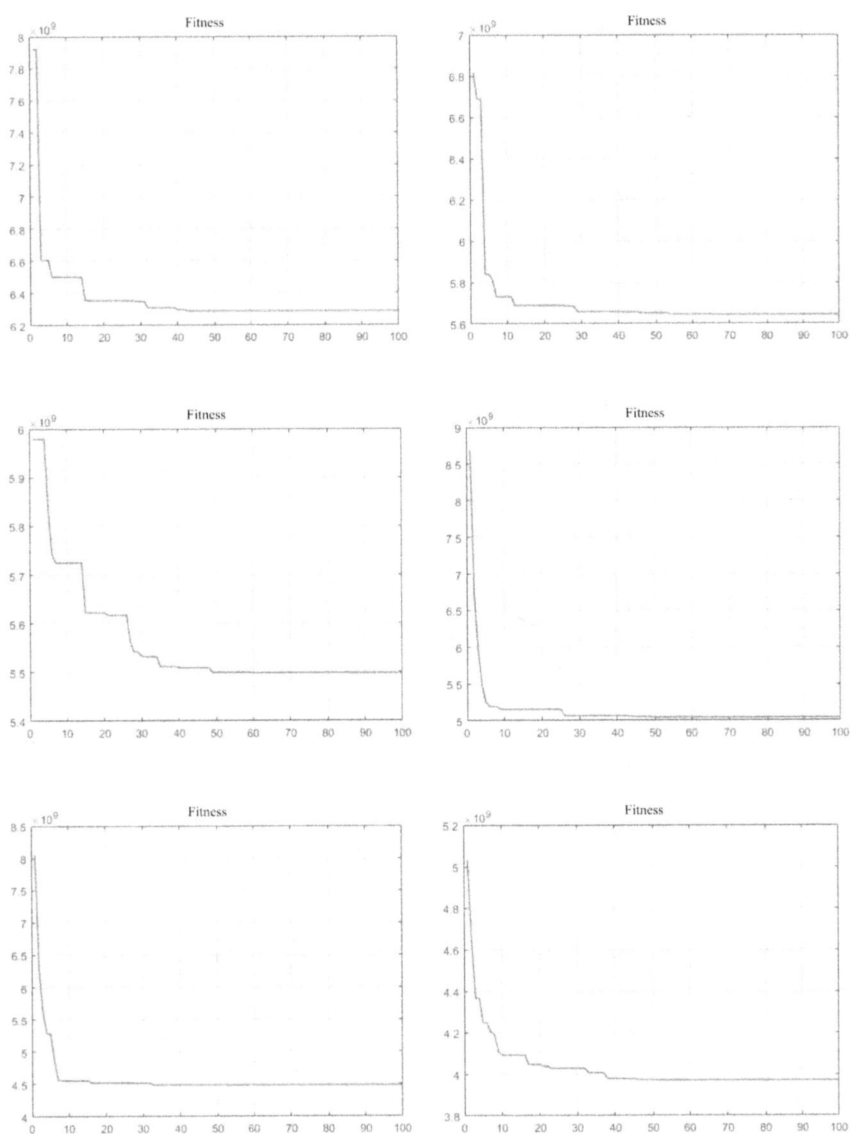

Figure 6.10 Iterative convergence curves corresponding to different constraint possibility degrees.

Table 6.7 Solution results of importance factors of different objective functions

μ	*Optimal design vector*	*Range of the objective function*	*Adaptation degree (10^9)*
0	[10.51,4.09,0.14]	[143.21,148.53]	3.37
1/3	[10.83,4.64,0.15]	[123.81,128.95]	3.79
2/3	[11.14,5.34,0.15]	[103.75,109.01]	4.14
1	[11.53,6.35,0.15]	[71.86,77.21]	4.54

setting different objective function importance factors μ. When the importance factor μ of the objective function is larger, the decision-making process has a stronger preference for the lower bound of the interval. When μ is smaller, the decision-making process has a stronger preference for the upper bound of the interval. When the importance factor μ of the objective function is closer to 0.5, the decision-making process has a stronger preference for the interval's midpoint. As shown in Figure 6.11, the convergence curves of the solution process corresponding to the importance factors of different objective functions are given.

Through the above two cases, the constraint possibility and the importance factor of the objective function mainly depend on the needs of the actual problem in the selection and setting process. If the overall reliability requirement of the system is low, a smaller constraint possibility value can be set. On the contrary, a larger constraint possibility value needs to be set. The setting of the objective function's importance factor depends on the engineering designer's preference for the upper and lower bounds of the interval and the actual engineering problem requirements. This section discusses various possibilities and important factors of the objective function, and effective optimization results are obtained. The algorithm is verified by mathematical examples, which provides ideas for solving subsequent engineering cases.

6.2.2 UBMDO method based on possibility theory

In MDO, considering the cognitive uncertainty of design variables and parameters has become one of the hot topics in MDO research. This section will study the Possibility-Based MDO (PBMDO) problem, which mainly includes when continuous variables and parameters only contain cognitive uncertainty, the uncertainty analysis model in the MDO environment, and the optimization method of the PBMDO problem, as shown in Figure 6.12.

In order to meet the requirements of high reliability and safety of complex coupled systems, RBMDO has attracted more and more attention.

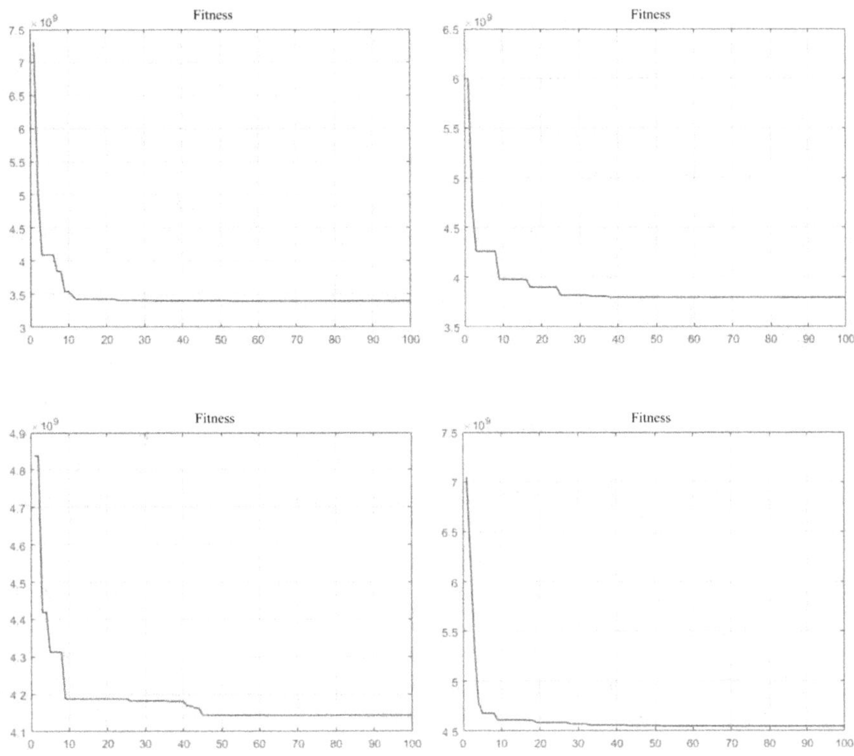

Figure 6.11 The iterative convergence curves corresponding to the importance factors of different objective functions.

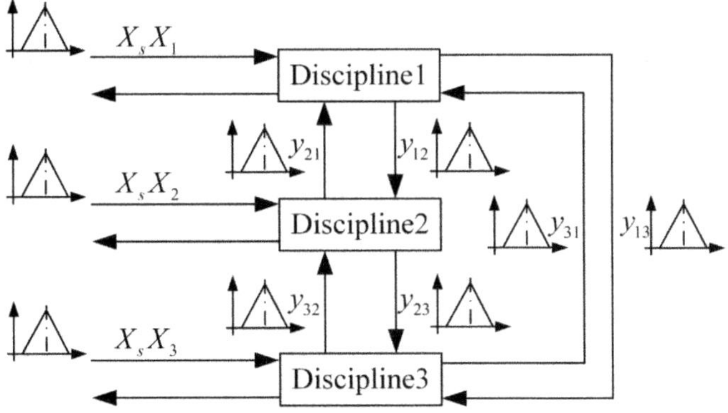

Figure 6.12 The propagation of cognitive uncertainty in MDO.

However, in the initial design stage of complex coupled systems, due to time, environment, manpower, and other reasons, there is a lack of sufficient information and data so that the probability distribution of variables and parameters cannot be accurately established, which is the basis of RBMDO. When limited data cannot establish the probability distribution of variables and parameters, RBMDO may obtain an unsafe design scheme. In the literature [30] (single subject), 50 sets of samples were randomly generated from two continuous normal variables with an assumed known probability distribution, and the probability distribution of these two random variables was obtained by fitting technique. When the performance function is linear, the reliability analysis shows that the 92% reliability obtained by fitting distribution is only equivalent to 77% reliability when the probability distribution is known. System nonlinearity and lack of data will increase the difference between the two. When the exact probability distribution cannot be obtained from limited data, the probabilistic method may not be competent for structural analysis and optimization because unreasonable uncertainty modeling produces more random uncertainty than physical uncertainty [31, 32].

The situation will become worse in multidisciplinary system design. Taking an MDO system with three disciplines as an example (as shown in Figure 6.12), the cognitive uncertainty in discipline 1 will be transmitted to disciplines 2 and 3 through coupling variables y_{12} and y_{13}, respectively, and will have an impact on the design constraints and output coupling variables in these two disciplines; at the same time, the uncertainties in disciplines 2 and 3 will be transmitted to discipline 1 through y_{21} and y_{31}, respectively, which will affect the design constraints and output coupling variables in discipline 1. Suppose the cognitive uncertainty in a certain discipline is not reasonably considered. In that case, it will affect the design of this discipline and the design of other disciplines through the coupling relationship between disciplines. Therefore, it is necessary to reasonably characterize cognitive uncertainty in multidisciplinary systems.

Possibility theory is a powerful tool for cognitive uncertainty [30–38]. The possibility method can deal with the uncertainty of insufficient information because it defines a fuzzy variable corresponding to the limited available data [34]. The main advantage of possibility analysis is that it uses the membership function of physical variables to retain the inherent random nature of these variables; the fuzzy operation is more straightforward than the probability method, especially in the case of more variables. Furthermore, when the data are insufficient, a more conservative design will be provided by the possibility method [30, 31, 33]. Therefore, PBMDO provides an alternative method for designing complex coupled systems when the probability distribution of variables and parameters cannot be accurately established due to insufficient data.

6.2.2.1 Possibility theory

The theory of possibility was first proposed by Zadeh [39], and then developed. The following definitions are derived from Reference [40] except for continuous and discrete fuzzy variables from Reference [41].

1. Ample domain

The set of sample space subsets is an Ample field, denoted by A, if it satisfies:

1. $\Omega \in \mathrm{A}$
2. If event $A \in \mathrm{A}$, then the complement $A^c \in \mathrm{A}$ of A
3. If $A_\gamma \in \mathrm{A}, \gamma \in \Gamma$, then $\bigcup_{\gamma \in \Gamma} A_\gamma \in \mathrm{A}$

Among them, Γ is the index set, which can be finite, countable or uncountable.

2. Possibility measure

A is an Ample field of the sample space Ω. A possibility measure is a function Π on A such that:

1. For all $A_\gamma \in \mathrm{A}$, $\Pi(A) \geq 0$
2. $\Pi(\varnothing) \geq 0$ ($\varnothing$ is an empty set), $\Pi(\Omega) = 1$
3. If $A_\gamma \in \mathrm{A}, \gamma \in \Gamma$, then $\Pi\left(\bigcup_{\gamma \in \Gamma} A_\gamma\right) = \sup_{\gamma \in \Gamma} \Pi(A_\gamma)$

Sample space Ω, Ample field A and possibility measure Π constitute the possibility space $(\Omega, \mathrm{A}, \Pi)$.

3. Fuzzy variables and fuzzy vectors

Fuzzy variable X is a function which maps from possibility space $(\Omega, \mathrm{A}, \Pi)$ to real number. The fuzzy variable X is characterized by the membership function $\Pi_X(x)$ corresponding to each point in the real number field $\mathbb{R}$. When X takes value x, $\Pi_X(x)$ is called the grade of membership of x in X.

Continuous fuzzy variables: When the membership function $\Pi_X(x)$ of a fuzzy variable is a continuous function, it is called a continuous fuzzy variable.

Discrete fuzzy variable: When there is a countable sequence $\{x_1, x_2, \ldots\}$ such that $\Pi\{X \neq x_1, X \neq x_2, \ldots\} = 0$.

The membership function $\Pi_X(x)$ of a continuous fuzzy variable X is bounded: For any $\alpha \in (0,1]$, the set $\{x: \Pi_X(x) \geq \alpha\}$ is bounded.

The membership function $\Pi_X(x)$ of a continuous fuzzy variable X is convex: For any $\alpha \in (0,1]$, the set $\{x : \Pi_X(x) \geq \alpha\}$ is convex. The membership function $\Pi_X(x)$ is convex if and only if for all x_1, x_2 and $\lambda \in [0,1]$, $\Pi_X(\lambda x_1 + (1-\lambda)x_2) \geq \min\{\Pi_X(x_1), \Pi_X(x_2)\}$.

The unity property of the membership function $\Pi_X(x)$ of a fuzzy variable X means that there is and only one point x^M such that $\Pi_X(x^M) = 1$, which is called the maximum grade point.

n-dimensional fuzzy vector $\mathbf{X}$ is a function that maps from the possibility space $(\Omega, \mathrm{A}, \Pi)$ to the real number field $\mathbb{R}^n$.

4. **Non-interactive nature**

Fuzzy variable X_1, X_2 and its membership function are $\Pi_{X_1}(x_1)$ and $\Pi_{X_2}(x_2)$ are non-interactive, respectively, if their joint membership function $\Pi_{X_1,X_2}(x_1, x_2)$ satisfies $\Pi_{X_1,X_2}(x_1, x_2) = \min\{\Pi_{X_1}(x_1), \Pi_{X_2}(x_2)\}$.

Fuzzy variables $X_1, X_2, \cdots, X_n$ are non-interactive if their joint membership functions satisfy $\Pi_{X_1,\ldots,X_n}(x_1, \ldots, x_n) = \min\{\Pi_{X_1}(x_1), \ldots, \Pi_{X_n}(x_n)\}$.

6.2.2.2 PBMDO optimization model

The PBMDO optimization model is

$$
\begin{aligned}
&\min_{DV} f(\mathbf{d}_{s,c}, \mathbf{d}_c, \mathbf{X}^M_{s,fc}, \mathbf{X}^M_{fc}, \mathbf{P}^M_{fc}, \mathbf{Y}^M) \\
&s.t.\ \ \Pi(G^{(i)}(\mathbf{d}_{s,c}, \mathbf{d}_{i,c}, \mathbf{X}_{s,fc}, \mathbf{X}_{i,fc}, \mathbf{P}_{i,fc}, \mathbf{Y}_{\bullet i}) > 0) \leq \alpha_t \\
&\qquad g^{(i)}(\mathbf{d}_{s,c}, \mathbf{d}_{i,c}, \mathbf{X}^M_{s,fc}, \mathbf{X}^M_{i,fc}, \mathbf{P}^M_{i,fc}, \mathbf{Y}^M_{\bullet i}) \leq 0 \\
&\qquad i = 1, 2, \cdots, nd \\
&\qquad \mathbf{d}^L_{s,c} \leq \mathbf{d}_{s,c} \leq \mathbf{d}^U_{s,c}, \mathbf{d}^L_c \leq \mathbf{d}_c \leq \mathbf{d}^U_c \\
&\qquad \mathbf{X}^{M,L}_{s,fc} \leq \mathbf{X}^M_{s,fc} \leq \mathbf{X}^{M,U}_{s,fc}, \mathbf{X}^{M,L}_{fc} \leq \mathbf{X}^M_{fc} \leq \mathbf{X}^{M,U}_{fc} \\
&DV = \{\mathbf{d}_{s,c}, \mathbf{d}_c, \mathbf{X}^M_{s,fc}, \mathbf{X}^M_{fc}\}
\end{aligned}
\tag{6.61}
$$

where the subscript 's,i' represents the shared variable, the local variable or parameter of discipline i, 'c,fc' refers to the continuous variable, the continuous fuzzy variable or parameter, and the superscript '(i)' refers to the discipline i. $\mathbf{d}_{s,c}$ is a continuous shared deterministic design variable. $\mathbf{d}_c = \{\mathbf{d}_{1,c}, \mathbf{d}_{2,c}, \cdots, \mathbf{d}_{nd,c}\}$ is composed of continuous local deterministic variables of all sub-disciplines, nd is the total number of sub-disciplines. $\mathbf{d}_{i,c}$ is a continuous local deterministic design variable in discipline i. $\mathbf{X}_{s,fc}$ is a continuous shared fuzzy variable; $\mathbf{X}_{fc} = \{\mathbf{X}_{1,fc}, \mathbf{X}_{2,fc}, \cdots, \mathbf{X}_{nd,fc}\}$ is composed of continuous local fuzzy variables of all sub-disciplines, $\mathbf{X}_{fc}^{M}$ is its maximum membership point. $\mathbf{X}_{i,fc}$ is a continuous local fuzzy variable in discipline i. $\mathbf{P}_{fc} = \{\mathbf{P}_{1,fc}, \mathbf{P}_{2,fc}, \cdots, \mathbf{P}_{nd,fc}\}$ is composed of continuous local fuzzy parameters of all sub-disciplines, $\mathbf{P}_{fc}^{M}$ is its maximum membership point, $\mathbf{P}_{i,fc}$ is a continuous local fuzzy parameter in discipline i. $f(\cdot)$ is the objective function, $\Pi[G^{(i)}(\cdot) > 0] \le \alpha_t$ is the possibility constraint in discipline i, where $G^{(i)}(\cdot)$ is the performance function, and the failure mode is defined as $G^{(i)}(\cdot) > 0$; $g^{(i)}$ is the deterministic constraint in discipline i. 'L' and 'U' are the lower and upper boundaries, respectively. $\mathbf{Y} = \{\mathbf{Y}_{1\bullet}, \ldots, \mathbf{Y}_{nd\bullet}\}$ is the set of all coupling variables, $\mathbf{Y}_{i\bullet}$ is the coupling variable output from discipline i to other disciplines, and $\mathbf{Y}_{\bullet i}$ is the coupling variable input from other disciplines to discipline i. $\mathbf{Y}^{M}$ is the value of the coupling variable at the design point.

The relationship between coupling variables, shared variables, local variables, and parameters is

$$\begin{aligned} &\mathbf{Y}_{i\bullet} = \mathbf{Y}_{i\bullet}(\mathbf{d}_{s,c}, \mathbf{d}_{i,c}, \mathbf{X}_{s,fc}, \mathbf{X}_{i,fc}, \mathbf{P}_{i,fc}, \mathbf{Y}_{\bullet i}) \\ &\mathbf{Y}_{i\bullet} = \{y_{ij}; j = 1 \sim nd, j \ne i\} \\ &i = 1 \sim nd \end{aligned} \tag{6.62}$$

This paper assumes that continuous fuzzy variables and parameters satisfy the non-interactive property, and their membership functions satisfy unity, strong convexity, and boundedness. The direct solution of PBMDO involves three nested loops: The outer loop optimizes the objective function and obtains the design point; the middle ring analyses the possibility of this design point; the internal ring performs discipline consistency analysis. Next, we discuss the uncertainty analysis (possibility analysis) model in PBMDO.

6.2.2.3 Uncertainty analysis (possibility analysis) model

When a design point is obtained, an uncertainty analysis (possibility analysis) is required to test whether the possibility constraints are satisfied at the design point. The possibility analysis consists of two steps:

1. The continuous fuzzy variables and parameters that satisfy the non-interactive property are transformed into standard fuzzy variables and parameters $\mathbf{V}$, and the transformation method is

$$v = \begin{cases} \Pi_X(x) - 1, & x \le X^M \\ 1 - \Pi_X(x), & x > X^M \end{cases} \tag{6.63}$$

where $\Pi_X(x)$ is the membership function of continuous fuzzy variable X, and X^M is its maximum membership point.

2. Solving the optimization problem

$$\begin{aligned} &\max_{\mathbf{V}} G(\mathbf{V}) \\ &s.t. \ \| \mathbf{V} \|_{\infty} \le 1 - \alpha_t \end{aligned} \tag{6.64}$$

The optimal solution of the above equation is the Most Probable Point (MPP) $\mathbf{V}^*$ and the function value $G(\mathbf{V}^*)$ of the performance function at MPP. If $G(\mathbf{V}^*) \le 0$, then the possibility constraint $\Pi(G > 0) \le \alpha_t$ is satisfied; and vice versa.

In MDO, when continuous variables and parameters contain cognitive uncertainty, the cognitive uncertainty of a discipline will be transmitted to other disciplines through coupling variables, and the cognitive uncertainty of other disciplines will also be transmitted to this discipline through the coupling relationship between disciplines. This means that the local fuzzy variables and parameters of other disciplines will affect the possibility constraints of this discipline through Eq. 6.62. It can be seen from the performance function $G^{(i)}(\mathbf{d}_{s,c}, \mathbf{d}_{i,c}, \mathbf{X}_{s,fc}, \mathbf{X}_{i,fc}, \mathbf{P}_{i,fc}, \mathbf{Y}_{\bullet i})$ and Eq. 6.62 of the possibility constraint in discipline *i* that the performance function implicitly contains all deterministic design variables, continuous fuzzy variables, and parameters in the multidisciplinary system because $\mathbf{Y}_{\bullet i}$ is a coupling variable input from other disciplines to discipline *i*, and it itself is a function of deterministic design variables, continuous local fuzzy variables and parameters in other disciplines.

In the possibility analysis, the continuous fuzzy variables and parameters $\mathbf{X}_{s,fc}$, $\mathbf{X}_{fc} = \{\mathbf{X}_{1,fc}, \ldots, \mathbf{X}_{nd,fc}\}$, and $\mathbf{P}_{fc} = \{\mathbf{P}_{1,fc}, \ldots, \mathbf{P}_{nd,fc}\}$ are first transformed into standard fuzzy variables and parameters $\mathbf{V}_{s,fc}$, $\mathbf{V}_{fc} = \{\mathbf{V}_{1,fc}, \ldots, \mathbf{V}_{nd,fc}\}$, and $\mathbf{VP}_{fc} = \{\mathbf{VP}_{1,fc}, \ldots, \mathbf{VP}_{nd,fc}\}$ (where $\mathbf{V}, \mathbf{VP}$ is standard fuzzy variable and standard fuzzy parameter, respectively). Based on Eq. 6.64, the possibility analysis model of possibility constraint in PBMDO discipline *i* is

$$\begin{aligned}
&\max_{DV} G^{(i)}(\mathbf{d}_{s,c},\mathbf{d}_{i,c},\mathbf{V}^{(i)}_{s,fc},\mathbf{V}^{(i)}_{i,fc},\mathbf{VP}^{(i)}_{i,fc},\mathbf{Y}^{(i)}_{\bullet i})\\
&s.t.\ \|(\mathbf{V}^{(i)}_{s,fc},\mathbf{V}^{(i)}_{fc},\mathbf{VP}^{(i)}_{fc})\|_{\infty}\leq 1-\alpha_t\\
&\quad y^{(i)}_{jm}=y_{jm}(\mathbf{d}_{s,c},\mathbf{d}_{j,c},\mathbf{V}^{(i)}_{s,fc},\mathbf{V}^{(i)}_{j,fc},\mathbf{VP}^{(i)}_{j,fc},\mathbf{Y}^{(i)}_{\bullet j})\\
&\quad j,m=1\sim nd;m\neq j\\
&DV=\{\mathbf{V}^{(i)}_{s,fc},\mathbf{V}^{(i)}_{fc},\mathbf{VP}^{(i)}_{fc},\mathbf{Y}^{(i)}\}
\end{aligned}\tag{6.65}$$

where the superscript (i) refers to the subject i. $\mathbf{d}_{s,c},\mathbf{d}_{i,c},\mathbf{d}_{j,c}$ is the corresponding deterministic design variable value in the obtained design point. $\mathbf{Y}^{(i)}$ is the coupling variable, which corresponds to the performance function $G^{(i)}$ of the possibility constraint in discipline i. In the above formula, the first design constraint contains all the transformed standard fuzzy variables and parameters because $G^{(i)}$ implicitly contains all the fuzzy variables and parameters. Using the idea of Individual Discipline Feasible (IDF), the coupling variable is used as an additional design variable, and the inter-disciplinary consistency requirement is used as an additional design constraint.

The optimization results are the MPP $\mathbf{V}^{*,(i)}_{s,fc},\mathbf{V}^{*,(i)}_{fc},\mathbf{VP}^{*,(i)}_{fc}$ in V space, the value $\mathbf{Y}^{*,(i)}$ of the coupling variable at MPP, and the function value $G^{(i)}$ of the performance function at MPP. The MPP $\mathbf{X}^{*,(i)}_{s,fc},\mathbf{X}^{*,(i)}_{fc},\mathbf{P}^{*,(i)}_{fc}$ in X space can be obtained by the inverse process of Eq. 6.63. If the function value $G^{(i)}\leq 0$, the possibility constraint $\Pi(G^{(i)}(\cdot)>0)\leq\alpha_t$ will be satisfied at the design point; otherwise, the possibility constraint will not be satisfied, and the design point will not be feasible.

From the above discussion, it can be seen that the possibility constraint $\Pi(G^{(i)}(\mathbf{d}_{s,c},\mathbf{d}_{i,c},\mathbf{X}_{s,fc},\mathbf{X}_{i,fc},\mathbf{P}_{i,fc},\mathbf{Y}_{\bullet i})>0)\leq\alpha_t$ is equivalent to the certainty constraint $G^{(i)}(\mathbf{d}_{s,c},\mathbf{d}_{i,c},\mathbf{X}^{*,(i)}_{s,fc},\mathbf{X}^{*,(i)}_{i,fc},\mathbf{P}^{*,(i)}_{i,fc},\mathbf{Y}^{*,(i)}_{\bullet i})\leq 0$. Therefore, Eq. 6.61 is equivalent to

$$\begin{aligned}
&\min_{DV} f(\mathbf{d}_{s,c},\mathbf{d}_{c},\mathbf{X}^{M}_{s,fc},\mathbf{X}^{M}_{fc},\mathbf{P}^{M}_{fc},\mathbf{Y}^{M})\\
&s.t.\ \ G^{(i)}(\mathbf{d}_{s,c},\mathbf{d}_{i,c},\mathbf{X}^{*,(i)}_{s,fc},\mathbf{X}^{*,(i)}_{i,fc},\mathbf{P}^{*,(i)}_{i,fc},\mathbf{Y}^{*,(i)}_{\bullet i})\leq 0\\
&\quad g^{(i)}(\mathbf{d}_{s,c},\mathbf{d}_{i,c},\mathbf{X}^{M}_{s,fc},\mathbf{X}^{M}_{i,fc},\mathbf{P}^{M}_{i,fc},\mathbf{Y}^{M}_{\bullet i})\leq 0\\
&\quad i=1,2,\cdots,nd\\
&\quad \mathbf{d}^{L}_{s,c}\leq\mathbf{d}_{s,c}\leq\mathbf{d}^{U}_{s,c},\mathbf{d}^{L}_{c}\leq\mathbf{d}_{c}\leq\mathbf{d}^{U}_{c}\\
&\quad \mathbf{X}^{M,L}_{s,fc}\leq\mathbf{X}^{M}_{s,fc}\leq\mathbf{X}^{M,U}_{s,fc},\mathbf{X}^{M,L}_{fc}\leq\mathbf{X}^{M}_{fc}\leq\mathbf{X}^{M,U}_{fc}\\
&DV=\{\mathbf{d}_{s,c},\mathbf{d}_{c},\mathbf{X}^{M}_{s,fc},\mathbf{X}^{M}_{fc}\}
\end{aligned}\tag{6.66}$$

where $\mathbf{X}_{s,fc}^{*,(i)}, \mathbf{X}_{i,fc}^{*,(i)}, \mathbf{P}_{i,fc}^{*,(i)}$ is the MPP of the performance function of the possibility constraint in the subject i in the $\mathbf{X}$ space, which corresponds to $\mathbf{X}_{s,fc}, \mathbf{X}_{i,fc}, \mathbf{P}_{i,fc}$, respectively; $\mathbf{Y}_{\bullet i}^{*,(i)}$ is the value of the coupling variable at MPP.

In order to solve PBMDO effectively and efficiently, two optimization methods of the PBMDO problem are proposed based on SORA and Safety Factor Assessment (SFA): PBMDO in the framework of SORA (PBMDO-SORA); PBMDO in the framework of SFA (PBMDO-SFA). Next, we first discuss PBMDO-SORA.

6.2.2.4 PBMDO-SORA

This section will discuss the PBMDO-SORA strategy, steps, and mathematical model in detail and give an example to verify.

6.2.2.4.1 PBMDO-SORA strategy

In order to effectively solve the PBMDO problem, the following two key technologies are used in PBMDO-SORA:

1. Performance Measure Approach (PMA). In RBDO and PBDO, the feasibility of using the PMA method to analyze reliability constraints and possibility constraints has significantly improved the computational efficiency
2. SORA. Using the basic idea of SORA, the possibility analysis is separated from the optimization, and the three-nested solution form of PBMDO is transformed into sequence-determined MDO and possibility analysis. The MPP obtained in the previous loop is used to modify the deterministic design constraints in this loop to improve the feasibility of the design point

6.2.2.4.2 PBMDO-SORA step

PBMDO-SORA includes the following steps:

Step 1: Set the initial value of the design variable $\mathbf{d}_{s,c}^{(0)}, \mathbf{d}_{c}^{(0)}, \mathbf{X}_{s,fc}^{M,(0)}, \mathbf{X}_{fc}^{M,(0)}$, where $k = 1$.

Step 2: Figure out the determined MDO to obtain the design point $\mathbf{d}_{s,c}^{(k)}, \mathbf{d}_{c}^{(k)}, \mathbf{X}_{s,fc}^{M,(k)}, \mathbf{X}_{fc}^{M,(k)}$. In the first loop, since the possibility analysis has not been carried out, the MPP value of each function is set to $\mathbf{X}_{s,fc}^{M,(0)},\mathbf{X}_{fc}^{M,(0)},\mathbf{P}_{fc}^{M}$; beginning with the second ring, the deterministic constraints in the determined MDO are modified using the MPP obtained in the previous ring.

Step 3: Possibility analysis. Prior to conducting a possibility analysis, all continuous fuzzy variables and parameters are converted into standard fuzzy variables and parameters by using Eq. 6.55. In the mathematical model of possibility analysis, interdisciplinary consistency is considered an additional design constraint.

Step 4: Test the convergence and feasibility. If all the possibility constraints are satisfied and the objective function value converges ($G^{(i),k} \le 0, i = 1 \sim nd; | f(k) - f(k-1) | \le \varepsilon$, ε is an arbitrarily small positive number), then stop; otherwise $k = k+1$, go to **Step 2**.

The mathematical models in **Steps 2** and **3** will be discussed and given later. When the possibility constraint is not satisfied, the following discussion is about using the SORA idea to modify the design constraints in the determined MDO.

When the possibility constraint $\Pi(G^{(i)}(\mathbf{d}_{s,c},\mathbf{d}_{i,c},\mathbf{X}_{s,fc},\mathbf{X}_{i,fc},\mathbf{P}_{i,fc},\mathbf{Y}_{\bullet i}) > 0) \le \alpha_t$ is not satisfied in the (k – 1)th ring, which means that the performance function has a function value $G^{(i),(k-1)} > 0$ at its MPP ($G^{(i)}(\mathbf{d}_{s,c},\mathbf{d}_{i,c},\mathbf{X}_{s,fc}^{*,(i)},\mathbf{X}_{i,fc}^{*,(i)},\mathbf{P}_{i,fc}^{*,(i)},\mathbf{Y}_{\bullet i}^{*,(i)}) \le 0$), then the MPP obtained by the possibility analysis of the (k – 1)th ring will be used to modify the MDO determined by the (k – 1)th ring. It can be seen from the discussion in Section 6.2.2.3 that satisfying the possibility constraint is equivalent to satisfying the deterministic constraint $G^{(i)}(\mathbf{d}_{s,c},\mathbf{d}_{i,c},\mathbf{X}_{s,fc}^{*,(i)},\mathbf{X}_{i,fc}^{*,(i)},\mathbf{P}_{i,fc}^{*,(i)},\mathbf{Y}_{\bullet i}^{*,(i)}) \le 0$, which means that the function value of the performance function at MPP is on bigger than zero, that is, the corresponding MPP of this performance function must be in the determined feasible region ($G^{(i)} \le 0$). Since the MPP in the (k – 1)th ring is not in a certain feasible region at this time, in order to improve the feasibility of the design point in the (k – 1)th ring, **S** is defined as the moving vector and the idea of SORA design movement is adopted

$$\begin{aligned}
&\mathbf{S}_{s,fc}^{(i),k} = \mathbf{X}_{s,fc}^{M,(k-1)} - \mathbf{X}_{s,fc}^{*,(i),(k-1)} \\
&\mathbf{S}_{j,fc}^{(i),k} = \mathbf{X}_{j,fc}^{M,(k-1)} - \mathbf{X}_{j,fc}^{*,(i),(k-1)} \\
&j = 1 \sim nd
\end{aligned} \tag{6.67}$$

Modify the design constraints in the MDO determined by the kth ring. Among them, $\mathbf{S}_{s,fc}^{(i),k},\mathbf{S}_{j,fc}^{(i),k}$ is the moving vector corresponding to $\mathbf{X}_{s,fc},\mathbf{X}_{j,fc}$ in discipline i. $\mathbf{X}_{s,fc}^{M,(k-1)},\mathbf{X}_{j,fc}^{M,(k-1)}; j=1\sim nd$ is the maximum membership degree point of the fuzzy variable in the design point obtained by the $(k-1)$th ring, $\mathbf{X}_{s,fc}^{*,(i),(k-1)},\mathbf{X}_{j,fc}^{*,(i),(k-1)}; j=1\sim nd$ is the MPP obtained by the possibility analysis of the $(k-1)$th ring.

Since the maximum membership point of the continuous fuzzy parameter is known, fixed and not a variable, the MPP of the continuous fuzzy parameter directly replaces $\mathbf{P}$ in the performance function $G^{(i)}(\cdot)$. The design constraint in the MDO determined by the kth ring is modified to

$$G_{\Pi}^{(i)}(\mathbf{d}_{s,c},\mathbf{d}_{i,c},\mathbf{X}_{s,fc}^{M}-\mathbf{S}_{s,fc}^{(i),k},\mathbf{X}_{i,fc}^{M}-\mathbf{S}_{i,fc}^{(i),k},\mathbf{P}_{i,fc}^{*,(i),(k-1)},\mathbf{Y}_{\bullet i}^{*,(i)})\le 0$$

6.2.2.4.3 *Mathematical model in PBMDO-SORA*

The mathematical models of deterministic MDO and possibility analysis mentioned in the previous section will be discussed and given in this section.

6.2.2.4.3.1 *THE MDO DETERMINED IN THE kth RING*

The MDO mathematical model determined in the kth ring is

$$\begin{aligned}
&\min_{DV} f(\mathbf{d}_{s,c}^{k},\mathbf{d}_{c}^{k},\mathbf{X}_{s,fc}^{M,k},\mathbf{X}_{fc}^{M,k},\mathbf{P}_{fc}^{M},\mathbf{Y}^{M,k})\\
&s.t.\ \ G_{\Pi}^{(i)}(\mathbf{d}_{s,c}^{k},\mathbf{d}_{i,c}^{k},\mathbf{X}_{s,fc}^{M,k}-\mathbf{S}_{s,fc}^{(i),k},\mathbf{X}_{i,fc}^{M,k}-\mathbf{S}_{i,fc}^{(i),k},\mathbf{P}_{i,fc}^{*,(i),(k-1)},\mathbf{Y}_{\bullet i}^{*,(i)})\le 0\\
&\qquad g^{(i)}(\mathbf{d}_{s,c}^{k},\mathbf{d}_{i,c}^{k},\mathbf{X}_{s,fc}^{M,k},\mathbf{X}_{i,fc}^{M,k},\mathbf{P}_{i,fc}^{M},\mathbf{Y}_{\bullet i}^{M,k})\le 0\\
&\qquad y_{jm}^{*,(i)}=y_{jm}(\mathbf{d}_{s,c}^{k},\mathbf{d}_{j,c}^{k},\mathbf{X}_{s,fc}^{M,k}-\mathbf{S}_{s,fc}^{(i),k},\mathbf{X}_{j,fc}^{M,k}-\mathbf{S}_{j,fc}^{(i),k},\mathbf{P}_{j,fc}^{*,(i),(k-1)},\mathbf{Y}_{\bullet j}^{*,(i)})\\
&\qquad i,j,m=1\sim nd; m\ne j\\
&\qquad y_{ij}^{M,k}=y_{ij}(\mathbf{d}_{s,c}^{k},\mathbf{d}_{i,c}^{k},\mathbf{X}_{s,fc}^{M,k},\mathbf{X}_{i,fc}^{M,k},\mathbf{P}_{i,fc}^{M},\mathbf{Y}_{\bullet i}^{M,k})\ \ i,j=1\sim nd; j\ne i\\
&\qquad \mathbf{d}_{s,c}^{L}\le\mathbf{d}_{s,c}^{k}\le\mathbf{d}_{s,c}^{U},\mathbf{d}_{c}^{L}\le\mathbf{d}_{c}^{k}\le\mathbf{d}_{c}^{U}\\
&\qquad \mathbf{X}_{s,fc}^{M,L}\le\mathbf{X}_{s,fc}^{M,k}\le\mathbf{X}_{s,fc}^{M,U},\mathbf{X}_{fc}^{M,L}\le\mathbf{X}_{fc}^{M,k}\le\mathbf{X}_{fc}^{M,U}\\
&DV=\{\mathbf{d}_{s,c}^{k},\mathbf{d}_{c}^{k},\mathbf{X}_{s,fc}^{M,k},\mathbf{X}_{fc}^{M,k},\mathbf{Y}^{M,k},\mathbf{Y}^{*}\}
\end{aligned}\tag{6.68}$$

where, the superscript 'k' is in the kth ring. $\mathbf{Y}^{*}=\{\mathbf{Y}^{*,(i)},i=1\sim nd\}$; $\mathbf{Y}^{*,(i)}$ is the value of the coupling variable at MPP, corresponding to the possibility constraint in discipline i. $G_{\Pi}^{(i)}$ is a modified deterministic design constraint corresponding to the possibility constraint in discipline i. $\mathbf{S}_{s,fc}^{(i),k},\mathbf{S}_{i,fc}^{(i),k},\mathbf{S}_{j,fc}^{(i),k}$ are the moving vectors of discipline i corresponding to $\mathbf{X}_{s,fc},\mathbf{X}_{i,fc},\mathbf{X}_{j,fc}$ in the

kth ring, respectively. $\mathbf{P}_{i,fc}^{*,(i),(k-1)}, \mathbf{P}_{j,fc}^{*,(i),(k-1)}$ is the MPP of the continuous fuzzy parameter $\mathbf{P}_{i,fc}, \mathbf{P}_{j,fc}$ obtained by discipline i in the $(k-1)$th ring.

In Eq. 6.68, the equality constraints for obtaining interdisciplinary consistency also need to be modified using MPP ($\mathbf{X}_{s,fc}^{*,(i),(k-1)}, \mathbf{X}_{j,fc}^{*,(i),(k-1)}, \mathbf{P}_{j,fc}^{*,(i),(k-1)}$)

$$\begin{aligned}
& y_{jm}^{*,(i)} = y_{jm}(\mathbf{d}_{s,c}^k, \mathbf{d}_{j,c}^k, \mathbf{X}_{s,fc}^{M,k} - \mathbf{S}_{s,fc}^{(i),k}, \mathbf{X}_{j,fc}^{M,k} - \mathbf{S}_{j,fc}^{(i),k}, \mathbf{P}_{j,fc}^{*,(i),(k-1)}, \mathbf{Y}_{\bullet j}^{*,(i)}) \\
& i, j, m = 1 \sim nd; m \neq j
\end{aligned}$$

where $\mathbf{X}_{j,fc}^{*,(i),(k-1)}, \mathbf{P}_{j,fc}^{*,(i),(k-1)}$ is the MPP of $\mathbf{X}_{j,fc}, \mathbf{P}_{j,fc}$ obtained by subject i in the $(k-1)$th ring.

6.2.2.4.3.2 POSSIBILITY ANALYSIS IN kth RING From the discussion of uncertainty analysis in Section 6.2.2.3, it can be seen that when the design point $\mathbf{d}_{s,c}^k, \mathbf{d}_c^k, \mathbf{X}_{s,fc}^{M,k}, \mathbf{X}_{fc}^{M,k}$ is obtained, the continuous fuzzy variables and parameters $\mathbf{X}_{s,fc}$, $\mathbf{X}_{fc} = \{\mathbf{X}_{1,fc}, \ldots, \mathbf{X}_{nd,fc}\}$, and $\mathbf{P}_{fc} = \{\mathbf{P}_{1,fc}, \ldots, \mathbf{P}_{nd,fc}\}$ are first transformed into standard fuzzy variables and parameters $\mathbf{V}_{s,fc}$, $\mathbf{V}_{fc} = \{\mathbf{V}_{1,fc}, \ldots, \mathbf{V}_{nd,fc}\}$, and $\mathbf{VP}_{fc} = \{\mathbf{VP}_{1,fc}, \ldots, \mathbf{VP}_{nd,fc}\}$, respectively, and then the optimization problem (the mathematical model of possibility analysis in the kth ring) is solved

$$\begin{aligned}
& \max_{DV} G^{(i)}(\mathbf{d}_{s,c}^k, \mathbf{d}_{i,c}^k, \mathbf{V}_{s,fc}^{(i),k}, \mathbf{V}_{i,fc}^{(i),k}, \mathbf{VP}_{i,fc}^{(i),k}, \mathbf{Y}_{\bullet i}^{(i)}) \\
& s.t. \; \| (\mathbf{V}_{s,fc}^{(i),k}, \mathbf{V}_{fc}^{(i),k}, \mathbf{VP}_{fc}^{(i),k}) \|_{\infty} \leq 1 - \alpha_t \\
& \quad y_{jm}^{(i)} = y_{jm}(\mathbf{d}_{s,c}^k, \mathbf{d}_{j,c}^k, \mathbf{V}_{s,fc}^{(i),k}, \mathbf{V}_{j,fc}^{(i),k}, \mathbf{VP}_{j,fc}^{(i),k}, \mathbf{Y}_{\bullet j}^{(i)}) \\
& \quad j, m = 1 \sim nd; m \neq j \\
& DV = \{\mathbf{V}_{s,fc}^{(i),k}, \mathbf{V}_{fc}^{(i),k}, \mathbf{VP}_{fc}^{(i),k}, \mathbf{Y}^{(i)}\} \\
& i = 1 \sim nd
\end{aligned} \tag{6.69}$$

where, the superscript 'k' is in the kth ring. $\mathbf{d}_{s,c}^k, \mathbf{d}_{i,c}^k, \mathbf{d}_{j,c}^k$ is the $\mathbf{d}_{s,c}, \mathbf{d}_{i,c}, \mathbf{d}_{j,c}$ value obtained after solving the MDO determined in the kth ring. $\mathbf{Y}^{(i)} = \{y_{jm}^{(i)}; j, m = 1 \sim nd, m \neq j\}$ is the value of the coupling variable at MPP.

The results of the possibility analysis are MPP ($\mathbf{V}_{s,fc}^{*,(i),k}, \mathbf{V}_{fc}^{*,(i),k}, \mathbf{VP}_{fc}^{*,(i),k}$) in the standard fuzzy space, the value of the coupling variable $\mathbf{Y}^{*,(i)}$ and $G^{(i),k}$,

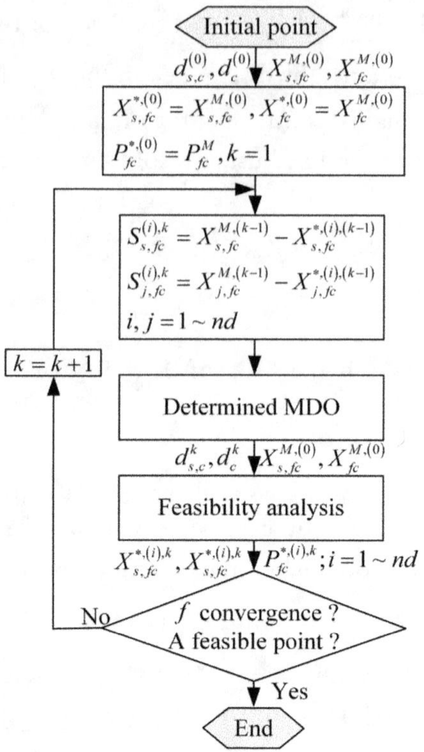

Figure 6.13 PBMDO-SORA flowchart.

$i = 1 \sim nd$ (the function value of the performance function in the MPP). MPP ($\mathbf{X}_{s,fc}^{*,(i),k}, \mathbf{X}_{fc}^{*,(i),k}, \mathbf{P}_{fc}^{*,(i),k}$) in $\mathbf{X}$ space can be obtained by the inverse process of Eq. 6.55. If the function value $G^{(i),k} \le 0$, the possibility which constraints $\Pi(G^{(i)}(\cdot) > 0) \le \alpha_t$ will be satisfied at the design point; otherwise, the possibility constraint is not satisfied, and the design point is not feasible. When the possibility constraints are not fully satisfied, and the objective function value does not converge, the design constraints in the determined MDO will be modified by using MPP. The whole PBMDO-SORA flow diagram is shown in Figure 6.13.

6.2.2.4.4 *Example*

Heart dipole problem

The heart dipole problem is a classic MDO verification example. The heart dipole problem involves using two artificial dipoles to determine the synthetic dipole distance of the human heart in a disk containing electrolytes. The magnitude, direction, and position of the two independent dipoles in

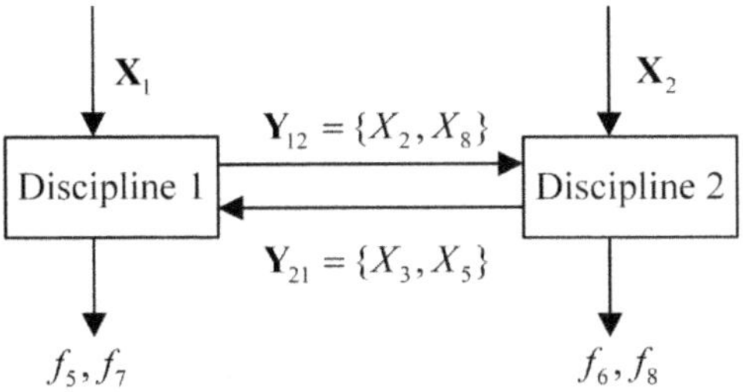

Figure 6.14 The heart dipole multidisciplinary problem.

the disk are solved by using the Gabor-Nelson equations. Taking these two independent dipoles as two sub-disciplines, the problem is transformed into an MDO optimization problem. The heart dipole multidisciplinary question is shown in Figure 6.14.

In this paper, the mathematical model of the heart dipole multidisciplinary problem is modified to highlight the coupling relationship between disciplines, and the original MDO problem is modified to a PBMDO problem.

Discipline 1: Continuous fuzzy variable $\mathbf{X}_1 = \{X_1, X_4\}$, deterministic design variable X_6, X_7, continuous fuzzy parameter d_{mx}, deterministic parameter d_a, d_c, d_e, input coupling variable $\mathbf{Y}_{21} = \{X_3, X_5\}$, output coupling variable $\mathbf{Y}_{12} = \{X_2, X_8\}$, output f_5, f_7.

where,

$$X_2 = d_{mx} - X_1$$

$$\begin{aligned} f_5 = {} & X_1(X_5^2 - X_7^2) - 2X_3X_5X_7 \\ & + (d_{mx} - X_1)\left[X_6^2 - \left(\frac{X_5X_1 + X_6(d_{mx} - X_1) - X_7X_3 - d_a}{X_4}\right)^2\right] \\ & - 2X_4X_6\left(\frac{X_5X_1 + X_6(d_{mx} - X_1) - X_7X_3 - d_a}{X_4}\right) - d_c \end{aligned}$$

$$\begin{aligned} f_7 = {} & X_1X_5(X_5^2 - 3X_7^2) + X_3X_7(X_7^2 - 3X_5^2) \\ & + (d_{mx} - X_1)X_6\left[X_6^2 - 3\left(\frac{X_5X_1 + X_6(d_{mx} - X_1) - X_7X_3 - d_a}{X_4}\right)^2\right] \\ & + X_4\left(\frac{X_5X_1 + X_6(d_{mx} - X_1) - X_7X_3 - d_a}{X_4}\right)\left[\left(\frac{X_5X_1 + X_6(d_{mx} - X_1) - X_7X_3 - d_a}{X_4}\right)^2 - 3X_6^2\right] \\ & - d_e \end{aligned}$$

The possibility constraint is

$$\begin{aligned} &\Pi[(-f_5) > 0] \le \alpha_t \\ &\Pi[(-f_7) > 0] \le \alpha_t \end{aligned} \tag{6.70}$$

Discipline 2: Continuous fuzzy variable $\mathbf{X}_2 = \{X_1, X_4\}$, deterministic design variable X_6, X_7, continuous fuzzy parameter d_{my}, deterministic parameter d_b, d_d, d_f, input coupling variable $\mathbf{Y}_{12} = \{X_2, X_8\}$, output coupling variable $\mathbf{Y}_{21} = \{X_3, X_5\}$, output is f_6, f_8.
where,

$$X_3 = d_{my} - X_4$$

$$\begin{aligned} f_6 = (d_{my} - X_4)&\left[\left(-\frac{X_7X_1 + X_8X_2 + X_6X_4 - d_b}{d_{my} - X_4}\right)^2 - X_7^2\right] \\ &+ 2X_1\left(-\frac{X_7X_1 + X_8X_2 + X_6X_4 - d_b}{d_{my} - X_4}\right)X_7 \\ &+ X_4(X_6^2 - X_8^2) + 2X_2X_6X_8 - d_d \end{aligned}$$

$$\begin{aligned} f_8 = (d_{my} - X_4)&\left(-\frac{X_7X_1 + X_8X_2 + X_6X_4 - d_b}{d_{my} - X_4}\right) \\ &\left[\begin{matrix}\left(-\dfrac{X_7X_1 + X_8X_2 + X_6X_4 - d_b}{d_{my} - X_4}\right)^2 \\ -3X_7^2\end{matrix}\right] \\ &- X_1X_7\left[X_7^2 - 3\left(-\frac{X_7X_1 + X_8X_2 + X_6X_4 - d_b}{d_{my} - X_4}\right)^2\right] \\ &+ X_4X_6(X_6^2 - 3X_8^2) - X_2X_8(X_8^2 - 3X_6^2) - d_f \end{aligned}$$

The possibility constraint is

$$\begin{aligned} &\Pi[(-f_6) > 0] \le \alpha_t \\ &\Pi[(-f_8) > 0] \le \alpha_t \end{aligned} \tag{6.71}$$

Table 6.8 Continuous fuzzy variables and parameters in heart dipole

Variables/ Parameters	*The largest vassal*	*Deviation*	*Type of membership function*	*Lower bounds of design variables*	*The upper bound of design variables*
X_1		0.03	Triangular	0.1	399.9
X_4		0.03	Triangular	0.1	399.9
d_{mx}	0.05	0.003	Triangular		
d_{my}	0.05	0.003	Triangular		

Table 6.9 The heart dipole problem PBMDO conventional optimization and PBMDO-SORA optimization results

	X_1^M	X_4^M	X_6	X_7	*Objective function values*	n_1	n_2	k
Regular optimization methods	1.0821	6.8173	0.5006	0.9812	0.2736	93101	93101	
PBMDO-SORA	27.9386	2.3497	0.3416	0.6349	0.0600	8613	8613	5

The objective is to minimize the value of $f_5 + f_6 + f_7 + f_8$ at the design point. The lower and upper bounds of deterministic design variables X_6 and X_7 are 0.1 and 399.9, and the deterministic parameter $d_a, d_b, d_c, d_d, d_e, d_f$ is 0.01. Table 6.8 is the description of the uncertainty of variables and parameters.

Table 6.9 lists the conventional optimization methods for the heart dipole problem PBMDO problem and the optimal solutions, objective function values, and number of disciplinary analyses of PBMDO-SORA.

The solution form of the nested loop takes the discipline consistency requirement as an additional design constraint in the possibility analysis. In order to further improve the computational efficiency, in the process of using PBMDO-SORA, the optimal solution of each optimization problem in the previous ring is taken as the starting point of the corresponding optimization problem in this ring. This strategy is not adopted since the nested ring solution form is not sequential. The conventional optimization method and PBMDO-SORA solution are solved from the same starting point, and the same optimization algorithm is used. It can be seen from the objective function values in Table 6.9 that the optimal solution obtained

Table 6.10 Function values of each performance function at MPP after conventional optimization and PBMDO-SORA optimization

	$-f_5$	$-f_7$	$-f_6$	$-f_8$
Regular optimization methods	−0.0601	−0.0438	−0.0003	−0.1410
PBMDO-SORA	−0.0180	5.3066×10^{-9}	7.9413×10^{-10}	−0.0328

by PBMDO-SORA is better than the optimal solution of the conventional optimization method. Moreover, the number of disciplinary analyses in PBMDO-SORA is much less than that of conventional optimization methods, significantly improving the solution efficiency. Table 6.10 lists the function values of each function at its corresponding MPP. In PBMDO-SORA, if the function values of the performance functions $-f_7$ and $-f_6$ are approximately zero, it means that the corresponding possibility constraint is the action constraint. These possibility constraints are satisfied if the remaining function values are less than zero.

6.3 UBMDO METHOD WITH MIXED UNCERTAINTY

In the actual multidisciplinary system design, variables, and parameters often contain both random and interval uncertainties, which makes the research of MDO under random and interval uncertainties become one of the focuses of MDO research. This section studies Mixed Variables (random and fuzzy variables) MDO (MVMDO) under aleatory and epistemic uncertainties. It mainly includes when the continuous variables and parameters contain random and interval uncertainties, and the uncertainty analysis model in the MDO environment is established as the optimization method of the MVMDO problem.

6.3.1 Multidisciplinary uncertainty analysis method considering aleatory and interval uncertainties

In traditional UBMDO, uncertain design variables are often regarded as random variables with probability distribution. However, in engineering applications, the probability distribution of random variables is often unknown due to the limited and missing data samples. At the same time, those uncertain design variables that do not have random uncertainty cannot be forced to be attached to the probability distribution. For a variable whose value is in a certain interval, if its distribution information cannot be obtained, its uncertainty should be described by interval theory. Compared with UBMDO considering only random variables, UBMDO considering interval variables will face more severe computational pressure in reliability analysis. In order to solve this problem, this chapter proposes a Magnetic Resonance Angiography (MRA) method based on the worst reliability of interval variables within their value range.

There will be interval variables when considering the mixed uncertainty in multidisciplinary engineering. Specifically, since $\boldsymbol{Y}$ is used to represent the coupling variables in the multidisciplinary design, the interval variables in this section can no longer be represented by Y, but represent the interval variables by using the superscript 'I,' and use a specific interval $\left[x^{\mathrm{I}}_{\mathrm{lower}}, x^{\mathrm{I}}_{\mathrm{upper}}\right]$ to represent these interval variables $\left(\boldsymbol{X}^{\mathrm{I}}_i, \boldsymbol{X}^{\mathrm{I}}_{\mathrm{S}}\right)$. Therefore, the UBMDO optimization model with random variables and interval mixed variables is

$$\begin{cases} \min\limits_{\mathrm{DV}} f\left(\boldsymbol{d}_i, \boldsymbol{d}_{\mathrm{S}}, \boldsymbol{X}^{\mathrm{R,M}}_i, \boldsymbol{X}^{\mathrm{R,M}}_{\mathrm{S}}, \boldsymbol{Y}^{\mathrm{R,M}}_{ji}, \bar{\boldsymbol{X}}^{\mathrm{I}}_i, \bar{\boldsymbol{X}}^{\mathrm{I}}_{\mathrm{S}}, \bar{\boldsymbol{Y}}^{\mathrm{I}}_{ji}\right) \\ \text{s.t.} \quad \Pr\left\{\boldsymbol{g}_i\left(\boldsymbol{d}_i, \boldsymbol{d}_{\mathrm{S}}, \boldsymbol{X}^{\mathrm{R}}_i, \boldsymbol{X}^{\mathrm{R}}_{\mathrm{S}}, \boldsymbol{Y}^{\mathrm{R}}_{ji}, \boldsymbol{X}^{\mathrm{I}}_i, \boldsymbol{X}^{\mathrm{I}}_{\mathrm{S}}, \boldsymbol{Y}^{\mathrm{I}}_{ji}\right) \geq 0\right\} \geq [R] = \Phi(\beta), \\ \qquad i, j = 1 \sim n,\ i \neq j, \\ \mathrm{DV} = \{\boldsymbol{d}, \boldsymbol{X}\} \end{cases} \tag{6.72}$$

where $\bar{\bullet}$ represents the numerical midpoint of the upper and lower limits of the interval variable $\boldsymbol{X}^{\mathrm{I}}$ range.

In addition, the consistency analysis between different disciplines in the internal cycle can be expressed as

$$\boldsymbol{Y}_{ij} = \boldsymbol{Y}_{ij}\left(\boldsymbol{d}_i, \boldsymbol{d}_{\mathrm{S}}, \boldsymbol{X}^{\mathrm{R}}_i, \boldsymbol{X}^{\mathrm{R}}_{\mathrm{S}}, \boldsymbol{Y}^{\mathrm{R}}_{ji}, \boldsymbol{X}^{\mathrm{I}}_i, \boldsymbol{X}^{\mathrm{I}}_{\mathrm{S}}, \boldsymbol{Y}^{\mathrm{I}}_{ji}\right) \tag{6.73}$$

where $\boldsymbol{Y}_{ij} = \left\{\boldsymbol{Y}^{\mathrm{R}}_{ij}, \boldsymbol{Y}^{\mathrm{I}}_{ij}\right\}$.

For the uncertainty constraints in Eq. 6.72, due to the existence of random and interval uncertainties, the random and interval uncertainty evaluation of the performance function should be carried out in the process of solving UBMDO. In order to maximize structural reliability and safety, reliability can be defined as the probability of the worst case. In addition, this section develops a double-loop optimization strategy. A three-loop optimization strategy is proposed for the mixed aleatory and interval uncertainty analysis in UBMDO.

The optimization model in the outer loop is to find the worst reliability caused by interval uncertainty, as shown in Eq. 6.74

$$\begin{cases} \min\limits_{\mathrm{DV}} \ \beta = \left\| \boldsymbol{U}^{*}\left(\boldsymbol{X}_{i}^{\mathrm{I}}, \boldsymbol{X}_{\mathrm{S}}^{\mathrm{I}}, \boldsymbol{Y}_{ji}^{\mathrm{I}}\right) \right\| \\ \text{s.t.} \ \ x_{\text{lower}}^{\mathrm{I}} \le \boldsymbol{X}_{i}^{\mathrm{I}}, \boldsymbol{X}_{\mathrm{S}}^{\mathrm{I}} \le x_{\text{upper}}^{\mathrm{I}}, \\ \qquad i, j = 1 \sim n, \ i \ne j, \\ \mathrm{DV} = \left\{ \boldsymbol{X}_{i}^{\mathrm{I}}, \boldsymbol{X}_{\mathrm{S}}^{\mathrm{I}} \right\} \end{cases} \tag{6.74}$$

The optimization model of the middle loop and the inner loop is to find the MPP while maintaining the multidisciplinary balance, as shown in Eq. 6.75

$$\begin{cases} \min\limits_{\mathrm{DV}} \ \beta = \left\| \boldsymbol{U}^{\mathrm{R}} \right\| \\ \text{s.t.} \ \ g_{i}\left(\boldsymbol{d}_{i}, \boldsymbol{d}_{\mathrm{S}}, \boldsymbol{U}_{i}^{\mathrm{R}}, \boldsymbol{U}_{\mathrm{S}}^{\mathrm{R}}, \boldsymbol{Y}_{ji}^{\mathrm{R}}, \boldsymbol{X}_{i}^{\mathrm{I}}, \boldsymbol{X}_{\mathrm{S}}^{\mathrm{I}}, \boldsymbol{Y}_{ji}^{\mathrm{I}}\right) = 0, \\ \qquad \boldsymbol{Y}_{ij} = \boldsymbol{Y}_{ij}\left(\boldsymbol{d}_{i}, \boldsymbol{d}_{\mathrm{S}}, \boldsymbol{U}_{i}^{\mathrm{R}}, \boldsymbol{U}_{\mathrm{S}}^{\mathrm{R}}, \boldsymbol{Y}_{ji}^{\mathrm{R}}, \boldsymbol{X}_{i}^{\mathrm{I}}, \boldsymbol{X}_{\mathrm{S}}^{\mathrm{I}}, \boldsymbol{Y}_{ji}^{\mathrm{I}}\right), \\ \mathrm{DV} = \left\{ \boldsymbol{U}_{i}^{\mathrm{R}}, \boldsymbol{U}_{\mathrm{S}}^{\mathrm{R}} \right\}, \ i, j = 1 \sim n, \ i \ne j \end{cases} \tag{6.75}$$

It is noticeable that that the optimization model in Eq. 6.75 is embedded in the optimization model in Eq. 6.74. The solution in Eq. 6.74 is the worst-case scenario combination $\boldsymbol{X}_{\text{worst}}^{\mathrm{I}} = \left(\boldsymbol{X}_{i}^{\mathrm{I}}, \boldsymbol{X}_{\mathrm{S}}^{\mathrm{I}}\right)$, and the solution in Eq. 6.75 is the worst-case MPP $\boldsymbol{U}_{\text{worst}}^{\mathrm{R},*} = \left(\boldsymbol{U}_{i}^{\mathrm{R},*}, \boldsymbol{U}_{\mathrm{S}}^{\mathrm{R},*}\right)$.

UBMDO based on the PMA is usually more efficient than the original UBMDO. Therefore, the three-loop optimization strategy using PMA can be given as follows.

In the outer ring, the model in Eq. 6.74 can be changed to

$$\begin{cases} \min\limits_{\mathrm{DV}} \ g\left(\boldsymbol{d}_{i}, \boldsymbol{d}_{\mathrm{S}}, \boldsymbol{U}_{i}^{\mathrm{R},*}, \boldsymbol{U}_{\mathrm{S}}^{\mathrm{R},*}, \boldsymbol{Y}_{ji}^{\mathrm{R}}, \boldsymbol{X}_{i}^{\mathrm{I}}, \boldsymbol{X}_{\mathrm{S}}^{\mathrm{I}}, \boldsymbol{Y}_{ji}^{\mathrm{I}}\right) \\ \text{s.t.} \ \ x_{\text{lower}}^{\mathrm{I}} \le \boldsymbol{X}_{i}^{\mathrm{I}}, \boldsymbol{X}_{\mathrm{S}}^{\mathrm{I}} \le x_{\text{upper}}^{\mathrm{I}}, \\ \qquad i, j = 1 \sim n, \ i \ne j, \\ \mathrm{DV} = \left\{ \boldsymbol{X}_{i}^{\mathrm{I}}, \boldsymbol{X}_{\mathrm{S}}^{\mathrm{I}} \right\} \end{cases} \tag{6.76}$$

In the middle and inner rings, the model in Eq. 6.75 can be changed to

$$\begin{cases} \min\limits_{\mathrm{DV}} \ g\left(\boldsymbol{d}_{i}, \boldsymbol{d}_{\mathrm{S}}, \boldsymbol{U}_{i}^{\mathrm{R}}, \boldsymbol{U}_{\mathrm{S}}^{\mathrm{R}}, \boldsymbol{Y}_{ji}^{\mathrm{R}}, \boldsymbol{X}_{i}^{\mathrm{I}}, \boldsymbol{X}_{\mathrm{S}}^{\mathrm{I}}, \boldsymbol{Y}_{ji}^{\mathrm{I}}\right) \\ \text{s.t.} \ \ \left\| \boldsymbol{U} \right\| = \Phi^{-1}\left(\left[R\right]\right), \\ \qquad \boldsymbol{Y}_{ij} = \boldsymbol{Y}_{ij}\left(\boldsymbol{d}_{i}, \boldsymbol{d}_{\mathrm{S}}, \boldsymbol{U}_{i}^{\mathrm{R}}, \boldsymbol{U}_{\mathrm{S}}^{\mathrm{R}}, \boldsymbol{Y}_{ji}^{\mathrm{R}}, \boldsymbol{X}_{i}^{\mathrm{I}}, \boldsymbol{X}_{\mathrm{S}}^{\mathrm{I}}, \boldsymbol{Y}_{ji}^{\mathrm{I}}\right), \\ \mathrm{DV} = \left\{ \boldsymbol{U}_{i}^{\mathrm{R}}, \boldsymbol{U}_{\mathrm{S}}^{\mathrm{R}} \right\}, \ i, j = 1 \sim n, \ i \ne j \end{cases} \tag{6.77}$$

where $\left[R\right]$ denotes the required reliability.

The solution of the model in Eqs. 6.76 and 6.77 is the combination of interval variables $\boldsymbol{X}^{\mathrm{I}}_{\text{worst}} = \left(\boldsymbol{X}^{\mathrm{I}}_i, \boldsymbol{X}^{\mathrm{I}}_{\mathrm{S}}\right)$ and MPP $\boldsymbol{U}^{\mathrm{R},*,[R]}_{\text{worst}}$ corresponding to the worst case of $[R]$, respectively. Then, the worst-case performance metric $G^{[R]}_{\text{MPP,worst}}$ can be calculated as

$$G^{[R]}_{\text{MPP,worst}} = g\left(\boldsymbol{d}_i, \boldsymbol{d}_{\mathrm{S}}, \boldsymbol{U}^{\mathrm{R},*,[R]}_{\text{worst}}, \boldsymbol{Y}^{\mathrm{R}}_{ji}, \boldsymbol{X}^{\mathrm{I}}_{\text{worst}}, \boldsymbol{Y}^{\mathrm{I}}_{ji}\right) \tag{6.78}$$

where $\boldsymbol{U}^{\mathrm{R},*,[R]}_{\text{worst}} = \left(\boldsymbol{U}^{\mathrm{R},*,[R]}_i, \boldsymbol{U}^{\mathrm{R},*,[R]}_{\mathrm{S}}\right)$, $\boldsymbol{X}^{\mathrm{I}}_{\text{worst}} = \left(\boldsymbol{X}^{\mathrm{I}}_{i,\text{worst}}, \boldsymbol{X}^{\mathrm{I}}_{\mathrm{S,worst}}\right)$.

It can be found that the uncertainty analysis in UBMDO includes a three-loop optimization process. If the system optimization is considered, the whole UBMDO process will include a four-loop optimization process. In order to simplify the UBMDO process as much as possible and reduce the amount of calculation, the three-loop optimization process of uncertainty analysis can be transformed into a double-loop optimization process by combining Eqs. 6.74 and 6.75, which is given as follows:

$$\begin{cases} \min\limits_{\mathrm{DV}} \ \beta = \left\| \boldsymbol{U}^*\left(\boldsymbol{X}^{\mathrm{I}}_i, \boldsymbol{X}^{\mathrm{I}}_{\mathrm{S}}, \boldsymbol{Y}^{\mathrm{I}}_{ji}\right) \right\| \\ \text{s.t.} \ \ g_i\left(\boldsymbol{d}_i, \boldsymbol{d}_{\mathrm{S}}, \boldsymbol{U}^{\mathrm{R}}_i, \boldsymbol{U}^{\mathrm{R}}_{\mathrm{S}}, \boldsymbol{Y}^{\mathrm{R}}_{ji}, \boldsymbol{X}^{\mathrm{I}}_i, \boldsymbol{X}^{\mathrm{I}}_{\mathrm{S}}, \boldsymbol{Y}^{\mathrm{I}}_{ji}\right) = 0, \\ \qquad \boldsymbol{Y}_{ij} = \boldsymbol{Y}_{ij}\left(\boldsymbol{d}_i, \boldsymbol{d}_{\mathrm{S}}, \boldsymbol{U}^{\mathrm{R}}_i, \boldsymbol{U}^{\mathrm{R}}_{\mathrm{S}}, \boldsymbol{Y}^{\mathrm{R}}_{ji}, \boldsymbol{X}^{\mathrm{I}}_i, \boldsymbol{X}^{\mathrm{I}}_{\mathrm{S}}, \boldsymbol{Y}^{\mathrm{I}}_{ji}\right), \\ \qquad x^{\mathrm{I}}_{\text{lower}} \le \boldsymbol{X}^{\mathrm{I}}_i, \boldsymbol{X}^{\mathrm{I}}_{\mathrm{S}} \le x^{\mathrm{I}}_{\text{upper}}, \\ \qquad i, j = 1 \sim n, \ i \ne j, \\ \mathrm{DV} = \left\{\boldsymbol{U}^{\mathrm{R}}_i, \boldsymbol{U}^{\mathrm{R}}_{\mathrm{S}}, \boldsymbol{X}^{\mathrm{I}}_i, \boldsymbol{X}^{\mathrm{I}}_{\mathrm{S}}\right\} \end{cases} \tag{6.79}$$

In addition, based on the same strategy, Eqs. 6.76 and 6.77 can be combined to obtain the worst-case performance metric, which is given as follows:

$$\begin{cases} \min\limits_{\mathrm{DV}} \ g\left(\boldsymbol{d}_i, \boldsymbol{d}_{\mathrm{S}}, \boldsymbol{U}^{\mathrm{R}}_i, \boldsymbol{U}^{\mathrm{R}}_{\mathrm{S}}, \boldsymbol{Y}^{\mathrm{R}}_{ji}, \boldsymbol{X}^{\mathrm{I}}_i, \boldsymbol{X}^{\mathrm{I}}_{\mathrm{S}}, \boldsymbol{Y}^{\mathrm{I}}_{ji}\right) \\ \text{s.t.} \ \ \|\boldsymbol{U}\| = \Phi^{-1}\left([R]\right), \\ \qquad x^{\mathrm{I}}_{\text{lower}} \le \boldsymbol{X}^{\mathrm{I}}_i, \boldsymbol{X}^{\mathrm{I}}_{\mathrm{S}} \le x^{\mathrm{I}}_{\text{upper}}, \\ \qquad \boldsymbol{Y}_{ij} = \boldsymbol{Y}_{ij}\left(\boldsymbol{d}_i, \boldsymbol{d}_{\mathrm{S}}, \boldsymbol{U}^{\mathrm{R}}_i, \boldsymbol{U}^{\mathrm{R}}_{\mathrm{S}}, \boldsymbol{Y}^{\mathrm{R}}_{ji}, \boldsymbol{X}^{\mathrm{I}}_i, \boldsymbol{X}^{\mathrm{I}}_{\mathrm{S}}, \boldsymbol{Y}^{\mathrm{I}}_{ji}\right), \\ \qquad i, j = 1 \sim n, \ i \ne j, \\ \mathrm{DV} = \left\{\boldsymbol{U}^{\mathrm{R}}_i, \boldsymbol{U}^{\mathrm{R}}_{\mathrm{S}}, \boldsymbol{X}^{\mathrm{I}}_i, \boldsymbol{X}^{\mathrm{I}}_{\mathrm{S}}\right\} \end{cases} \tag{6.80}$$

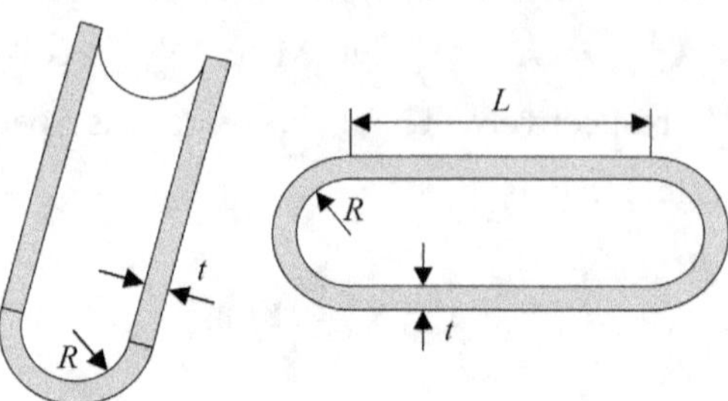

Figure 6.15 Pressure vessel structure diagram.

Table 6.11 Random uncertainty description of pressure vessel

Variable	*Mean value*	*Standard deviation*	*Dispersion pattern*	*The lower bound of the mean*	*The upper bound of the mean*
R	R^M	$0.01R^M$	Normality	0.1	36
T	T^M	$0.01T^M$	Normality	0.5	6.0
L	L^M	$0.01L^M$	Normality	0.1	140
S_t	40	4	Normality		

Table 6.12 Pressure vessel interval uncertainty description

Variable	*The upper bound of the interval*	*The lower bound of the interval*
P	2.723	5.057

Example: Pressure vessel design

Figure 6.15 shows the structure of the pressure vessel. The variables and parameters of the design optimization problem have random and interval uncertainties.

The radius R, length L, thickness T (unit is in), and the allowable tensile stress S_t of the material of the pressure vessel have random uncertainty, and the uncertainty information is expressed in Table 6.11. The internal pressure P (unit is klb) has interval uncertainty, and its uncertainty information is expressed in Table 6.12. In this chapter, the reliability is set to 0.99.

Three performance indexes need to be considered in the design optimization of pressure vessels: Volume, structural stress, and structural size. In this optimization problem, the design variables are the pressure vessel's radius R, length L, and thickness T. The objective functions are to maximize the internal volume $V_{\text{internal capacity}}$ while minimizing the weight (minimizing the weight is to minimize the volume $V_{\text{pressure vessel}}$ of the pressure vessel), maximize $V_{\text{internal capacity}}$, and minimize $V_{\text{pressure vessel}}$, respectively. The constraint function is the residual performance index.

The objective function volume can be expressed as

$$V = V_{\text{pressure vessel}} - V_{\text{internal capacity}} \tag{6.81}$$

The volume of the pressure vessel in the objective function can be expressed as

$$V_{\text{pressure vessel}} = \frac{4}{3}\pi(R+t)^3 + \pi(R+t)^2 L - \left[\frac{4}{3}\pi R^3 + \pi R^2 L\right] \tag{6.82}$$

The internal volume of the pressure vessel in the objective function can be expressed as

$$V_{\text{internal capacity}} = \frac{4}{3}\pi R^3 + \pi R^2 L \tag{6.83}$$

The stress constraint of the pressure vessel can be expressed as

$$g_1 = \frac{PR}{t} - S_t \le 0 \tag{6.84}$$

The size constraint of the pressure vessel can be expressed as

$$\begin{cases} g_2 = 5t - R \le 0, \\ g_3 = R + t - 40 \le 0, \\ g_4 = 2R + L + 2t - 150 \le 0 \end{cases} \tag{6.85}$$

The MDO problem has two disciplines: Internal volume and pressure vessel volume. The relationship between the disciplines is shown in Figure 6.16.

In the MDO problem, the design optimization problem's design variables are the radius R, length L, and thickness T of the pressure vessel. The objective function is to maximize the internal volume while minimizing the weight structure weight $M_{\text{pressure vessel}}$ or volume $V_{\text{pressure vessel}}$, and the constraint function is the residual performance index.

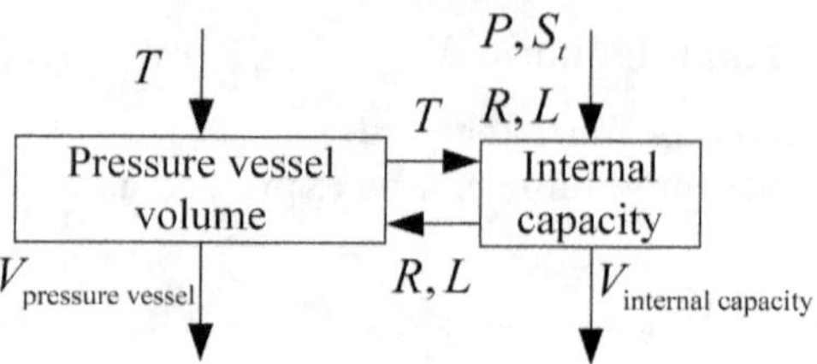

Figure 6.16 Pressure vessel discipline relationship.

Table 6.13 Optimization results

	R	T	L	V	$V_{\text{pressure vessel}}$	$V_{\text{interior volume}}$
Only consider random	34.71	5.25	69.99	-2.61×10^5	1.78×10^5	4.41×10^5
The mixture of random and interval	33.23	6.00	71.30	-2.04×10^5	1.97×10^5	4.01×10^5

Considering the uncertainty information of each variable, the UBMDO model of the pressure vessel is

$$\begin{cases} \min \quad V = V_{\text{pressure vessel}} - V_{\text{internal capacity}} \\ \qquad = \dfrac{4}{3}\pi(R^M + T^M)^3 + \pi(R^M + T^M)^2 L^M \\ \qquad -2\left[\dfrac{4}{3}\pi\left(R^M\right)^3 + \pi\left(R^M\right)^2 L^M\right] \\ \text{s.t.} \quad \Pr\left\{g_1 = \dfrac{PR}{T} - S_t \le 0\right\} \ge 0.99 \\ \qquad \Pr\left\{g_2 = 5T - R \le 0\right\} \ge 0.99 \\ \qquad \Pr\left\{g_3 = R + T - 40 \le 0\right\} \ge 0.99 \\ \qquad \Pr\left\{g_4 = 2R + L + 2T - 150 \le 0\right\} \ge 0.99 \end{cases} \tag{6.86}$$

In this section, the design scheme considering only random uncertainty is compared with the design scheme considering random and interval uncertainty in this chapter. The comparison results are shown in Table 6.13. The optimization iteration process is shown in Figure 6.17.

Table 6.13 shows that although the optimization results obtained by the optimization method in this chapter are more conservative than the results

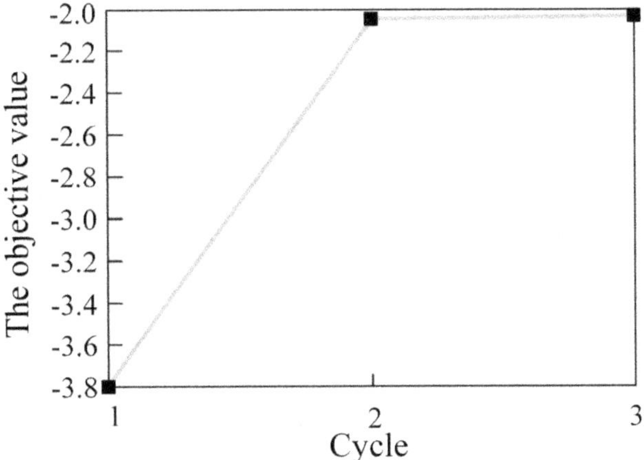

Figure 6.17 Calculating the convergence process.

only considering random uncertainty, it also considers the influence of random and interval uncertainty. Compared with the traditional optimization method, which only considers random uncertainty, it is more in line with the actual situation and has higher reality and feasibility.

6.3.2 UBMDO method based on probability theory and possibility theory

6.3.2.1 MVMDO optimization model

The MVMDO optimization model is

$$\begin{aligned}
&\min_{DV} f(\mathbf{d}_{s,c}, \mathbf{d}_{c}, \mathbf{X}^{M}_{s,c}, \mathbf{X}^{M}_{c}, \mathbf{P}^{M}_{c}, \mathbf{Y}^{M}) \\
&s.t.\ \ \Pi(G^{(i)}(\mathbf{d}_{s,c}, \mathbf{d}_{i,c}, \mathbf{X}_{s,c}, \mathbf{X}_{i,c}, \mathbf{P}_{i,c}, \mathbf{Y}_{\bullet i}) > 0) \leq \alpha_{t} \\
&\quad\ \ g^{(i)}(\mathbf{d}_{s,c}, \mathbf{d}_{i,c}, \mathbf{X}^{M}_{s,c}, \mathbf{X}^{M}_{i,c}, \mathbf{P}^{M}_{i,c}, \mathbf{Y}^{M}_{\bullet i}) \leq 0 \\
&\quad\ \ i = 1, 2, \cdots, nd \\
&\quad\ \ \mathbf{d}^{L}_{s,c} \leq \mathbf{d}_{s,c} \leq \mathbf{d}^{U}_{s,c}, \mathbf{d}^{L}_{c} \leq \mathbf{d}_{c} \leq \mathbf{d}^{U}_{c} \\
&\quad\ \ \mathbf{X}^{M,L}_{s,c} \leq \mathbf{X}^{M}_{s,c} \leq \mathbf{X}^{M,U}_{s,c}, \mathbf{X}^{M,L}_{c} \leq \mathbf{X}^{M}_{c} \leq \mathbf{X}^{M,U}_{c} \\
&DV = \{\mathbf{d}_{s,c}, \mathbf{d}_{c}, \mathbf{X}^{M}_{s,c}, \mathbf{X}^{M}_{c}\}
\end{aligned} \tag{6.87}$$

where the subscript s,i represents the shared variable, the local variable or parameter of discipline i, c is the continuous design variable, the continuous random or fuzzy variable/parameter, and the superscript (i)

refers to discipline i. $\mathbf{d}_{s,c}$ is a continuous shared deterministic design variable; $\mathbf{d}_c = \{\mathbf{d}_{1,c}, \mathbf{d}_{2,c}, \cdots, \mathbf{d}_{nd,c}\}$ is composed of continuous local deterministic design variables of all sub-disciplines, nd is the total number of sub-disciplines. $\mathbf{d}_{i,c}$ is a continuous local deterministic design variable in discipline i. $\mathbf{X}_{s,c} = \{\mathbf{X}_{s,rc}, \mathbf{X}_{s,fc}\}$ is a continuous shared variable with uncertainty, including a continuous shared random variable $\mathbf{X}_{s,rc}$ and a continuous shared fuzzy variable $\mathbf{X}_{s,fc}$, $\mathbf{X}_{s,c}^{M}$ is its mean or maximum membership point. $\mathbf{X}_c = \{\mathbf{X}_{rc}, \mathbf{X}_{fc}\} = \{(\mathbf{X}_{1,rc}, \ldots, \mathbf{X}_{nd,rc}), (\mathbf{X}_{1,fc}, \ldots, \mathbf{X}_{nd,fc})\}$ is composed of continuous local random and fuzzy variables in all sub-disciplines, and $\mathbf{X}_c^M$ is its mean or maximum membership point.

$\mathbf{X}_{i,c} = \{\mathbf{X}_{i,rc}, \mathbf{X}_{i,fc}\}$ is a continuous variable with uncertainty in subject i, including continuous local random variable $\mathbf{X}_{i,rc}$ and continuous fuzzy variable $\mathbf{X}_{i,fc}$, $\mathbf{X}_{i,c}^{M}$ is its mean or maximum membership point. $\mathbf{P}_c = \{\mathbf{P}_{rc}, \mathbf{P}_{fc}\} = \{(\mathbf{P}_{1,rc}, \ldots, \mathbf{P}_{nd,rc}), (\mathbf{P}_{1,fc}, \ldots, \mathbf{P}_{nd,fc})\}$ is composed of continuous local random and fuzzy parameters in all sub-disciplines, and $\mathbf{P}_c^M$ is its mean or maximum membership point. $\mathbf{P}_{i,c} = \{\mathbf{P}_{i,rc}, \mathbf{P}_{i,fc}\}$ is a continuous parameter with uncertainty in subject i, including continuous local random parameter $\mathbf{P}_{i,rc}$ and continuous fuzzy parameter $\mathbf{P}_{i,fc}$, $\mathbf{P}_{i,c}^{M}$ is its mean or maximum membership point. $f(\cdot)$ is the objective function; $\Pi[G^{(i)}(\cdot) > 0] \le \alpha_t$ is the possibility constraint in discipline i, $G^{(i)}(\cdot)$ is the performance function, and the failure mode is defined as $G^{(i)}(\cdot) > 0$; $g^{(i)}$ is the deterministic constraint in discipline i. 'L' and 'U' are the lower and upper boundaries, respectively. $\mathbf{Y} = \{\mathbf{Y}_{1\bullet}, \ldots, \mathbf{Y}_{nd\bullet}\}$ is the set of all coupling variables, $\mathbf{Y}_{i\bullet}$ is the coupling variable output from discipline i to other disciplines, and $\mathbf{Y}_{\bullet i}$ is the coupling variable input from other disciplines to discipline i. $\mathbf{Y}^M$ is the value of the coupling variable at the design point.

The relationship between coupling variables, shared variables, local variables, and parameters is

$$
\begin{aligned}
&\mathbf{Y}_{i\bullet} = \mathbf{Y}_{i\bullet}(\mathbf{d}_{s,c}, \mathbf{d}_{i,c}, \mathbf{X}_{s,c}, \mathbf{X}_{i,c}, \mathbf{P}_{i,c}, \mathbf{Y}_{\bullet i}) \\
&\mathbf{Y}_{i\bullet} = \{y_{ij}; j = 1 \sim nd, j \ne i\} \\
&i = 1 \sim nd
\end{aligned}
\tag{6.88}
$$

The direct solution of MVMDO involves three nested loops: The outer loop optimizes the objective function and obtains the design point; the intermediate ring performs uncertainty analysis on this design point; the internal ring performs discipline consistency analysis. Next, we discuss the probability/possibility analysis model in the MVMDO.

6.3.2.2 Uncertainty analysis model

In a single discipline, when variables and parameters contain both random and epistemic uncertainties, the concept of conditional failure probability is used, and the value of conditional failure probability is equal to that of conditional probability. Using the possibility theory, the PMA method for uncertainty analysis is proposed. Reviewing the introduction of single-disciplinary MVDO, the method includes two steps:

1. Using Rosenblatt transformation, continuous random variables, and parameters are transformed into standard normal variables and parameters in standard normal space (U-space). The continuous fuzzy variables and parameters satisfying the non-interactive property are transformed into standard fuzzy variables and parameters in the standard fuzzy space (V-space). The transformation method is

$$v = \begin{cases} \Pi_X(x) - 1, & x \le X^M \\ 1 - \Pi_X(x), & x > X^M \end{cases} \tag{6.89}$$

   where $\Pi_X(x)$ is the membership function of continuous fuzzy variable X, and X is its maximum membership point.

2. Solving the optimization problem

$$\begin{aligned} &\max_{\mathbf{U},\mathbf{V}} G(\mathbf{U},\mathbf{V}) \\ &s.t.\ \ \|\mathbf{U}\|_2 \le -\Phi^{-1}(\alpha_t) \\ &\qquad \|\mathbf{V}\|_\infty \le 1-\alpha_t \end{aligned} \tag{6.90}$$

The optimal solution of the above formula is Most Probable/Possible Point (MPPP) $(\mathbf{U}^*,\mathbf{V}^*)$, and the function value $G(\mathbf{U}^*,\mathbf{V}^*)$ of the performance function at MPPP. If $G(\mathbf{U}^*,\mathbf{V}^*) \le 0$, then the possibility constraint $\Pi(G>0) \le \alpha_t$ is satisfied; and vice versa.

It can be seen from the discussion in the previous chapter that the performance function $G^{(i)}(\mathbf{d}_{s,c},\mathbf{d}_{i,c},\mathbf{X}_{s,c},\mathbf{X}_{i,c},\mathbf{P}_{i,c},\mathbf{Y}_{\bullet i})$ of the possibility constraint in discipline i ($i = 1 \sim nd$) implicitly contains all deterministic design variables, continuous random variables and parameters, and continuous fuzzy variables and parameters in the multidisciplinary system.

When a design point (deterministic design variable, the mean of continuous random variables, the maximum membership point of continuous fuzzy variables) is obtained, in the uncertainty analysis of MVMDO, the continuous random variables and parameters $\mathbf{X}_{s,rc}$, $\mathbf{X}_{rc}=\{\mathbf{X}_{1,rc},\ldots,\mathbf{X}_{nd,rc}\}$ and $\mathbf{P}_{rc}=\{\mathbf{P}_{1,rc},\ldots,\mathbf{P}_{nd,rc}\}$ are first converted into standard normal variables and parameters $\mathbf{U}_{s,rc}$, $\mathbf{U}_{rc}=\{\mathbf{U}_{1,rc},\ldots,\mathbf{U}_{nd,rc}\}$, and $\mathbf{UP}_{rc}=\{\mathbf{UP}_{1,rc},\ldots,\mathbf{UP}_{nd,rc}\}$ (where $\mathbf{U},\mathbf{UP}$ is standard random variable and standard random parameter, respectively). The continuous fuzzy variables and parameters $\mathbf{X}_{s,fc}$, $\mathbf{X}_{fc}=\{\mathbf{X}_{1,fc},\ldots,\mathbf{X}_{nd,fc}\}$, and $\mathbf{P}_{fc}=\{\mathbf{P}_{1,fc},\ldots,\mathbf{P}_{nd,fc}\}$ are transformed into standard fuzzy variables and parameters $\mathbf{V}_{s,fc}$, $\mathbf{V}_{fc}=\{\mathbf{V}_{1,fc},\ldots,\mathbf{V}_{nd,fc}\}$, and $\mathbf{VP}_{fc}=\{\mathbf{VP}_{1,fc},\ldots,\mathbf{VP}_{nd,fc}\}$ (where $\mathbf{V},\mathbf{VP}$ is standard fuzzy variable and standard fuzzy parameter, respectively). Based on Eq. 6.90, the probability/possibility analysis model of possibility constraint in MVMDO discipline ith is

$$\begin{aligned}
&\max_{DV} G^{(i)}(\mathbf{d}_{s,c},\mathbf{d}_{i,c},\mathbf{U}^{(i)}_{s,rc},\mathbf{V}^{(i)}_{s,fc},\mathbf{U}^{(i)}_{i,rc},\mathbf{V}^{(i)}_{i,fc},\mathbf{UP}^{(i)}_{i,rc},\mathbf{VP}^{(i)}_{i,fc},\mathbf{Y}^{(i)}_{\bullet i})\\
&s.t.\ \|(\mathbf{U}^{(i)}_{s,rc},\mathbf{U}^{(i)}_{rc},\mathbf{UP}^{(i)}_{rc})\|_2\le -\Phi^{-1}(\alpha_t)\\
&\quad \|(\mathbf{V}^{(i)}_{s,fc},\mathbf{V}^{(i)}_{fc},\mathbf{VP}^{(i)}_{fc})\|_\infty\le 1-\alpha_t\\
&\quad y^{(i)}_{jm}=y_{jm}(\mathbf{d}_{s,c},\mathbf{d}_{j,c},\mathbf{U}^{(i)}_{s,rc},\mathbf{V}^{(i)}_{s,fc},\mathbf{U}^{(i)}_{j,rc},\mathbf{V}^{(i)}_{j,fc},\mathbf{UP}^{(i)}_{j,rc},\mathbf{VP}^{(i)}_{j,fc},\mathbf{Y}^{(i)}_{\bullet j})\\
&\quad j,m=1\sim nd;m\ne j\\
&DV=\{\mathbf{U}^{(i)}_{s,rc},\mathbf{V}^{(i)}_{s,fc},\mathbf{U}^{(i)}_{rc},\mathbf{V}^{(i)}_{fc},\mathbf{UP}^{(i)}_{rc},\mathbf{VP}^{(i)}_{fc},\mathbf{Y}^{(i)}\}
\end{aligned} \tag{6.91}$$

where the superscript (i) refers to the subject i. $\mathbf{d}_{s,c},\mathbf{d}_{i,c},\mathbf{d}_{j,c}$ is the corresponding deterministic design variable value in the obtained design point. $\mathbf{Y}^{(i)}$ is the coupling variable, which corresponds to the performance function $G^{(i)}$ of the possibility constraint in discipline i. In the above formula, the first design constraint contains all transformed standard normal variables and parameters, and the second design constraint contains all transformed standard fuzzy variables and parameters because $G^{(i)}$ implicitly contains all random, fuzzy variables and parameters. Using the idea of IDF, the coupling variable is used as an additional design variable, and the inter-disciplinary consistency requirement is used as an additional design constraint.

The optimization results are the MPPP $\mathbf{U}^{*,(i)}_{s,rc},\mathbf{V}^{*,(i)}_{s,fc},\mathbf{U}^{*,(i)}_{rc},\mathbf{V}^{*,(i)}_{fc},\mathbf{UP}^{*,(i)}_{rc},\mathbf{VP}^{*,(i)}_{fc}$, the value $\mathbf{Y}^{*,(i)}$ of the coupling variable at MPPP, and the function value $G^{(i)}$ of the performance function at MPPP. The MPPP $\mathbf{X}^{*,(i)}_{s,rc},\mathbf{X}^{*,(i)}_{s,fc},\mathbf{X}^{*,(i)}_{rc},\mathbf{X}^{*,(i)}_{fc},\mathbf{P}^{*,(i)}_{rc},\mathbf{P}^{*,(i)}_{fc}$

in X space can be obtained by the inverse process of Rosenblatt's transformation and Eq. 6.89. If the function value $G^{(i)} \leq 0$, then the possibility constraint $\Pi(G^{(i)}(\cdot) > 0) \leq \alpha_t$ will be satisfied at the design point; on the contrary, the possibility constraint is not satisfied, and the design point is not feasible.

In order to solve MVMDO effectively and efficiently, based on SORA and SFA ideas, two optimization methods for the MVMDO problem are proposed: MVMDO in the framework of SORA (MVMDO-SORA); MVMDO in the framework of SFA (MVMDO-SFA).

6.3.2.3 MVMDO-SORA

This section will discuss the MVMDO-SORA strategy, steps, and mathematical model and give an example to verify.

6.3.2.3.1 MVMDO-SORA strategy

In order to effectively solve the MVMDO problem, the following two key technologies are used in MVMDO-SORA:

1. Performance evaluation method PMA. In RBDO, PBDO, and MVDO, the feasibility of using the PMA method to analyze reliability and possibility constraints has significantly improved the computational efficiency
2. SORA. Using the basic idea of SORA, the uncertainty analysis is separated from the optimization, and the three-nested solution form of MVMDO is transformed into sequence-determined MDO and uncertainty analysis. In each loop, the deterministic design variables, the mean of continuous random variables, and the maximum membership point of continuous fuzzy variables are obtained by solving the determined MDO. Then, the uncertainty analysis is carried out. When the possibility constraint is not satisfied, the MPPP obtained by the uncertainty analysis is used to modify the deterministic design constraints in the MDO determined by the next loop to improve the feasibility of the design point. In this way, multidisciplinary optimization and uncertainty analysis in each loop are sequential rather than nested, improving the solution efficiency significantly

6.3.2.3.2 MVMDO-SORA steps

MVMDO-SORA includes the following steps:

Step 1. Set the initial value of the design variable $\mathbf{d}_{s,c}^{(0)}, \mathbf{d}_c^{(0)}, \mathbf{X}_{s,c}^{M,(0)}, \mathbf{X}_c^{M,(0)}$, where $k = 1$.

Step 2. Solve the determined MDO to obtain the design point $\mathbf{d}_{s,c}^{(k)},\mathbf{d}_{c}^{(k)},\mathbf{X}_{s,c}^{M,(k)},\mathbf{X}_{c}^{M,(k)}$. In the first loop, since the uncertainty analysis has not been carried out, the MPPP value of each function is set to $\mathbf{X}_{s,c}^{M,(0)},\mathbf{X}_{c}^{M,(0)},\mathbf{P}_{c}^{M}$; beginning with the second ring, using the MPPP obtained in the previous ring to modify the deterministic constraints in the determined MDO.

Step 3. Uncertainty analysis. Before uncertainty analysis, Rosenblatt transformation is used to transform all continuous random variables and parameters into standard normal variables and parameters, and continuous fuzzy variables and parameters are transformed into standard fuzzy variables and parameters by using Eq. 6.89. In the mathematical model of uncertainty analysis, the interdisciplinary consistency requirement is used as an additional design constraint.

Step 4. Test the convergence and feasibility. If all the possibility constraints are satisfied and the objective function value converges ($G^{(i),k} \le 0, i = 1 \sim nd; | f(k) - f(k-1) | \le \varepsilon$, ε is an arbitrarily small positive number), then stop; otherwise $k = k+1$, go to **Step 2.**

The mathematical models in **Steps 2** and **3** will be discussed and given later. When the possibility constraint is not satisfied, the following discussion is about using the SORA idea to modify the design constraints in the determined MDO.

When the possibility constraint $\Pi(G^{(i)}(\mathbf{d}_{s,c},\mathbf{d}_{i,c},\mathbf{X}_{s,c},\mathbf{X}_{i,c},\mathbf{P}_{i,c},\mathbf{Y}_{\bullet i}) > 0) \le \alpha_t$ is not satisfied in the $(k-1)$th ring, which means that the performance function has a function value $G^{(i),(k-1)} > 0$ at its MPPP ($\mathbf{X}_{s,c}^{*,(i),(k-1)},\mathbf{X}_{c}^{*,(i),(k-1)},\mathbf{P}_{c}^{*,(i),(k-1)}$), then the MPPP obtained by the possibility analysis of the (k – 1)th ring will be used to modify the MDO determined in the (k – 1)th ring. From the discussion in Section 6.3.2.2, it can be seen that the determination of possibility constraint satisfaction means that the function value of the performance function at its MPPP is less than or equal to zero. That is, the corresponding MPPP of the performance function must be in the determined feasible region ($G^{(i)} \le 0$). Since the MPPP in the (k – 1)th ring is not in the certain feasible region at this time, in order to improve the feasibility of the design point in the (k – 1)th ring, **S** is defined as the moving vector, and the idea of SORA design movement is adopted

$$\begin{aligned}
&\mathbf{S}_{s,c}^{(i),k} = \mathbf{X}_{s,c}^{M,(k-1)} - \mathbf{X}_{s,c}^{*,(i),(k-1)} \\
&\mathbf{S}_{j,c}^{(i),k} = \mathbf{X}_{j,c}^{M,(k-1)} - \mathbf{X}_{j,c}^{*,(i),(k-1)} \qquad (6.92)\\
&i,j = 1 \sim nd
\end{aligned}$$

Modify the design constraints in the MDO determined by the kth ring. Among them, $\mathbf{S}_{s,c}^{(i),k}, \mathbf{S}_{j,c}^{(i),k}$ is the moving vector corresponding to $\mathbf{X}_{s,c}, \mathbf{X}_{j,c}$ in discipline i. $\mathbf{X}_{s,c}^{M,(k-1)}, \mathbf{X}_{j,c}^{M,(k-1)}; j = 1 \sim nd$ is the mean value of continuous random variables and the maximum membership point of continuous fuzzy variables in the design point obtained in the $(k-1)$th ring, and $\mathbf{X}_{s,c}^{*,(i),(k-1)}, \mathbf{X}_{j,c}^{*,(i),(k-1)}; j = 1 \sim nd$ is the MPPP obtained by uncertainty analysis in the $(k-1)$th ring.

Since the mean value of continuous random parameters and the maximum membership point of continuous fuzzy parameters are known, fixed and not variables, the MPPP of continuous random parameters and continuous fuzzy parameters directly replaces $\mathbf{P}$ in the performance function $G^{(i)}(\cdot)$. The design constraint in the MDO determined by the kth ring is modified to

$$G_{\Pi}^{(i)}(\mathbf{d}_{s,c}, \mathbf{d}_{i,c}, \mathbf{X}_{s,c}^{M} - \mathbf{S}_{s,c}^{(i),k}, \mathbf{X}_{i,c}^{M} - \mathbf{S}_{i,c}^{(i),k}, \mathbf{P}_{i,c}^{*,(i),(k-1)}, \mathbf{Y}_{\bullet i}^{*,(i)}) \le 0$$

6.3.2.3.3 The mathematical model of MVMDO-SORA

The mathematical models of deterministic MDO and uncertainty analysis mentioned in the previous section will be discussed and given in this section.

6.3.2.3.3.1 THE MDO DETERMINED IN THE kth RING The MDO mathematical model determined in the kth ring is

$$\begin{aligned}
&\min_{DV} f(\mathbf{d}_{s,c}^{k}, \mathbf{d}_{c}^{k}, \mathbf{X}_{s,c}^{M,k}, \mathbf{X}_{c}^{M,k}, \mathbf{P}_{c}^{M}, \mathbf{Y}^{M,k}) \\
&s.t.\ \ G_{\Pi}^{(i)}(\mathbf{d}_{s,c}^{k}, \mathbf{d}_{i,c}^{k}, \mathbf{X}_{s,c}^{M,k} - \mathbf{S}_{s,c}^{(i),k}, \mathbf{X}_{i,c}^{M,k} - \mathbf{S}_{i,c}^{(i),k}, \mathbf{P}_{i,c}^{*,(i),(k-1)}, \mathbf{Y}_{\bullet i}^{*,(i)}) \le 0 \\
&\qquad g^{(i)}(\mathbf{d}_{s,c}^{k}, \mathbf{d}_{i,c}^{k}, \mathbf{X}_{s,c}^{M,k}, \mathbf{X}_{i,c}^{M,k}, \mathbf{P}_{i,c}^{M}, \mathbf{Y}_{\bullet i}^{M,k}) \le 0 \\
&\qquad y_{jm}^{*,(i),k} = y_{jm}(\mathbf{d}_{s,c}^{k}, \mathbf{d}_{j,c}^{k}, \mathbf{X}_{s,c}^{M,k} - \mathbf{S}_{s,c}^{(i),k}, \mathbf{X}_{j,c}^{M,k} - \mathbf{S}_{j,c}^{(i),k}, \mathbf{P}_{j,c}^{*,(i),(k-1)}, \mathbf{Y}_{\bullet j}^{*,(i)}) \\
&\qquad i,j,m = 1 \sim nd; m \ne j \qquad\qquad\qquad\qquad\qquad\qquad\qquad\qquad\qquad\qquad (6.93) \\
&\qquad y_{ij}^{M,k} = y_{ij}(\mathbf{d}_{s,c}^{k}, \mathbf{d}_{i,c}^{k}, \mathbf{X}_{s,c}^{M,k}, \mathbf{X}_{i,c}^{M,k}, \mathbf{P}_{i,c}^{M}, \mathbf{Y}_{\bullet i}^{M,k})\ \ i,j = 1 \sim nd; j \ne i \\
&\qquad \mathbf{d}_{s,c}^{L} \le \mathbf{d}_{s,c}^{k} \le \mathbf{d}_{s,c}^{U}, \mathbf{d}_{c}^{L} \le \mathbf{d}_{c}^{k} \le \mathbf{d}_{c}^{U} \\
&\qquad \mathbf{X}_{s,c}^{M,L} \le \mathbf{X}_{s,c}^{M,k} \le \mathbf{X}_{s,c}^{M,U}, \mathbf{X}_{c}^{M,L} \le \mathbf{X}_{c}^{M,k} \le \mathbf{X}_{c}^{M,U} \\
&DV = \{\mathbf{d}_{s,c}^{k}, \mathbf{d}_{c}^{k}, \mathbf{X}_{s,c}^{M,k}, \mathbf{X}_{c}^{M,k}, \mathbf{Y}^{M,k}, \mathbf{Y}^{*}\}
\end{aligned}$$

where, the superscript k is in the kth ring. $\mathbf{Y}^{*} = \{\mathbf{Y}^{*,(i)}, i = 1 \sim nd\}$; $\mathbf{Y}^{*,(i)}$ is the value of the coupling variable at MPPP, corresponding to the possibility constraint in discipline i. $G_{\Pi}^{(i)}$ is a modified deterministic design constraint corresponding to the possibility constraint in discipline i. $\mathbf{S}_{s,c}^{(i),k}, \mathbf{S}_{i,c}^{(i),k}, \mathbf{S}_{j,c}^{(i),k}$ are the moving vectors of discipline i corresponding to $\mathbf{X}_{s,c}, \mathbf{X}_{i,c}, \mathbf{X}_{j,c}$ in the kth ring, respectively. $\mathbf{P}_{i,c}^{*,(i),(k-1)}, \mathbf{P}_{j,c}^{*,(i),(k-1)}$ is the MPPP of the parameter $\mathbf{P}_{i,c}, \mathbf{P}_{j,c}$ obtained by discipline i in the $(k-1)$th ring, respectively.

In the Eq. 6.93, in order to obtain the equality constraint of interdisciplinary consistency, MPPP ($\mathbf{X}_{s,c}^{*,(i),(k-1)}, \mathbf{X}_{j,c}^{*,(i),(k-1)}, \mathbf{P}_{j,c}^{*,(i),(k-1)}$) is also needed to be modified to

$$
\begin{aligned}
& y_{jm}^{*,(i),k} = y_{jm}(\mathbf{d}_{s,c}^{k}, \mathbf{d}_{j,c}^{k}, \mathbf{X}_{s,c}^{M,k} - \mathbf{S}_{s,c}^{(i),k}, \mathbf{X}_{j,c}^{M,k} - \mathbf{S}_{j,c}^{(i),k}, \mathbf{P}_{j,c}^{*,(i),(k-1)}, \mathbf{Y}_{\bullet j}^{*,(i)}) \\
& i,j,m = 1 \sim nd; m \neq j
\end{aligned}
$$

where $\mathbf{X}_{j,c}^{*,(i),(k-1)}, \mathbf{P}_{j,c}^{*,(i),(k-1)}$ is the MPPP of $\mathbf{X}_{j,c}, \mathbf{P}_{j,c}$ obtained by discipline i in the $(k-1)$th ring.

6.3.2.3.3.2 *UNCERTAINTY ANALYSIS IN THE kth RING* From the discussion of uncertainty analysis in Section 6.3.2.2, it can be seen that when the design point $\mathbf{d}_{s,c}^{k}, \mathbf{d}_{c}^{k}, \mathbf{X}_{s,c}^{M,k}, \mathbf{X}_{c}^{M,k}$ is obtained, the continuous random variables and parameters $\mathbf{X}_{s,rc}$, $\mathbf{X}_{rc} = \{\mathbf{X}_{1,rc}, \ldots, \mathbf{X}_{nd,rc}\}$, and $\mathbf{P}_{rc} = \{\mathbf{P}_{1,rc}, \ldots, \mathbf{P}_{nd,rc}\}$ are first converted into standard normal variables and parameters $\mathbf{U}_{s,rc}$, $\mathbf{U}_{rc} = \{\mathbf{U}_{1,rc}, \ldots, \mathbf{U}_{nd,rc}\}$, and $\mathbf{UP}_{rc} = \{\mathbf{UP}_{1,rc}, \ldots, \mathbf{UP}_{nd,rc}\}$, respectively; the continuous fuzzy variables and parameters $\mathbf{X}_{s,fc}$, $\mathbf{X}_{fc} = \{\mathbf{X}_{1,fc}, \ldots, \mathbf{X}_{nd,fc}\}$, and $\mathbf{P}_{fc} = \{\mathbf{P}_{1,fc}, \ldots, \mathbf{P}_{nd,fc}\}$ are transformed into standard fuzzy variables and parameters $\mathbf{V}_{s,fc}$, $\mathbf{V}_{fc} = \{\mathbf{V}_{1,fc}, \ldots, \mathbf{V}_{nd,fc}\}$, and $\mathbf{VP}_{fc} = \{\mathbf{VP}_{1,fc}, \ldots, \mathbf{VP}_{nd,fc}\}$, respectively, and then the optimization problem (the mathematical model of uncertainty analysis in the kth ring) is solved

$$
\begin{aligned}
& \max_{DV} G^{(i)}(\mathbf{d}_{s,c}^{k}, \mathbf{d}_{i,c}^{k}, \mathbf{U}_{s,rc}^{(i),k}, \mathbf{V}_{s,fc}^{(i),k}, \mathbf{U}_{i,rc}^{(i),k}, \mathbf{V}_{i,fc}^{(i),k}, \mathbf{UP}_{i,rc}^{(i),k}, \mathbf{VP}_{i,fc}^{(i),k}, \mathbf{Y}_{\bullet i}^{(i)}) \\
& s.t. \; \| (\mathbf{U}_{s,rc}^{(i),k}, \mathbf{U}_{rc}^{(i),k}, \mathbf{UP}_{rc}^{(i),k}) \|_2 \leq -\Phi^{-1}(\alpha_t) \\
& \quad \| (\mathbf{V}_{s,fc}^{(i),k}, \mathbf{V}_{fc}^{(i),k}, \mathbf{VP}_{fc}^{(i),k}) \|_\infty \leq 1 - \alpha_t \\
& \quad y_{jm}^{(i)} = y_{jm}(\mathbf{d}_{s,c}^{k}, \mathbf{d}_{j,c}^{k}, \mathbf{U}_{s,rc}^{(i),k}, \mathbf{V}_{s,fc}^{(i),k}, \mathbf{U}_{j,rc}^{(i),k}, \mathbf{V}_{j,fc}^{(i),k}, \mathbf{UP}_{j,rc}^{(i),k}, \mathbf{VP}_{j,fc}^{(i),k}, \mathbf{Y}_{\bullet j}^{(i)}) \\
& \quad j,m = 1 \sim nd; m \neq j \\
& DV = \{\mathbf{U}_{s,rc}^{(i),k}, \mathbf{V}_{s,fc}^{(i),k}, \mathbf{U}_{rc}^{(i),k}, \mathbf{V}_{fc}^{(i),k}, \mathbf{UP}_{rc}^{(i),k}, \mathbf{VP}_{fc}^{(i),k}, \mathbf{Y}^{(i)}\} \\
& i = 1 \sim nd
\end{aligned}
\tag{6.94}
$$

where, the superscript k is in the kth ring. $\mathbf{d}_{s,c}^{k}, \mathbf{d}_{i,c}^{k}, \mathbf{d}_{j,c}^{k}$ is the $\mathbf{d}_{s,c}, \mathbf{d}_{i,c}, \mathbf{d}_{j,c}$ value obtained after solving the MDO determined in the kth ring. $\mathbf{Y}^{(i)} = \{y_{jm}^{(i)}; j, m = 1 \sim nd, m \neq j\}$ is the value of the coupling variable at MPPP.

The results of uncertainty analysis are MPPP ($\mathbf{U}_{s,rc}^{*,(i),k}, \mathbf{V}_{s,fc}^{*,(i),k}, \mathbf{U}_{rc}^{*,(i),k}, \mathbf{V}_{fc}^{*,(i),k}, \mathbf{UP}_{rc}^{*,(i),k}, \mathbf{VP}_{fc}^{*,(i),k}$), the coupling variable value $\mathbf{Y}^{*,(i)}$ and the function value $G^{(i),k}$ of the performance function at MPPP, where $i = 1 \sim nd$. The MPPP ($\mathbf{X}_{s,c}^{*,(i),k} = \{\mathbf{X}_{s,rc}^{*,(i),k}, \mathbf{X}_{s,fc}^{*,(i),k}\}$, $\mathbf{X}_{c}^{*,(i),k} = \{\mathbf{X}_{rc}^{*,(i),k}, \mathbf{X}_{fc}^{*,(i),k}\}$, and $\mathbf{P}_{c}^{*,(i),k} = \{\mathbf{P}_{rc}^{*,(i),k}, \mathbf{P}_{fc}^{*,(i),k}\}$) in X space can be obtained by the inverse process of Rosenblatt's transformation Eq. 6.89. If the function value is $G^{(i),k} \leq 0$, then the possibility constraint $\Pi(G^{(i)}(\cdot) > 0) \leq \alpha_t$ will be satisfied at the design point; otherwise, the possibility constraint is not satisfied, and the design point is infeasible. When the possibility constraints are not fully satisfied and the objective function value does not converge, the design constraints are modified in the identified MDO using MPPP.

The entire MVMDO-SORA process diagram is shown in Figure 6.18.

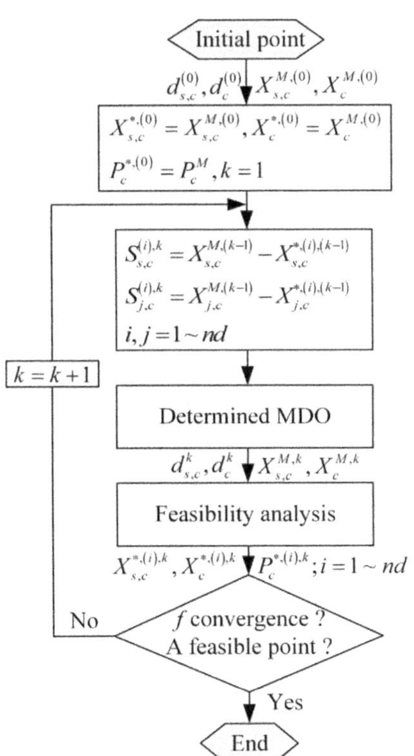

Figure 6.18 MVMDO-SORA flowchart.

6.3.2.3.4 *Example*

Conceptual ship design

Conceptual ship design example [42, 43] contains six design variables: Length (L), beam (B), depth (D), draft (T), block coefficient (C_B), and speed in knots (V_k). The three objective functions are to minimize Transportation cost (C_T), maximize Annual cargo (AC), and minimize Lightship weight. Based on the two disciplines of Cargo and Cost, this paper modifies it into a multidisciplinary optimization problem. Conceptual ship design Multidisciplinary issues are shown in Figure 6.19.

In Figure 6.19, DWT , DWT_C, DC , and C_C represent deadweight, cargo deadweight, daily consumption and capital costs, respectively, which are the coupling variables transmitted from Cargo to Cost. $RTPA$ are round trips per year, which is the coupling variable transmitted from the discipline Cost to the discipline Cargo. The detailed model of the example can be seen in Reference [42, 43].

In this example, the uncertainty of variables is described in Table 6.14. Among them, length (L), ship width (B), ship depth (D), and draft (T) are continuous random variables, and square coefficient (C_B), and speed (V_k) are continuous fuzzy variables.

Table 6.15 lists the conventional optimization and MVMDO-SORA optimization results of the conceptual ship design MVMDO problem, as well as the optimal solution, objective function value, and solution loop number of the PBMDO problem (where the PBMDO optimization results come from the previous section and are solved by PBMDO-SORA). The conventional optimization method refers to the nested loop solution form, but the discipline consistency requirement is used as an additional design constraint in the possibility analysis. In order to further improve the computational

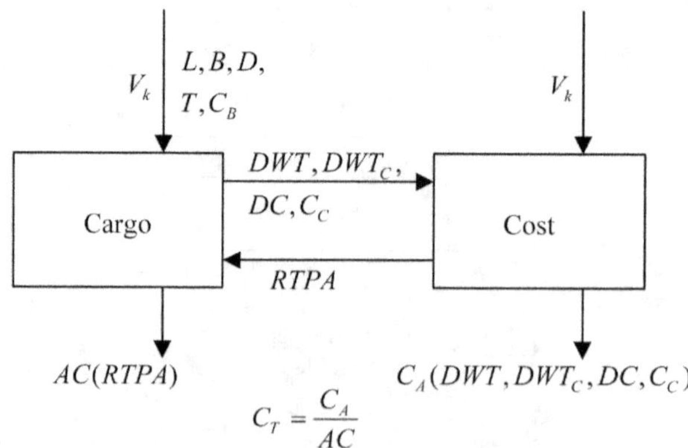

Figure 6.19 Multidisciplinary issues of conceptual ship design.

Table 6.14 Continuous random and fuzzy variables in conceptual ship design MVMDO problem

Random variable	*Mean value*	*Standard deviation*	*Dispersion pattern*	*Lower bounds of design variables*	*The upper bound of design variables*
L		0.5	Normal	151.5	272.82
B		0.15	Normal	20.45	31.86
D		0.075	Normal	13.225	24.775
T		0.05	Normal	10.15	11.56
Fuzzy variable	*The largest vassal*	*Deviation*	*Type of membership function*	*Lower bounds of design variables*	*The upper bound of design variables*
C_B		0.0105	Triangular	0.6405	0.7395
V_k		0.225	Triangular	11.225	19.775

Table 6.15 Conceptual ship design MVMDO conventional optimization and MVMDO-SORA optimization results

	L	B	D	T	C_B	V_k	C_T	k
Conventional optimization	194.0453	31.8224	15.8247	11.5598	0.7004	13.8171	8.4352	
MVMDO-SORA	203.0764	31.8600	16.5371	11.5097	0.7371	11.8917	8.3459	39
PBMDO-SORA	211.0736	31.8600	15.9530	11.5600	0.7260	11.2291	8.3579	3

efficiency in the MVMDO-SORA solution process, the optimal solution of each optimization problem in the previous loop is taken as the starting point of the corresponding optimization problem in this loop. The conventional optimization method and MVMDO-SORA solution are solved from the same starting point, and the same optimization algorithm is used.

From the objective function values in Table 6.15, the optimal solution obtained by MVMDO-SORA is better than the optimal solution of the conventional optimization method. Moreover, the number of disciplines analyses ($n_1 = 42462, n_2 = 42462$) in MVMDO-SORA is less than the number of analyses required by conventional optimization methods ($n_1 = 133337, n_2 = 133337$), improving the solution's efficiency. After MVMDO-SORA optimization, the function values of each function at MPPP are $-0.2737, -2.5276, -1.0888, -0.8908, -0.5487, -4.5190\times10^5$, $-4.1489\times10^4, -0.1803, -1.3156, -1.4113\times10^4, -4.3152\times10^5$, respectively. Each performance function value is less than zero, indicating that each possibility constraint is satisfied. For this MVMDO and PBMDO problem, MVMDO-SORA and PBMDO-SORA are solved from the same starting point, and the same optimization algorithm is used. From the value of the objective function, the optimal solution of PBMDO is more conservative than that of MVMDO.

6.3.2.4 MVMDO-SFA

This section will discuss the MVMDO-SFA strategy, steps, and mathematical model and give an example to verify.

6.3.2.4.1 MVMDO-SFA strategy

In order to effectively solve the MVMDO problem, the following two key technologies are used in the MVMDO-SFA:

1. Performance evaluation method PMA. In RBDO, PBDO, and MVDO, the feasibility of using the PMA method to analyze reliability and possibility constraints has significantly improved the computational efficiency
2. SFA. Using the basic idea of SFA, the uncertainty analysis is separated from the optimization, and the three-nested solution form of MVMDO is transformed into sequence-determined MDO and uncertainty analysis. In each loop, the deterministic design variables, the mean of continuous random variables, and the maximum membership point of continuous fuzzy variables are obtained by solving the determined MDO, and then the uncertainty analysis is carried out. When the possibility constraint is not satisfied, the performance function is used to modify the design constraints in the MDO determined by the next loop to improve the feasibility of the design point by using the function value at all MPPPs obtained by the performance function. In this way, multidisciplinary optimization and uncertainty analysis in each loop are sequential rather than nested, improving the solution efficiency significantly

6.3.2.4.2 MVMDO-SFA steps

MVMDO-SORA includes the following steps:

Step 1. Set the initial value of the design variable $\mathbf{d}_{s,c}^{(0)}, \mathbf{d}_{c}^{(0)}, \mathbf{X}_{s,c}^{M,(0)}, \mathbf{X}_{c}^{M,(0)}$, $k = 1$.

Step 2. Solve the determined MDO to obtain the design point $\mathbf{d}_{s,c}^{(k)}, \mathbf{d}_{c}^{(k)}, \mathbf{X}_{s,c}^{M,(k)}, \mathbf{X}_{c}^{M,(k)}$. In the first loop, since the uncertainty analysis has not been carried out, the constraint movement of each function is set to zero. Starting from the second ring, in the determined MDO, the deterministic constraint uses the performance function to modify the function value at all MPPPs it has obtained.

Step 3. Uncertainty analysis. Before uncertainty analysis, Rosenblatt transformation is used to transform all continuous random variables and parameters into standard normal variables and parameters, and Eq. 6.89 is used to transform continuous fuzzy variables and parameters into standard fuzzy variables and parameters. The interdisciplinary consistency requirement is used as an additional design constraint in the mathematical uncertainty analysis model.

Step 4. Test the convergence and feasibility. If all the possibility constraints are satisfied and the objective function value converges ($G^{(i),k} \le 0, i = 1 \sim nd; | f(k) - f(k-1) | \le \varepsilon$), then stop; otherwise $k = k+1$, go to **Step 2.**

The mathematical models in **Step 2** and **3** will be discussed and given later. When the possibility constraint is not satisfied, the following discussion is about using the SFA idea to modify the design constraints in the determined MDO.

When the possibility constraint $\Pi(G^{(i)}(\mathbf{d}_{s,c}, \mathbf{d}_{i,c}, \mathbf{X}_{s,c}, \mathbf{X}_{i,c}, \mathbf{P}_{i,c}, \mathbf{Y}_{\bullet i}) > 0) \le \alpha_t$ is not satisfied in the $(k - 1)$th ring, the performance function has a function value $G^{(i),(k-1)} > 0$ at its MPPP ($\mathbf{X}_{s,c}^{*,(i),(k-1)}, \mathbf{X}_{c}^{*,(i),(k-1)}, \mathbf{P}_{c}^{*,(i),(k-1)}$). From the discussion in Section 6.3.2.2, it can be seen that the determination of possibility constraint satisfaction means that the function value of the performance function at its MPPP is less than or equal to zero. That is, the MPPP of the performance function must be in the determined feasible region ($G^{(i)} \le 0$). In order to improve the feasibility of the design point in the kth ring, G^S is defined as the constraint movement and the idea of SFA constraint movement is adopted

$$G^{S,(i),k} = G^{S,(i),(k-1)} + G^{(i),(k-1)} \tag{6.95}$$

Modify the design constraints in the MDO determined by the kth ring. Here, $G^{S,(i),k}$ is the constraint shift of the performance function $G^{(i)}(\cdot)$ in the kth ring, and $G^{S,(i),(k-1)}$ is the constraint shift in the $(k - 1)$th ring. The design constraint in the MDO determined by the kth ring is modified to

$$G^{(i)}(\mathbf{d}_{s,c}, \mathbf{d}_{i,c}, \mathbf{X}_{s,c}^{M}, \mathbf{X}_{i,c}^{M}, \mathbf{P}_{i,c}^{M}, \mathbf{Y}_{\bullet i}^{M}) + G^{S,(i),k} \le 0$$

6.3.2.4.3 The mathematical model of MVMDO-SFA

The mathematical models of deterministic MDO and uncertainty analysis mentioned in the previous section will be discussed and given in this section. The MDO mathematical model determined in the kth ring is

$$\begin{aligned}
&\min_{DV} f(\mathbf{d}_{s,c}^{k},\mathbf{d}_{c}^{k},\mathbf{X}_{s,c}^{M,k},\mathbf{X}_{c}^{M,k},\mathbf{P}_{c}^{M},\mathbf{Y}^{M,k})\\
&s.t.\ \ G^{(i)}(\mathbf{d}_{s,c}^{k},\mathbf{d}_{i,c}^{k},\mathbf{X}_{s,c}^{M,k},\mathbf{X}_{i,c}^{M,k},\mathbf{P}_{i,c}^{M},\mathbf{Y}_{\bullet i}^{M,k})+G^{S,(i),k}\le 0\\
&\quad\ \ g^{(i)}(\mathbf{d}_{s,c}^{k},\mathbf{d}_{i,c}^{k},\mathbf{X}_{s,c}^{M,k},\mathbf{X}_{i,c}^{M,k},\mathbf{P}_{i,c}^{M},\mathbf{Y}_{\bullet i}^{M,k})\le 0\\
&\quad\ \ i=1\sim nd\\
&\quad\ \ y_{ij}^{M,k}=y_{ij}(\mathbf{d}_{s,c}^{k},\mathbf{d}_{i,c}^{k},\mathbf{X}_{s,c}^{M,k},\mathbf{X}_{i,c}^{M,k},\mathbf{P}_{i,c}^{M},\mathbf{Y}_{\bullet i}^{M,k})\ \ i,j=1\sim nd;j\ne i\\
&\quad\ \ \mathbf{d}_{s,c}^{L}\le\mathbf{d}_{s,c}^{k}\le\mathbf{d}_{s,c}^{U},\mathbf{d}_{c}^{L}\le\mathbf{d}_{c}^{k}\le\mathbf{d}_{c}^{U}\\
&\quad\ \ \mathbf{X}_{s,c}^{M,L}\le\mathbf{X}_{s,c}^{M,k}\le\mathbf{X}_{s,c}^{M,U},\mathbf{X}_{c}^{M,L}\le\mathbf{X}_{c}^{M,k}\le\mathbf{X}_{c}^{M,U}\\
&DV=\{\mathbf{d}_{s,c}^{k},\mathbf{d}_{c}^{k},\mathbf{X}_{s,c}^{M,k},\mathbf{X}_{c}^{M,k},\mathbf{Y}^{M,k}\}
\end{aligned}\tag{6.96}$$

where the design variable DV includes the coupling variable $\mathbf{Y}^{M}$ at the design point in addition to the continuous deterministic design variable, the mean value of the continuous random variable and the maximum membership point of the continuous fuzzy variable. The interdisciplinary consistency requirement at $\mathbf{Y}^{M}$ is taken as an additional design constraint.

Compared with the MDO (Eq. 6.93) determined in MVMDO-SORA and the MDO (Eq. 6.96) determined in MVMDO-SFA, only the equality constraints added at $\mathbf{Y}^{M}$ to meet the discipline consistency requirements appear in Eq. 6.96, and the corresponding $\mathbf{Y}^{*}$ (the value of the coupling variable at MPPP) of the performance function in each possibility constraint and the equality constraints added to meet the discipline consistency requirements do not appear. The reason is that MVMDO-SFA moves the deterministic constraint boundary $G^{(i)}(\mathbf{d}_{s,c},\mathbf{d}_{i,c},\mathbf{X}_{s,c}^{M},\mathbf{X}_{i,c}^{M},\mathbf{P}_{i,c}^{M},\mathbf{Y}_{\bullet i}^{M})=0$ as a whole through the constraint movement to improve the feasibility of the design point in the next loop; MVMDO-SORA moves the deterministic boundary by design movement, which is essentially the moving design constraint $G_{\Pi}^{(i)}(\cdot)\le 0$ approximates $G^{(i)}(\mathbf{d}_{s,c},\mathbf{d}_{i,c},\mathbf{X}_{s,c}^{*,(i)},\mathbf{X}_{i,c}^{*,(i)},\mathbf{P}_{i,c}^{*,(i)},\mathbf{Y}_{\bullet i}^{*,(i)})\le 0$ in use Eq. 6.93, which makes MVMDO-SORA need to obtain the corresponding $\mathbf{Y}_{\bullet i}^{*,(i)}$, $i=1\sim nd$, of each performance function $G^{(i)}(\cdot)$. Therefore, in Eq. 6.93, the corresponding $\mathbf{Y}^{*}$ of the performance function in each possibility constraint (and as an additional design variable), as well as the equality constraints are added to meet the requirements of interdisciplinary consistency. Due to the above reasons, when the number of coupling variables and possibility constraints is large, the MDO determined in MVMDO-SFA has a

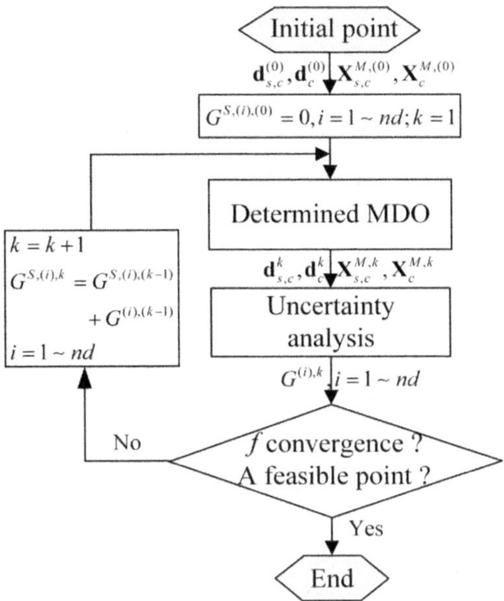

Figure 6.20 MVMDO-SFA flowchart.

much smaller solution scale than the MDO determined in MVMDO-SORA, so the requirements for the algorithm are relatively low. Moreover, it is not necessary to carry out disciplinary analysis to obtain $\mathbf{Y}^*$, so it is very beneficial to improve the solution's efficiency.

The mathematical model of the uncertainty analysis in the *k*th ring is the same as the uncertainty analysis model in Section 6.3.2.3. The flowchart of MVMDO-SFA is shown in Figure 6.20.

6.3.2.4.4 Examples

Conceptual ship design

For the conceptual ship design MVMDO problem in Section 6.3.2.3 (all conditions are the same), the MVMDO-SFA is used to solve it. Table 6.16 lists the conventional optimization method of conceptual ship design MVMDO problem, the optimal solution of MVMDO-SORA, MVMDO-SFA, and PBMDO, the objective function value and the number of solving cycles (PBMDO optimization results are solved by PBMDO-SFA). The conventional optimization method refers to the nested loop solution form, but the discipline consistency requirement is used as an additional design constraint in the possibility analysis. In the MVMDO-SFA solution process, the optimal solution of each optimization problem in the previous loop is

Table 6.16 Different optimization methods conceptual ship design MVMDO optimization results

	L	B	D	T	C_B	V_k	C_T	k
Conventional optimization	194.0453	31.8224	15.8247	11.5598	0.7004	13.8171	8.4352	
MVMDO-SORA	203.0764	31.8600	16.5371	11.5097	0.7371	11.8917	8.3459	39
MVMDO-SFA	194.7338	31.8600	15.8250	11.5600	0.6975	13.7861	8.4306	89
PBMDO-SFA	194.1986	31.6673	15.9530	11.5600	0.7036	13.7642	8.4464	27

taken as the starting point of the corresponding optimization problem in this loop. In the conventional optimization method, MVMDO-SORA and MVMDO-SFA, the solution is solved from the same starting point, and the same optimization algorithm is used.

From the objective function values in Table 6.16, the optimal solution obtained by MVMDO-SFA is superior to the conventional optimization method, and the optimal solution obtained by MVMDO-SORA is the best of the three methods. The number of disciplinary analyses ($n_1 = 37440, n_2 = 37440$) in MVMDO-SFA is less than the number of analyses ($n_1 = 133337, n_2 = 133337$) required by conventional optimization methods, so it improves the solution's efficiency. After MVMDO-SFA optimization, the function values of each function at its MPPP are $-0.0150, -2.4930, -1.8974, -0.4808, 2.1987\times10^{-11}, -4.5613\times10^{5}$, $-$ 3.7502 $\times$ 10^4, and $-0.1551, -1.6682, -2.8564\times10^{4}, -2.4193\times10^{5}$, respectively; the performance function corresponding to the fifth possibility constraint is approximately zero at its MPPP, indicating that the possibility constraint is an action constraint. The remaining function values are all less than zero, indicating that these possibility constraints are satisfied. For this MVMDO and PBMDO problem, MVMDO-SFA and PBMDO-SFA are solved from the same starting point, and the same optimization algorithm is used. From the value of the objective function, the optimal solution of PBMDO is more conservative than that of MVMDO.

MVMDO-SFA obtains the optimal solution after 89 cycles, while MVMDO-SORA obtains the optimal solution after 39 cycles, and the optimal solution obtained by MVMDO-SORA is better than that of MVMDO-SFA. The number of disciplinary analyses in MVMDO-SFA is less than that required by MVMDO-SORA. Compared with the conventional optimization method, both MVMDO-SORA and MVMDO-SFA improve the solution efficiency and obtain better optimal solutions.

REFERENCES

[1] Daniels H. E. (1954). Saddlepoint approximations in statistics. The Annals of Mathematical Statistics, 25(4): 631–650.
[2] Barndorff-Nielsen O. E., Cox D. R. (1994). Inference and Asymptotics. New York: Chapman & Hall.
[3] Wood A. T. A., Booth J. G., Butler R. W. (1993). Saddlepoint approximations to the CDF of some statistics with nonnormal limit distributions. Journal of the American Statistical Association, 88(422): 680–686.
[4] Barndorff-Nielsen O. E., Cox D. R. (1979). Edgeworth and saddle-point approximation with statistical application. Journal of the Royal Statistical Society, 41(3): 279–312.
[5] Du X., Sudjianto A. (2004). First order saddlepoint approximation for reliability analysis. AAIA Journal, 42(6): 1199–1207.
[6] Sudjianto A., Du X., Chen W. (2005). Probabilistic sensitivity analysis in engineering design using uniform sampling and saddlepoint approximation. SAE Transactions, 114(6): 267–276.
[7] Goutis C., Casella G. (1999). Explaining the saddlepoint approximation. The American Statistician, 53(3): 216–224.
[8] Huzurbazar S. (1999). Practical saddlepoint approximations. The American Statistician, 53(3): 225–232.
[9] Lugannani R., Rice S. O. (1980). Saddlepoint approximation for the distribution of the sum of independent random variables. Advances in Applied Probability, 12(2): 475–490.
[10] Gatto R., Ronchetti E. (1996). General saddlepoint approximations of marginal densities and tail probabilities. Journal of the American Statistical Association, 91(434): 666–673.
[11] Jensen J. L. (1995). Saddlepoint Approximations. Clarendon, Jamaica: Oxford Science Publications.
[12] Daniels H. E. (1987). Tail probability approximations. International Statistical Review, 55(1): 37–48.
[13] Lu H., Cao S., Zhu Z., Zhang Y. (2020). An improved high order moment-based saddlepoint approximation method for reliability analysis. Applied Mathematical Modelling, 82: 836–847.
[14] Du X., Sudjianto A. (2004). A saddlepoint approximation method for uncertainty analysis. 2004 ASME International Design Engineering Technical Conferences and the Computers and Information in Engineering Conference, Salt Lake City, UT, 445–452.
[15] Huang B., Du X. (2006). Uncertainty analysis by dimension reduction integration and saddlepoint approximations. Journal of Mechanical Design, 128(1): 26–33.
[16] Tvedt L. (1989). Second Order Reliability by an Exact Integral. Berlin Heidelberg: Springer, 377–384.
[17] Yuen K. V., Wang J., Au S. K. (2007). Application of saddlepoint approximation in reliability analysis of dynamic systems. Earthquake Engineering and Engineering Vibration, 6(45): 391–400.

[18] XIAO N. C. (2012). Research on structural reliability methods under stochastic and epistemic uncertainties. PhD thesis of University of Electronic Science and Technology of China (in Chinese), Chengdu, China.
[19] Barndorff-Nielsen O. E. (1986). Inference on full or partial parameters based on the standardized signed log likelihood ratio. Biometrika, 73(2): 307–322.
[20] Du X., Sudjianto A. (2004). First order saddlepoint approximation for reliability analysis. AIAA Journal, 42(6): 1199–1207.
[21] Yan Y. (2012). Research on numerical simulation method of structural reliability analysis. Master thesis of Tsinghua University (in Chinese), Beijing, China.
[22] Au S. K., Beck J. L. (2001). Estimation of small failure probabilities in high dimensions by subset simulation. Probabilistic Engineering Mechanics, 16(4): 263–277.
[23] Au S. K., Beck J. L. (2003). Subset simulation and its application to seismic risk based on dynamic analysis. Journal of Engineering Mechanics, 129(8): 901–917.
[24] Kodiyalam S. Evaluation of methods for multidisciplinary design optimization (MDO), Phase I[OL]. www.cs.odu.edu/~mln/ltrs-pdfs/NASA-98-cr208716.pdf.
[25] Jiang C. (2008). Interval-based uncertainty optimization theory and algorithm. PhD thesis of Hunan University (in Chinese), Changsha, China.
[26] Wang L., Xiong C., Hu J., Wang X., Qiu Z. (2018). Sequential MDO and reliability analysis under interval uncertainty. Aerospace Science and Technology, 80: 508–519.
[27] Nakahara Y., Sasaki M., Gen M. (1992). On the linear programming problems with interval coefficients. Computers and Industrial Engineering, 23(1–4): 301–304.
[28] Zhang C., Li D., Liang J. (2020). Interval-valued hesitant fuzzy multi-granularity three-way decisions in consensus processes with applications to multi-attribute group decision making. Information Sciences, 511: 192–211.
[29] Wang C. N., Ma M. X., Huang Z. J. (2009). Parameter mapping method for coupled multidisciplinary design systems. Journal of Mechanical Engineering, 45(4): 231–237.
[30] Choi K. K., Youn B. D., Du L. (2005). Integration of reliability- and possibility-based design optimization using performance measure approach. SAE Transactions, 114(6): 247–256.
[31] Du L., Choi K. K., Youn B. D. (2006). Inverse possibility analysis method for possibility-based design optimization. AIAA Journal, 44(11): 2682–2690.
[32] Du L., Choi K. K. (2008). An inverse analysis method for design optimization with both statistical and fuzzy uncertainties. Structural and Multidisciplinary Optimization, 37(2): 107–119.
[33] Nikolaidis E., Chen S., Cudney H., et al. (2004). Comparison of probability and possibility for design against catastrophic failure under uncertainty. Journal of Mechanical Design, 126(3): 386–394.
[34] Youn B. D. (2007). Integrated framework for design optimization under aleatory and/or epistemic uncertainties using adaptive-loop method. Journal of Risk and Reliability, 221(2): 107–119.

[35] Youn B. D., Choi K. K., Du L., et al. (2007). Integration of possibility-based optimization and robust design for epistemic uncertainty. Journal of Mechanical Design, 129(8): 876–882.
[36] Du L., Choi K. K., Youn B. D., et al. (2006). Possibility-based design optimization method for design problems with both statistical and fuzzy input data. Journal of Mechanical Design, 128(4): 928–935.
[37] Mourelatos Z. P., Zhou J. (2005). Reliability estimation and design with insufficient data based on possibility theory. AIAA Journal, 43(8): 1696–1705.
[38] Zhou J., Mourelatos Z. P. (2008). A sequential algorithm for possibility-based design optimization. Journal of Mechanical Design, 130(1): 011001-1–011001-10.
[39] Zadeh L. A. (1978). Fuzzy sets as a basis for a theory of possibility. Fuzzy Sets and Systems, 1(1): 3–28.
[40] Du L. (2006). Reliability- and Possibility-based Design Optimization Using Inverse Analysis Methods. The University of Iowa, Iowa, USA, 68(04).
[41] Ristic B., Gilliam C., Byrne M., Benavoli A. (2020). A tutorial on uncertainty modeling for machine reasoning. Information Fusion, 55: 30–44.
[42] Hart C. G. (2010). MDO of complex engineering systems for cost assessment under uncertainty. PhD thesis of University of Michigan.
[43] Parsons M. G., Scott R. L. (2004). Formulation of multicriterion design optimization problems for solution with scalar numerical optimization methods. Journal of Ship Research, 48(1): 61–76.

Chapter 7

Application of advanced UBMDO in aviation equipment

With the rapid development of aviation technology, the requirements for performance, safety, and reliability of aviation equipment are increasing daily. In this context, the Multidisciplinary Design Optimization (MDO) method, especially the Uncertainty-based Multidisciplinary Design Optimization (UBMDO) method, plays an increasingly important role in the design and development of aviation equipment. This chapter will discuss the application of advanced UBMDO in the aviation field through a specific case, namely the design of the airborne flip drive device.

The airborne flip drive device is the core component of the variant aircraft, which is responsible for accurately driving the aircraft to perform specific actions. However, in practical engineering applications, the existence of uncertainties may affect the performance stability and credibility of the drive device. Therefore, the design optimization process must use uncertainty-based design optimization technology to quantitatively consider the influence of these uncertainties and ensure the stability and reliability of the drive equipment as much as possible.

In addition, although the traditional serial design optimization method is widely used, it is time consuming, inefficient, and challenging to obtain the best solution for the system. By using the MDO method, the drive equipment can be divided into disciplines and multidisciplinary analysis according to the function. Moreover, the design optimization can be done from the whole system's perspective. In this way, the design optimization scheme can be closer to the optimal solution of the system while considering the interaction between components. In this chapter, two UBMDO methods will be used to discuss the UBMDO problem of the flip drive mechanism. At the same time, this chapter will analyze and explain the final design optimization results.

7.1 INTRODUCTION OF THE FLIP DRIVE MECHANISM

In order to cope with the continuous improvement of the comprehensive needs of aircraft, such as flight efficiency, maneuverability, and multi-task

DOI: 10.1201/9781003464792-7

adaptability, morphing aircraft has gradually become a new generation of development direction [1–4]. The common morphing technology of morphing aircraft mainly involves the folding of the wing, the torsion of the wing, the deformation of the leading and trailing edges, and the change of the sweep angle of the wing [5].

The flip drive mechanism installed inside the wing structure is integral to driving the deformation motion, which is very important for the morphing aircraft to complete the corresponding prescribed action. The flip drive mechanism refers to a very compact shape in the folding state, and the deployment process has the corresponding mechanism motion characteristics. Its fully expanded state is a large-span mechanism that can carry a certain load. The flip drive mechanism needs to have the following characteristics: A large stiffness-weight ratio, accurate motion trajectory, and fast response time.

In the design, the lightweight requirements of the flip drive mechanism must be met first. At the same time, the mechanical properties of each component in the mechanism must be considered. The mechanical properties will directly affect the motion accuracy of the mechanism and then affect the aerodynamics and strength of the wing. On the other hand, while considering the influence of mechanical properties on the motion accuracy of the flip drive mechanism, the structural form of the flip drive mechanism will determine the motion response time of the mechanism.

Applying the flip drive mechanism can change the area and aspect ratio of the airfoil to a certain extent. This deformation can give the shaper high battlefield survivability and a good attack effect on the enemy. The American Morphing Advanced Structures (MAS) program has done much work to expand the aircraft's flight envelope and improve the corresponding operational performance [5]. The foldable wing proposed by Lockheed Martin applies shape memory polymer-based variable stiffness skins in the folding drive mechanism. The maximum airfoil area can be doubled before and after the change of this mechanism [6–9]. The new unmanned carrier-based aircraft X-47B developed by the United States is the first generation of unmanned aircraft that can take off and land autonomously on aircraft carriers. The wing can still achieve good folding performance while equipped with multiple ailerons and highly integrated electronic and hydraulic systems. Such a structure makes the aircraft occupy a smaller space on the aircraft carrier, increasing the carrier's aircraft capacity to a certain extent.

In this chapter, the multidisciplinary analysis of the turnover drive mechanism is carried out, and the coupling relationship between the mechanical system components is considered. The mechanism's MDO model is established using the CO method. In order to study the influence of uncertain factors on the reliability of the flip drive mechanism and explore the

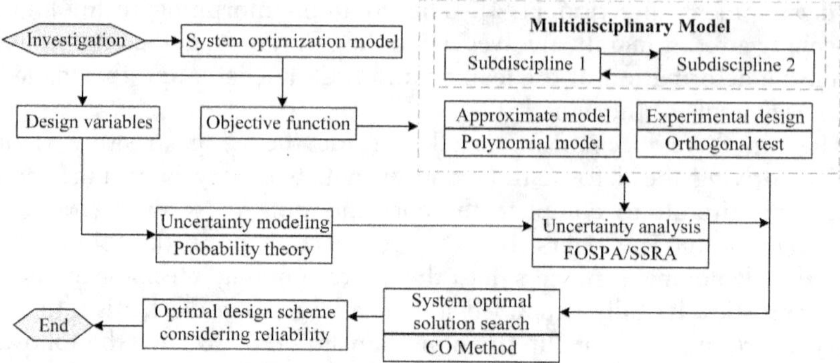

Figure 7.1 RBMDO flowchart of the flip drive mechanism.

transmission of uncertain factors between coupling disciplines, this chapter will use First-order Saddle Point Approximation (FOSPA)-MDO-Sequential Optimization and Reliability Assessment (SORA) and Subset Simulation Reliability Analysis (SSRA)-MDO-SORA to optimize the solution respectively. The UBMDO flowchart of the flip drive mechanism is shown in Figure 7.1.

The flip drive mechanism is a key component of the morphing aircraft. Next, an example of the flip drive mechanism of a morphing aircraft is given. The flipping drive mechanism of a morphing aircraft studied in this example is a variant of the crank slider mechanism, and the mechanism composition is simple. As shown in Figure 7.2, the flipping drive mechanism mainly comprises five parts: turnover section, flip actuator, slide rail, the slider, the connecting rod, and the triangular support.

The specific working process is as follows: The hydraulic rod in the flip actuator drives the slider and the connecting rod through hydraulic pressure, and the slider moves back and forth along the slide rail's direction, forming a moving pair. At the same time, the slider transmits power to drive the flip section to rotate around the fixed shaft A to form a rotary pair. The connecting rod rotates around the fixed shaft B to form a rotary pair. The rotary shaft C is the hinge point of the connecting rod, the slider, and the hydraulic rod in the flip actuator cylinder, forming a rotary pair. E is the hinge fulcrum of the flip actuator and the flip section, forming a rotary pair. The position of the E point is the final position of the unfolding state of the flip section, and the position of E'' point is the final position of the folding state of the flip section. The slide rail is fixedly connected with the flip section and can be considered a whole. All the shaft positioning points take the position of point A as the coordinate origin, as shown in Figure 7.3.

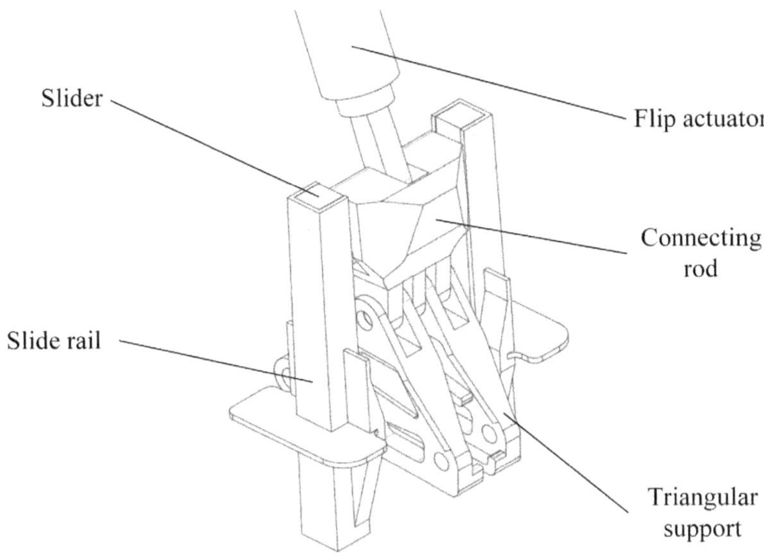

Figure 7.2 Schematic diagram of the flip drive mechanism.

7.2 DISCIPLINE DIVISION AND OPTIMIZATION PROBLEM DEFINITION OF FLIP DRIVE MECHANISM

The flip drive mechanism is a key component to adjust the morphing aircraft's geometric structure and aerodynamic shape. It must have lightweight and compact structure characteristics while considering the load-bearing and motion performance requirements. Specifically, the design optimization of the turnover drive mechanism needs to consider four performance indicators: Turnover torque T_t, mechanism weight M_t, turnover time t_t, and stress of each component in mechanism σ_t.

In the MDO, the overturning moment T_t is usually used as the objective function of the design optimization problem. Other performances are considered as constraints. Through investigation, it was determined that the design optimization problem contains seven local variables, three coupling variables, and one shared variable. The specific meanings are shown in Table 7.1 and Figure 7.4, respectively.

The design optimization problem of the flip drive mechanism mainly considers the layout position of each hinge point in the mechanism. It is mainly to meet the requirements of mechanism positioning size, mechanism motion performance, and output torque. At the same time, the influence of load on the stress distribution of the mechanism should also be considered. In MDO, the definition and division of disciplines, in addition to representing

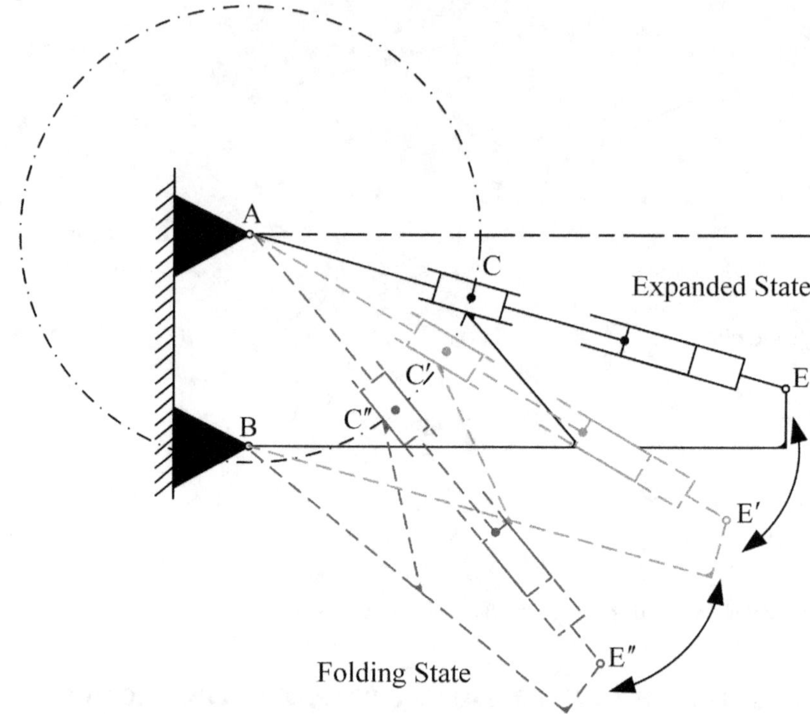

Figure 7.3 Schematic diagram of the working process of the flip drive mechanism.

Table 7.1 Variable information in the design optimization problem of the flipping drive mechanism

	Design variable	*Actual meaning*	*Lower bound*	*Upper bound*
Power input discipline local variables	x_E mm	The abscissa of the hinge fulcrum of the flip actuator and the flip section	0	954
	y_E mm	The ordinate of the hinge fulcrum of the flip actuator and the flip section	–208	0
	D mm	The piston diameter of the flip actuator cylinder	80	120
Local variables of power transmission discipline	x_B mm	Transverse coordinates of the hinged fulcrum of connecting rod and triangular support	0	954
	y_B mm	The ordinate of the hinge fulcrum of the connecting rod and the triangular support	–208	0
	l mm	Length of connecting rod	132	160
	α°	The angle between the initial position connecting the rod and the horizontal line	4.5	6.5

Table 7.1 (Cont.)

	Design variable	Actual meaning	Lower bound	Upper bound
Coupling variable	x_C	Transverse coordinates of hinge fulcrum of connecting rod and turnover actuator	0	954
	y_C	The ordinate of the hinge fulcrum of the connecting rod and the flip actuator cylinder	-208	0
	F N	Hydraulic axial force	0	12.4×10^5
Shared variable		The angle between the slide rail and horizontal line	4.5	6.5

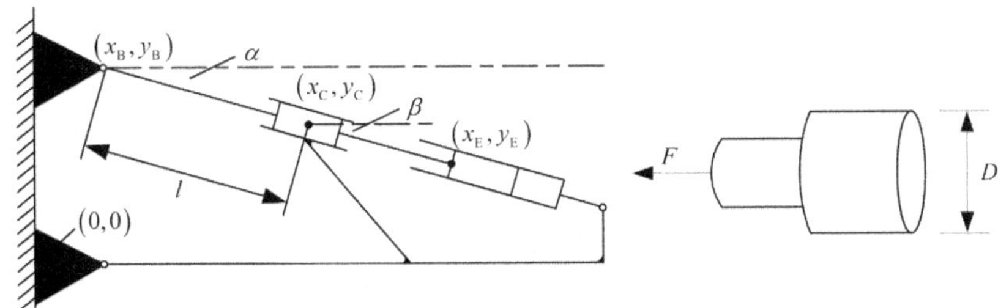

Figure 7.4 Schematic diagram of the actual meaning of design variables.

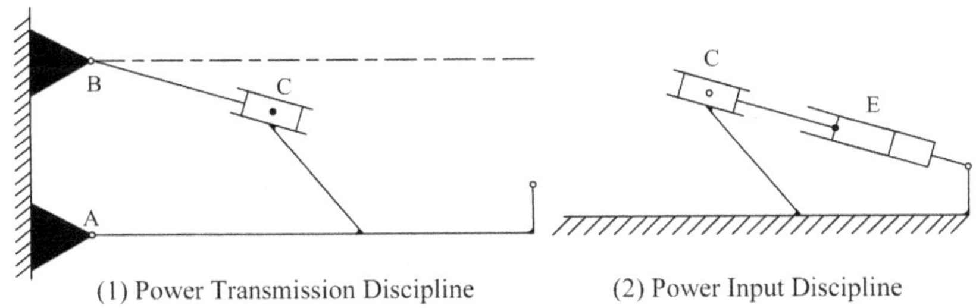

Figure 7.5 Discipline division of flip drive mechanism.

different fields, is a general term for different stages of product design, components and functions contained therein, etc. [10]. Therefore, according to the different functional forms of component composition, the flip drive mechanism is divided into power input discipline and power transmission discipline. As shown in Figure 7.5, the power input discipline consists of a flip actuator, a slide rail, and a slider; the power transmission discipline is composed of a connecting rod, triangular support, slide rail, and slider.

The coupling relationship between the two disciplines is shown in Figure 7.6. The power input discipline inputs the hydraulic axial force of

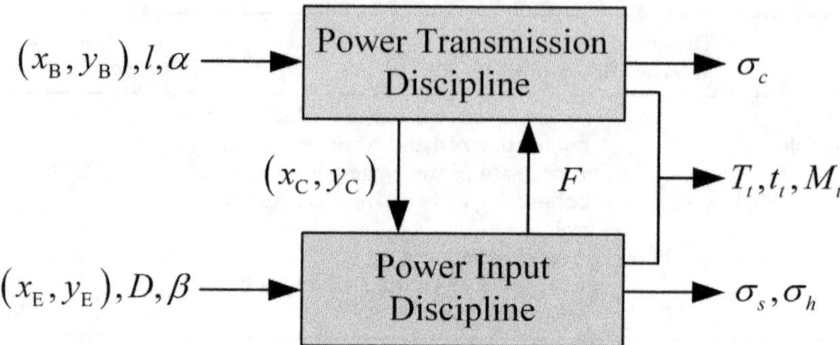

Figure 7.6 The discipline coupling relationship of the flip drive mechanism.

the flip actuator into the flip drive mechanism. The power transmission discipline drives the flip section to rotate around the fixed shaft A through the coupled input hydraulic axial force. When designing and optimizing the connecting rod, it is necessary to consider the influence of hydraulic axial force on the stress of the connecting rod. This may affect the determination of the C coordinate of the hinge point of the hydraulic rod in the connecting rod and the flip actuator. The power transmission discipline inputs the coordinates of the hinge connection point C of the connecting rod and the hydraulic rod in the flip actuator to the power input discipline. Then, the stress distribution is affected by the change in the length of the hydraulic rod. Finally, the size of the input hydraulic axial force is limited by it.

According to the above requirements discussion and analysis, the flip torque T_t is usually used as the objective function when designing and optimizing the flip drive mechanism. At the same time, it is also necessary to meet the conditions of mechanism weight, movement time, mechanism strength, and installation space limitation. Here, the design optimization model of the flip drive mechanism is determined as Eq. 7.1.

$$\begin{aligned}
&Max\ \ T_t = T_t\left(x_B, y_B, x_C, y_C, x_E, y_E, \beta, D\right) \\
&s.t.\ \ \sigma_c = \sigma_c\left(x_B, y_B, l, \alpha, F\right) \le 505, \\
&\qquad \sigma_h = \sigma_h\left(x_C, y_C, x_E, y_E, D\right) \le 505, \\
&\qquad \sigma_s = \sigma_s\left(\beta, x_C, y_C, x_E, y_E, D\right) \le 505, \\
&\qquad t_t = t_t\left(x_B, y_B, x_C, y_C, x_E, y_E, \beta, D\right) \le 35, \\
&\qquad M_t = M_t\left(x_B, y_B, x_C, y_C, x_E, y_E\right) \le 16.5, \\
&\qquad 0 \le x_B, x_C, x_E \le 954,
\end{aligned} \tag{7.1}$$

$$
\begin{aligned}
&208 \le y_B, y_C, y_E \le 0,\\
&132 \le l \le 160,\\
&4.5 \le \alpha, \beta \le 6.5,\\
&80 \le D \le 120,\\
&0 \le F \le 12.4 \times 10^5
\end{aligned}
\qquad (7.1)
$$

where $x_C = l \times \cos\alpha + x_B$, $y_C = l \times \sin\alpha + y_B$, and $F = P \times 10^6 \times \frac{\pi}{4} \times (D \times 10^{-4})^2$. P is the hydraulic pressure in the cylinder when turned over. Because the material used in the mechanism is carbon steel AISI1045, the yield strength limit is 505 MPa in the strength constraint.

7.3 UNCERTAINTY MODELING OF FLIP DRIVE MECHANISM

Before the uncertainty modeling, the flip drive mechanism's virtual prototype modeling and response surface modeling can be studied. This can help readers more intuitively understand the model of this flip drive mechanism and the specific basis for uncertainty modeling.

Virtual prototyping modeling technology has been widely used in aerospace, vehicle and ship, engineering machinery, and other fields [11]. Compared with physical prototypes, virtual prototypes can significantly reduce manufacturing costs and improve analysis efficiency through virtual analysis. Furthermore, with the deepening of the design, the virtual prototype model can be constantly revised. So, it can play a good role in the whole design process. In this case, ADAMS will be used to simulate and analyze the dynamic characteristics of the flip drive mechanism, and information on motion speed, peak load, and another mechanism will be obtained.

After establishing the three-dimensional simulation model of the flip drive mechanism, the model is first imported into the finite element analysis software for flexible body analysis. In order to facilitate the import of the mechanism simulation model in ADAMS, each component in the mechanism needs to be processed separately. Using the method of free meshing, solid185 and mass21 elements are used to complete the meshing and define key points, respectively. At the same time, the physical properties, such as elastic modulus and Poisson's ratio of the component material, are set. Selecting key points is crucial for the whole virtual prototype simulation process.

In ADAMS, the key point after adding the motion pair is the MARKER point used to simulate different motions. Inaccurate point position will lead to motion deviation in the later simulation process. Usually, when the component is meshed, the key points of the position of the motion pair are set, and the corresponding rigid domain is generated to analyze the velocity and load of the motion pair. As shown in Figure 7.7, taking the connecting rod in the flip drive mechanism as an example, the simulation model of the connecting rod is meshed. At the same time, the key points and the rigid domain marked as the green region are defined.

In order to simulate the movement process of the flip drive mechanism, this example adds the revolute joint and the sliding joint to the simulation model of the mechanism, respectively. Specifically, a revolute pair is added at the hinge support point connected to the connecting rod and the triangular support to simulate the flipping process. A sliding pair is added between the slide rail and the slider to simulate the reciprocating movement of the slider along the direction of the slide rail. A revolute pair is added at the hinge fulcrum of the slider and the connecting rod to simulate the rotation of the slider during the overall turnover of the mechanism. A rotary pair is added at the hinge fulcrum of the connecting rod and the hydraulic rod to simulate the rotation of the flipping actuator during the mechanism's overall flipping process. The driving force is added to the hydraulic rod of the flip actuator to simulate the input power of the flip drive mechanism. Figure 7.8 is the mechanism model after adding the corresponding kinematic pair.

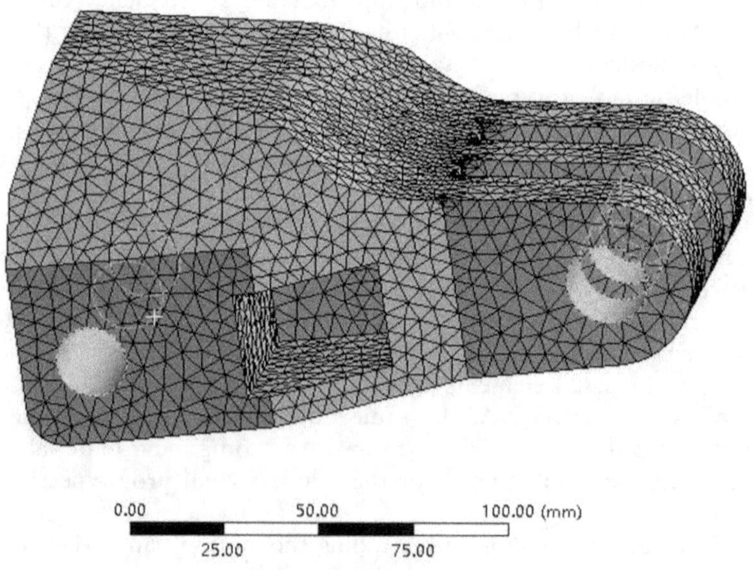

Figure 7.7 Grid division and rigid domain definition of connecting rod.

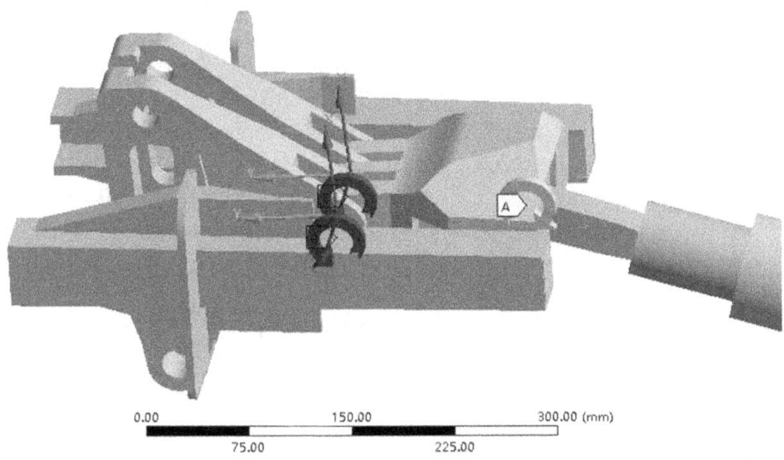

Figure 7.8 The addition of motion pairs in the flip drive mechanism.

Next, the response surface modeling of the flip drive mechanism based on the experimental design method is studied. As described in the previous section, alternative evaluation models can be established using low computational cost approximation techniques. It can be used for coupling discipline analysis, sensitivity calculation, etc. It is one of the key research directions of MDO. The response surface modeling based on the experimental design method is a common global approximation method, which mainly consists of two main areas of research:

7.3.1 Research on experimental design method

As mentioned in the previous chapter, the main process of the experimental design is how to reasonably and effectively arrange and implement a limited number of tests. The operation cost is still high even if the experimental design is based on a virtual prototype. Therefore, in the experimental design of the flip drive mechanism, from a statistical point of view, the functional relationship between the test factors and the response value is analyzed to the greatest extent within the number of affordable tests. In this section, the orthogonal test method is used to select representative data points from the global test factors, and these data points have uniform dispersion characteristics and neat comparability.

7.3.2 Research on approximation technology

After obtaining the corresponding sample points through the experimental design, the approximation technology mainly studies how to fit the original problem efficiently and accurately and then constructs the corresponding approximate model. This section uses the second-order polynomial response surface method in the polynomial response surface method. Taking the performance function of the overturning moment as an example, the coefficient φ of each item in Eq. 7.2 is determined by the least square regression method using the sample points obtained in the orthogonal test.

$$T_t(\mathbf{D}) \approx \varphi_0 + \sum_{i=1}^{8} \varphi_i \mathrm{D}_i + \sum_{i=1}^{8} \varphi_{ii} \mathrm{D}_i^2 + \sum_{i} \sum_{j>i} \varphi_{ij} \mathrm{D}_i \mathrm{D}_j \tag{7.2}$$

where **D** is the test factor of an orthogonal test when establishing the response surface of the overturning moment, $\mathbf{D} = (\mathrm{D}_i, i = 1 \sim 8) = (x_{\mathrm{B}}, y_{\mathrm{B}}, x_{\mathrm{C}}, y_{\mathrm{C}}, x_{\mathrm{E}}, y_{\mathrm{E}}, \beta, D)$. The whole process of experimental design and response surface construction is shown in Figure 7.9.

Then, the uncertainty analysis is carried out. There are multiple sources of uncertainty when dealing with the design optimization problem of the flip

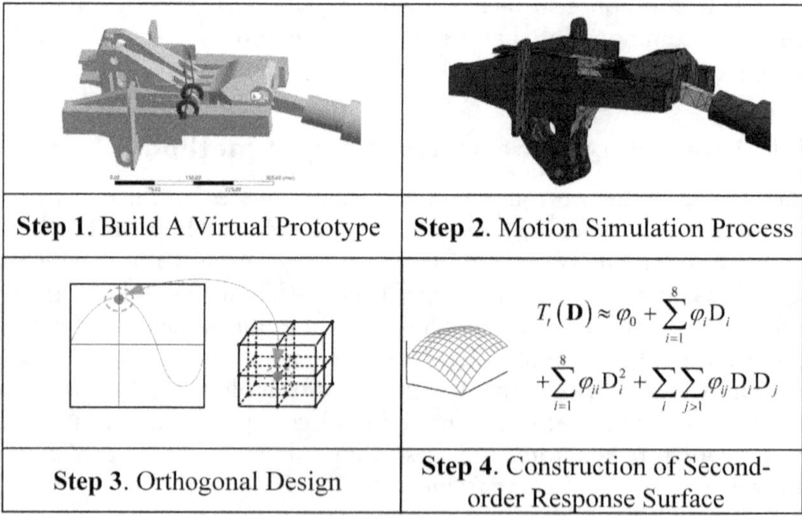

Figure 7.9 Orthogonal test process and response surface construction of flip drive mechanism.

Table 7.2 Uncertainty description of each variable in the design optimization problem of the flip drive mechanism

Variable	*Mean*	*Standard deviation*	*Distribution*
x_E	x_E^M	$0.01x_E^M$	Normal
y_E	y_E^M	$0.01y_E^M$	Normal
D	D^M	$0.001D^M$	Normal
x_B	x_B^M	$0.01x_B^M$	Normal
y_B	y_B^M	$0.01y_B^M$	Normal
l	l^M	$0.001l^M$	Normal
α	α^M	$0.01\alpha^M$	Normal
x_C	x_C^M	$0.01x_C^M$	Normal
y_C	y_C^M	$0.01y_C^M$	Normal
F	F^M	$0.01F^M$	Normal
β	β^M	$0.01\beta^M$	Normal

drive mechanism, such as parts processing dimension error, parts assembly error, material property error, load uncertainty, etc. This section will mainly consider random uncertainty factors, including part machining size errors and component assembly errors. Then, combined with the relevant technical data, the uncertainty information of each design variable is expressed in Table 7.2.

After considering the uncertainty information of each variable, the reliability-based design optimization model of the flip drive mechanism is

$$
\begin{aligned}
\text{Max}\ \ & T_t = T_t\left(x_B^M, y_B^M, x_C^M, y_C^M, x_E^M, y_E^M, \beta^M, D^M\right) \\
\text{s.t.}\ \ & \Pr\left[\sigma_c\left(x_B, y_B, l, \alpha, F\right) \le 505\right] \ge [R], \\
& \Pr\left[\sigma_b\left(x_C, y_C, x_E, y_E, D\right) \le 505\right] \ge [R], \\
& \Pr\left[\sigma_s\left(\beta, x_C, y_C, x_E, y_E, D\right) \le 505\right] \ge [R], \\
& t_t = t_t\left(x_B^M, y_B^M, x_C^M, y_C^M, x_E^M, y_E^M, \beta^M, D^M\right) \le 35, \\
& M_t = M_t\left(x_B^M, y_B^M, x_C^M, y_C^M, x_E^M, y_E^M\right) \le 16.5, \\
& 0 \le x_B^M, x_C^M, x_E^M \le 954, \\
& 208 \le y_B^M, y_C^M, y_E^M \le 0, \\
& 132 \le l^M \le 160, \\
& 4.5 \le \alpha^M, \beta^M \le 6.5, \\
& 80 \le D^M \le 120, \\
& 0 \le F^M \le 12.4 \times 10^5
\end{aligned}
\tag{7.3}
$$

where $x_{\mathrm{C}}^{\mathrm{M}} = l^{\mathrm{M}} \times \cos\alpha^{\mathrm{M}} + x_{\mathrm{B}}^{\mathrm{M}}$, $y_{\mathrm{C}}^{\mathrm{M}} = l^{\mathrm{M}} \times \sin\alpha^{\mathrm{M}} + y_{\mathrm{B}}^{\mathrm{M}}$, $F^{\mathrm{M}} = P \times 10^{6} \times \frac{\pi}{4} \times (D^{\mathrm{M}} \times 10^{-4})^{2}$, and $[R]$ is the specified reliability.

7.4 UBMDO IMPLEMENTATION AND RESULT ANALYSIS OF THE FLIP DRIVE MECHANISM

Considering that the flip drive mechanism is used in aviation, it needs to have high reliability. Therefore, in the design optimization problem, the allowable failure probability is $1-[R]=[p_f]=10^{-5}$. In this section, FOSPA-MDO-SORA and SSRA-MDO-SORA will be used for design optimization. Then, these two methods will be introduced.

7.4.1 FOSPA-MDO-SORA

This strategy can decouple reliability analysis and optimization design. It uses the Most Likely Point (MLP) in the previous loop operation to correct the constraints of the deterministic design optimization model in the narrow loop, gradually increase the reliability of the design point. Sequential deterministic MDO and reliability analyses determine the Reliability-based Multidisciplinary Design Optimization (RBMDO) problem.

In each loop of SORA, deterministic MDO is first performed, followed by reliability analysis. Specifically, the operation of each ring mainly includes the following four steps:

Step 1. The initial value of the optimal design variable in the kth ring operation is set to $\mathbf{d}_{\mathrm{s}}^{(k-1)}, \mathbf{d}^{(k-1)}, \mathbf{X}_{\mathrm{s}}^{\mathrm{M},(k-1)}, \mathbf{X}^{\mathrm{M},(k-1)}$, and the value of design $\mathbf{d}_{\mathrm{s}}^{(k)}, \mathbf{d}^{(k)}, \mathbf{X}_{\mathrm{s}}^{\mathrm{M},(k)}, \mathbf{X}^{\mathrm{M},(k)}$ is obtained. In the first loop operation, the MLP of each limit state function is set to $\mathbf{d}_{\mathrm{s}}^{(0)}, \mathbf{d}^{(0)}, \mathbf{X}_{\mathrm{s}}^{\mathrm{M},(0)}, \mathbf{X}^{\mathrm{M},(0)}$ because the reliability analysis has not been carried out. Starting from the second loop, in the deterministic MDO model, the MLP obtained modify each limit state function by the previous loop operation.

Step 2. Perform reliability analysis. First, the MLP of the limit state function is solved by Eq. 7.4.

$$\begin{aligned}
&\max_{\mathrm{X}_{\mathrm{R}}} \prod f_{\mathrm{X}_{\mathrm{s}}}(\mathrm{X}_{\mathrm{s}}) f_{\mathrm{X}}(\mathrm{X}) f_{\mathrm{P}}(\mathrm{P}) \\
&\text{s.t. } G_i(\mathbf{d}_{\mathrm{s}}, \mathbf{d}_i, \mathbf{X}_{\mathrm{R}}, \mathbf{Y}) = 0, \\
&\qquad \mathbf{Y}_{i\bullet} = \mathbf{Y}_{i\bullet}(\mathbf{d}_{\mathrm{s}}, \mathbf{d}_i, \mathbf{X}_{\mathrm{R}}, \mathbf{Y}_{\bullet i})
\end{aligned} \tag{7.4}$$

After obtaining the MLP, the limit state function $G_i\left(\mathbf{d}_s,\mathbf{d}_i,\mathbf{X}_s,\mathbf{X}_i,\mathbf{P},\mathbf{Y}_{\bullet i}\right)$ is expanded to the first-order Taylor approximation at MLP by Eq. 7.5.

$$\begin{aligned} G_i\left(\mathbf{d}_s,\mathbf{d}_i,\mathbf{X}_R,\mathbf{Y}_{\bullet i}\right) &\approx \widehat{G}_i\left(\mathbf{d}_s,\mathbf{d}_i,\mathbf{X}_R,\mathbf{Y}_{\bullet i}\right) \\ &= G_i\left(\mathbf{d}_s,\mathbf{d}_i,\mathbf{X}_R^*,\mathbf{Y}_{\bullet i}\right)+\nabla G_i\left(\mathbf{X}_R^*\right)\left(\mathbf{X}_R-\mathbf{X}_R^*\right)^T \end{aligned} \tag{7.5}$$

where $\nabla G_i\left(\mathbf{X}_R^*\right)$ is the gradient of $G_i\left(\mathbf{d}_s,\mathbf{d}_i,\mathbf{X}_R,\mathbf{Y}_{\bullet i}\right)$ at $\mathbf{X}_R^*$, as shown in Eq. 7.6.

$$\nabla G_i\left(\mathbf{X}_R^*\right)=\left.\left(\frac{\partial G_i}{\partial \mathrm{X}_s},\frac{\partial G_i}{\partial \mathrm{X}_i},\frac{\partial G_i}{\partial \mathrm{P}}\right)\right|_{\mathbf{X}_R^*} \tag{7.6}$$

Then, the limit state function value $G_i^{(1-\Phi(-\beta_t))}$ corresponds to the approximate limit state function, satisfying the specified reliability solved using the first-order approximate limit state function, as shown in Eq. 7.7.

$$\Pr\left[\widehat{G}_i \le G_i^{(1-\Phi(-\beta_t))}\right]=1-\Phi\left(-\beta_t\right) \tag{7.7}$$

The reliability calculation of Eq. 7.7 can be obtained by the saddle point approximation method algebra. The specific solution process of A can be seen in Reference [12].

Step 3. Test the feasibility and convergence. If all the reliability and deterministic constraints are satisfied, and the objective function value of the system converges, means $G_i^{(1-\Phi(-\beta_t))}\le 0$, $g_0\le 0$, $h_0=0$, $g_i\le 0$, $h_i=0$, and $\left\|f^{(k)}-f^{(k-1)}\right\|\le\varepsilon$. Where ε is an arbitrarily small positive number, $i=1\sim n$, then the optimization stops. Else, go to **Step 4.**

Step 4. Construct a new deterministic MDO model. First, solve the corresponding expected MLP value $\hat{\mathrm{X}}_R=\left(\hat{\mathrm{X}}_s,\hat{\mathrm{X}},\hat{\mathrm{P}}\right)$ when the limit state function value is $G_i^{(1-\Phi(-\beta_t))}$. The mathematical relationship between $G_i^{(1-\Phi(-\beta_t))}$ and $\hat{\mathrm{X}}_R=\left(\hat{\mathrm{X}}_s,\hat{\mathrm{X}},\hat{\mathrm{P}}\right)$ is

$$\Pr\left[\widehat{G}_i\left(\mathbf{d}_s,\mathbf{d}_i,\widehat{\mathbf{X}}_s,\widehat{\mathbf{X}},\widehat{\mathbf{P}},\widehat{\mathbf{Y}}_{\bullet i}\right)\le G_i^{(1-\Phi(-\beta_t))}\right]=1-\Phi\left(-\beta_t\right) \tag{7.8}$$

In order to improve the feasibility of the design point in the *k*th ring operation, the SORA idea is used to construct the moving vector **S** to modify the optimization constraints in the deterministic MDO model in the *k*th ring.

$$\begin{aligned} \mathbf{S}_{\mathrm{X}_s}^{(k)} &= \mathbf{X}_s^{\mathrm{M},(k-1)} - \widehat{\mathbf{X}}_s^{(k-1)} \\ \mathbf{S}_{\mathrm{X}_i}^{(k)} &= \mathbf{X}_i^{\mathrm{M},(k-1)} - \widehat{\mathbf{X}}_i^{(k-1)} \end{aligned} \tag{7.9}$$

where $i = 1 \sim n$, $\mathbf{X}_s^{\mathrm{M},(k-1)}$, and $\mathbf{X}_i^{\mathrm{M},(k-1)}$ are the design point mean of the deterministic MDO in the $(k-1)$th ring, respectively, $\hat{\mathbf{X}}_s^{(k-1)}$ and $\hat{\mathbf{X}}_i^{(k-1)}$ are the corresponding expected MLPs.

Since the MLP of the random design parameter is known and unchangeable, its MLP is directly involved in the construction of the deterministic MDO in the *k*th ring. The optimization constraint of the deterministic MDO model in the *k*th ring is

$$G_{\Pi i}\left(\mathbf{d}_s^{(k)}, \mathbf{d}_i^{(k)}, \mathbf{X}_s^{\mathrm{M},(k)} - \mathbf{S}_{\mathrm{X}_s}^{(k)}, \mathbf{X}_i^{\mathrm{M},(k)} - \mathbf{S}_{\mathrm{X}_i}^{(k)}, \hat{\mathbf{P}}^{(k-1)}, \widehat{\mathbf{Y}}_{\bullet i}^{(k-1)}\right) \le 0 \tag{7.10}$$

After obtaining the modified deterministic MDO model, go back to **Step 2**.

According to the above strategies and steps, the deterministic MDO model in the *k*th ring can be obtained as Eq. 7.11.

$$\begin{aligned} &\min_{\mathrm{DV}} f\left(\mathbf{d}_s^{(k)}, \mathbf{d}^{(k)}, \mathbf{X}_s^{\mathrm{M},(k)}, \mathbf{X}^{\mathrm{M},(k)}, \mathbf{P}^{\mathrm{M}}, \mathbf{Y}^{\mathrm{M},(k)}\right) \\ &\text{s.t. } G_{\Pi i}\left(\mathbf{d}_s^{(k)}, \mathbf{d}_i^{(k)}, \mathbf{X}_s^{\mathrm{M},(k)} - \mathbf{S}_{\mathrm{X}_s}^{(k)}, \mathbf{X}_i^{\mathrm{M},(k)} - \mathbf{S}_{\mathrm{X}_i}^{(k)}, \hat{\mathbf{P}}^{(k-1)}, \widehat{\mathbf{Y}}_{\bullet i}^{(k-1)}\right) \le 0, \\ &\quad g_0\left(\mathbf{d}_s^{(k)}, \mathbf{d}^{(k)}, \mathbf{X}_s^{\mathrm{M},(k)}, \mathbf{X}^{\mathrm{M},(k)}, \mathbf{P}^{\mathrm{M}}, \mathbf{Y}^{\mathrm{M},(k)}\right) \le 0, \\ &\quad h_0\left(\mathbf{d}_s^{(k)}, \mathbf{d}^{(k)}, \mathbf{X}_s^{\mathrm{M},(k)}, \mathbf{X}^{\mathrm{M},(k)}, \mathbf{P}^{\mathrm{M}}, \mathbf{Y}^{\mathrm{M},(k)}\right) = 0, \\ &\quad g_i\left(\mathbf{d}_s^{(k)}, \mathbf{d}_i^{(k)}, \mathbf{X}_s^{\mathrm{M},(k)}, \mathbf{X}_i^{\mathrm{M},(k)}, \mathbf{P}^{\mathrm{M}}, \mathbf{Y}_{\bullet i}^{\mathrm{M},(k)}\right) \le 0, \\ &\quad h_i\left(\mathbf{d}_s^{(k)}, \mathbf{d}_i^{(k)}, \mathbf{X}_s^{\mathrm{M},(k)}, \mathbf{X}_i^{\mathrm{M},(k)}, \mathbf{P}^{\mathrm{M}}, \mathbf{Y}_{\bullet i}^{\mathrm{M},(k)}\right) = 0, \\ &\quad \mathrm{DV} = \left\{\mathbf{d}_s, \mathbf{d}, \mathbf{X}_s^{\mathrm{M}}, \mathbf{X}^{\mathrm{M}}\right\} \end{aligned} \tag{7.11}$$

where $\hat{\mathbf{Y}}^{(k-1)}$ is the value of the coupling variable at the expected MLP in the $(k-1)$th ring operation. $G_{\Pi i}$ is a modified deterministic optimization constraint corresponding to the reliability constraint in the *i*th sub-discipline. $\mathbf{S}_{\mathrm{X}_s}^{(k)}$ and $\mathbf{S}_{\mathrm{X}_i}^{(k)}$ are the moving vectors corresponding to the design variables X_s and X_i in the *i*th sub-discipline, respectively. $\hat{\mathbf{P}}^{(k-1)}$ is the value of the random design parameter at the expected MLP in the $(k-1)$th ring operation. The mean value of coupling variables needs to meet the requirements of interdisciplinary consistency

$$\mathbf{Y}_{i\bullet}^{\mathrm{M},(k)} = y_{i\bullet}\left(\mathbf{d}_{\mathrm{s}}^{(k)}, \mathbf{d}_i^{(k)}, \mathbf{X}_{\mathrm{s}}^{\mathrm{M},(k)}, \mathbf{X}_i^{\mathrm{M},(k)}, \mathbf{P}^{\mathrm{M}}, \mathbf{Y}_{\bullet i}^{\mathrm{M},(k)}\right) \tag{7.12}$$

When the design point $\mathbf{d}_{\mathrm{s}}, \mathbf{d}, \mathbf{X}_{\mathrm{s}}^{\mathrm{M}}, \mathbf{X}^{\mathrm{M}}$ is obtained, the MLP and the expected MLP need to be solved. It is for the first-order approximation of the limit state function, subsequent reliability analysis, and a new loop deterministic MDO model correction.

$$\begin{aligned}
&\max_{\mathrm{DV}} \prod f_{\mathrm{X_s}}\left(\mathrm{X}_{\mathrm{s}}^{\mathrm{M},(k)}\right) f_{\mathrm{X}}\left(\mathrm{X}^{\mathrm{M},(k)}\right) f_{\mathrm{P}}\left(\mathrm{P}^{\mathrm{M},(k)}\right)\\
&\text{s.t. } G_i\left(\mathbf{d}_{\mathrm{s}}^{(k)}, \mathbf{d}_i^{(k)}, \mathbf{X}_{\mathrm{s}}^{\mathrm{M},(k)}, \mathbf{X}_i^{\mathrm{M},(k)}, \mathbf{P}^{\mathrm{M},(k)}, \mathbf{Y}_{\bullet i}^{\mathrm{M},(k)}\right) = 0\\
&\quad \mathbf{Y}_{i\bullet}^{\mathrm{M},(k)} = y_{i\bullet}\left(\mathbf{d}_{\mathrm{s}}^{(k)}, \mathbf{d}_i^{(k)}, \mathbf{X}_{\mathrm{s}}^{\mathrm{M},(k)}, \mathbf{X}_i^{\mathrm{M},(k)}, \mathbf{P}^{\mathrm{M},(k)}, \mathbf{Y}_{\bullet i}^{\mathrm{M},(k)}\right)\\
&\quad \mathrm{DV} = \left\{\mathbf{X}_{\mathrm{s}}^{\mathrm{M},(k)}, \mathbf{X}^{\mathrm{M},(k)}, \mathbf{P}^{\mathrm{M},(k)}\right\}
\end{aligned} \tag{7.13}$$

In the kth ring, MLP can be obtained by solving the following optimization model.

$$\begin{aligned}
&\max_{\mathrm{DV}} \prod f_{\mathrm{X_s}}\left(\mathrm{X}_{\mathrm{s}}^{\mathrm{M},(k)}\right) f_{\mathrm{X}}\left(\mathrm{X}^{\mathrm{M},(k)}\right) f_{\mathrm{P}}\left(\mathrm{P}^{\mathrm{M},(k)}\right)\\
&\text{s.t. } G_i\left(\mathbf{d}_{\mathrm{s}}^{(k)}, \mathbf{d}_i^{(k)}, \mathbf{X}_{\mathrm{s}}^{\mathrm{M},(k)} - \mathbf{S}_{\mathrm{X_s}}^{(k)}, \mathbf{X}_i^{\mathrm{M},(k)} - \mathbf{S}_{\mathrm{X}_i}^{(k)}, \mathbf{P}^{\mathrm{M},(k)}, \mathbf{Y}_{\bullet i}^{\mathrm{M},(k)}\right) = 0\\
&\quad \mathbf{Y}_{i\bullet}^{\mathrm{M},(k)} = y_{i\bullet}\left(\mathbf{d}_{\mathrm{s}}^{(k)}, \mathbf{d}_i^{(k)}, \mathbf{X}_{\mathrm{s}}^{\mathrm{M},(k)}, \mathbf{X}_i^{\mathrm{M},(k)}, \mathbf{P}^{\mathrm{M},(k)}, \mathbf{Y}_{\bullet i}^{\mathrm{M},(k)}\right)\\
&\quad \mathrm{DV} = \left\{\mathbf{X}_{\mathrm{s}}^{\mathrm{M},(k)}, \mathbf{X}^{\mathrm{M},(k)}, \mathbf{P}^{\mathrm{M},(k)}\right\}
\end{aligned} \tag{7.14}$$

It should be noted that the MLP obtained by Eq. 7.13 is the first-order linear approximation of the limit state function, and the expected MLP obtained by Eq. 7.14 is the initial point for constructing the moving vector. The specific meaning is shown in Figure 7.10.

7.4.2 SSRA-MDO-SORA

The SSRA-MDO-SORA strategy adopted in this section can also decouple reliability analysis and optimization design. The constraints of the deterministic design optimization model in the next loop are corrected by using the Simulation Most Probable Point (SMPP) obtained in the previous loop operation. It would gradually improve the reliability of the design point. A series of sequential deterministic MDO and reliability analyses can solve the RBMDO problem.

In each loop operation of SSRA-MDO-SORA, deterministic MDO is first performed, and then reliability analysis is performed. Specifically, the operation of each ring mainly includes the following six steps:

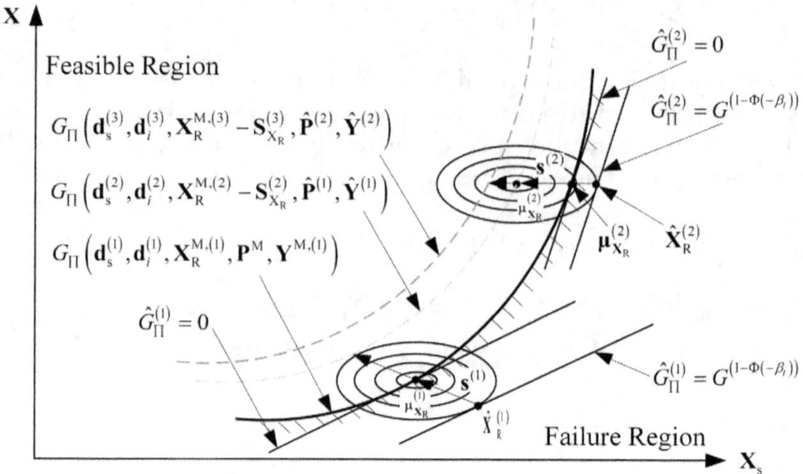

Figure 7.10 Schematic diagram of FOSPA-MDO-SORA.

Step 1. Solve deterministic MDO. The initial value of the optimal design variable in the first loop operation is set to $\mathbf{d}_s^{(0)},\mathbf{d}^{(0)},\mathbf{X}_s^{M,(0)},\mathbf{X}^{M,(0)}$. Using corresponding MDO method to obtain the value of design point $\mathbf{d}_s^{(1)},\mathbf{d}^{(1)},\mathbf{X}_s^{M,(1)},\mathbf{X}^{M,(1)}$. Since the reliability analysis has not been carried out in the first loop operation, the reliability of the obtained limit state function $G^{(1)}=G\left(\mathbf{d}_s^{(1)},\mathbf{d}^{(1)},\mathbf{X}_s^{M,(1)},\mathbf{X}^{M,(1)}\right)$ in the feasible region is about 0.5. Therefore, the Monte Carlo Simulation (MCS) method can conduct the reliability analysis in the first loop operation.

Step 2. Perform MCS reliability analysis. It is assumed that there are $N^{(1)}$ sample points, the reliability of the limit state function $G^{(1)}$ at the design points $\mathbf{X}_s^{M,(1)}$ and $\mathbf{X}^{M,(1)}$ can be obtained by Eq. 7.15.

$$\Pr[G(\mathbf{d}_s,\mathbf{d}_i,\mathbf{X}_R,\mathbf{P},\mathbf{Y})\le 0]\approx 1-p_f=1-\frac{1}{N}\sum_{i=1}^{N}I_G\left(x_{Ri}\right) \tag{7.15}$$

where $I_G(\bullet)$ is the indicator function, when $G\in F$, $I_G(\bullet)=1$; when $G\notin F$, $I_G(\bullet)=0$. The calculation error ξ of the MCS method for reliability evaluation can be calculated by Eq. 7.16.

$$\xi=\left(\frac{1-p_f}{Np_f}\right)^{\frac{1}{2}} \tag{7.16}$$

Step 3. Obtain the SMPP of the first ring. The limit state functions corresponding to all sample points are arranged in ascending order according to their values, as $G_1 < G_2 < \cdots < G_N$. In this sequence, take the value of the ith limit state function, and set $G_s^{(1)} = \left\{ G_i \middle| i = \text{int}\left(\left(1 - p_f\right) \times N^{(1)}\right) \right\}$. The random design point corresponding to the limit state function value is SMPP, $\mathbf{X}_{\text{SMPP}}^{(1)} = \left\{ \mathbf{X}_\text{R} \middle| G_i\left(\mathbf{d}, \mathbf{X}_\text{R}\right) = G_s^{(1)} \right\}$.

Step 4. Construct and solve a new deterministic MDO model. Let $k = k + 1$, by using the moving vector **S** in Eq. 7.17, the optimization constraints in the deterministic MDO model in the kth ring are modified. Solve the kth ring deterministic MDO problem to obtain the value of the design point $\mathbf{d}_\text{s}^{(k)}, \mathbf{d}^{(k)}, \mathbf{X}_\text{s}^{\text{M},(k)}, \mathbf{X}^{\text{M},(k)}$ for the kth ring.

$$\begin{aligned} \mathbf{S}_{\text{X}_\text{s}}^{(k)} &= \mathbf{X}_\text{s}^{\text{M},(k-1)} - \mathbf{X}_{\text{s,SMPP}}^{(k-1)}, \\ \mathbf{S}_{\text{X}_i}^{(k)} &= \mathbf{X}_i^{\text{M},(k-1)} - \mathbf{X}_{i,\text{SMPP}}^{(k-1)} \end{aligned} \tag{7.17}$$

Step 5. Define a series of failure domains for subset simulation reliability assessment. First, define m failure domains $F_j\left(j = 1 \sim m\right)$, including the final sub-failure domain F_m. And $P\left(F_1\right)$ can be calculated by Eq. 7.18.

$$P\left(F_1\right) \approx \frac{1}{N_1} \sum_{i=1}^{N_1} I_G\left(x_{\text{R}i}\right) \tag{7.18}$$

where $x_{\text{R}i}, i = 1 \sim N_1$ is the sample point of independent and identically distributed Probability Density Function (PDF) $f\left(x_\text{R}\right)$.

$P\left(F_{j+1} \middle| F_j\right)$ can be calculated by Eq. 7.19.

$$P\left(F_{j+1} \middle| F_j\right) \approx \frac{1}{N_j} \sum_{i=1}^{N_j} I_G\left(x_{\text{R}i}\right) \tag{7.19}$$

The failure probability of the limit state function $G^{(k)}$ at the design point $\mathbf{d}_\text{s}^{(k)}, \mathbf{d}^{(k)}, \mathbf{X}_\text{s}^{\text{M},(k)}, \mathbf{X}^{\text{M},(k)}$ is subsequently obtained using Eq. 7.20.

$$\begin{aligned} p_f &= P(F_m) = P\left(\bigcap_{j=1}^{m} F_j\right) = P\left(F_m \middle| \bigcap_{j=1}^{m-1} F_j\right) P\left(\bigcap_{j=1}^{m-1} F_j\right) \\ &= P(F_m | F_{m-1}) P\left(\bigcap_{j=1}^{m-1} F_j\right) = \cdots = P(F_1) \prod_{j=1}^{m-1} P(F_{j+1} | F_j) \end{aligned} \tag{7.20}$$

If the probability of failure is no bigger than the allowable probability $[p_f]$, the deterministic constraint is satisfied, and the objective function value is stable, which means $p_f^{(k)} \le [p_f]$, $g_0 \le 0$, $h_0 = 0$, $g_i \le 0$, $h_i = 0$, and $\|f^{(k)} - f^{(k-1)}\| \le \varepsilon$. ε is an arbitrarily small positive number, the optimization process stops. Otherwise, go to **Step 6**.

Step 6. Find the SMPP for the kth ring. First, define the final desired conditional failure probability $\hat{P}^{(k)}(F_m | F_{m-1})$ as

$$\begin{aligned} \hat{P}^{(k)}(F_m | F_{m-1}) &= \frac{[p_f]}{P^{(k)}(F_{m-1})} = \frac{[p_f]}{P^{(k)}(F_1) \prod_{j=1}^{m-1} P^{(k)}(F_{j+1} | F_j)} \\ &= P\left(G \le G_s^{(k)} \middle| G \le G_{m-1}\right) \end{aligned} \tag{7.21}$$

where the final expected conditional failure probability $\hat{P}^{(k)}(F_m | F_{m-1})$ satisfies $P^{(k)}(F_{m-1}) \times \hat{P}^{(k)}(F_m | F_{m-1}) = [p_f]$.

Same as in **Step 3**, the limit state function values corresponding to the $N_m^{(k)}$ sample points used for calculating $P^{(k)}(F_m | F_{m-1})$ are sorted in ascending order of size, as $G_1 < G_2 < \cdots < G_{N_m^{(k)}}$. The randomized design point corresponding to this limit state function value $G_s^{(k)}$ is the SMPP in the kth loop, which means $\mathbf{X}_{\text{SMPP}}^{(k)} = \left\{\mathbf{X}_{\text{R}} \middle| G_i(\mathbf{d}, \mathbf{X}_{\text{R}}) = G_s^{(k)}\right\}$, back to **Step 4**.

Based on the above strategy and steps, the deterministic MDO model in the kth ring can be obtained as Eq. 7.22.

$$\begin{aligned}
&\min_{\mathrm{DV}} f\left(\mathbf{d}_s^{(k)},\mathbf{d}^{(k)},\mathbf{X}_s^{\mathrm{M},(k)},\mathbf{X}^{\mathrm{M},(k)},\mathbf{P}^{\mathrm{M}},\mathbf{Y}^{\mathrm{M},(k)}\right)\\
&\text{s.t. } G_{\Pi i}^{(k)}\left(\mathbf{d}_s^{(k)},\mathbf{d}_i^{(k)},\mathbf{X}_s^{\mathrm{M},(k)}-\mathbf{S}_{\mathrm{X}_s}^{(k)},\mathbf{X}_i^{\mathrm{M},(k)}-\mathbf{S}_{\mathrm{X}_i}^{(k)},\mathbf{P}_{\mathrm{SMPP}}^{(k-1)},\mathbf{Y}_{\bullet i}^{\mathrm{M},(k-1)}\right)\le 0,\\
&\quad g_0\left(\mathbf{d}_s^{(k)},\mathbf{d}^{(k)},\mathbf{X}_s^{\mathrm{M},(k)},\mathbf{X}^{\mathrm{M},(k)},\mathbf{P}^{\mathrm{M}},\mathbf{Y}^{\mathrm{M},(k)}\right)\le 0,\\
&\quad h_0\left(\mathbf{d}_s^{(k)},\mathbf{d}^{(k)},\mathbf{X}_s^{\mathrm{M},(k)},\mathbf{X}^{\mathrm{M},(k)},\mathbf{P}^{\mathrm{M}},\mathbf{Y}^{\mathrm{M},(k)}\right)= 0,\\
&\quad g_i\left(\mathbf{d}_s^{(k)},\mathbf{d}_i^{(k)},\mathbf{X}_s^{\mathrm{M},(k)},\mathbf{X}_i^{\mathrm{M},(k)},\mathbf{P}^{\mathrm{M}},\mathbf{Y}_{\bullet i}^{\mathrm{M},(k)}\right)\le 0,\\
&\quad h_i\left(\mathbf{d}_s^{(k)},\mathbf{d}_i^{(k)},\mathbf{X}_s^{\mathrm{M},(k)},\mathbf{X}_i^{\mathrm{M},(k)},\mathbf{P}^{\mathrm{M}},\mathbf{Y}_{\bullet i}^{\mathrm{M},(k)}\right)= 0,\\
&\quad \mathrm{DV}=\left\{\mathbf{d}_s,\mathbf{d},\mathbf{X}_s^{\mathrm{M}},\mathbf{X}^{\mathrm{M}}\right\}
\end{aligned} \tag{7.22}$$

where $\mathbf{P}_{\mathrm{SMPP}}^{(k-1)}$ is the value of the stochastic design parameter in the $(k-1)$ th loop operation SMPP. $\mathbf{Y}_{\bullet i}^{(k-1)}$ is the value of the coupling variable in the $(k-1)$th loop operation that satisfies the interdisciplinary consistency requirement. $G_{\Pi i}$ is the modified deterministic optimization constraint corresponding to the reliability constraint in ith sub-discipline. $\mathbf{S}_{\mathrm{X}_s}^{(k)}$ and $\mathbf{S}_{\mathrm{X}_i}^{(k)}$ are the movement vectors corresponding to the design variables X_s and X_i in sub-discipline ith, respectively.

Coupled variable means need to satisfy inter-disciplinary consistency requirements

$$\mathbf{Y}_{i\bullet}^{\mathrm{M},(k)}=y_{i\bullet}\left(\mathbf{d}_s^{(k)},\mathbf{d}_i^{(k)},\mathbf{X}_s^{\mathrm{M},(k)},\mathbf{X}_i^{\mathrm{M},(k)},\mathbf{P}^{\mathrm{M}},\mathbf{Y}_{\bullet i}^{\mathrm{M},(k)}\right) \tag{7.23}$$

When the design point $\mathbf{d}_s,\mathbf{d},\mathbf{X}_s^{\mathrm{M}},\mathbf{X}^{\mathrm{M}}$ is obtained, the SMPP needs to be solved based on the random variable's distributional properties to construct the deterministic MDO model in the new loop. The process of finding the SMPP and constructing the transfer vectors in the kth ring is shown in Figure 7.11. The flowchart of SSRA-MDO-SORA is shown in Figure 7.12.

7.4.3 Optimization and result analysis of flip drive mechanism considering uncertainty

This section will optimize the design of the flip drive mechanism based on the above two methods. When using the SSRA-MDO-SORA method, an initial failure domain and four intermediate failure domains are divided as $F_j = 0.1, j = 1 \sim 5$. Since there is only one system design objective, the deterministic MDO under the SORA strategy adopts the CO method in Chapter 2. The system-level optimization model is shown in Eq. 7.24.

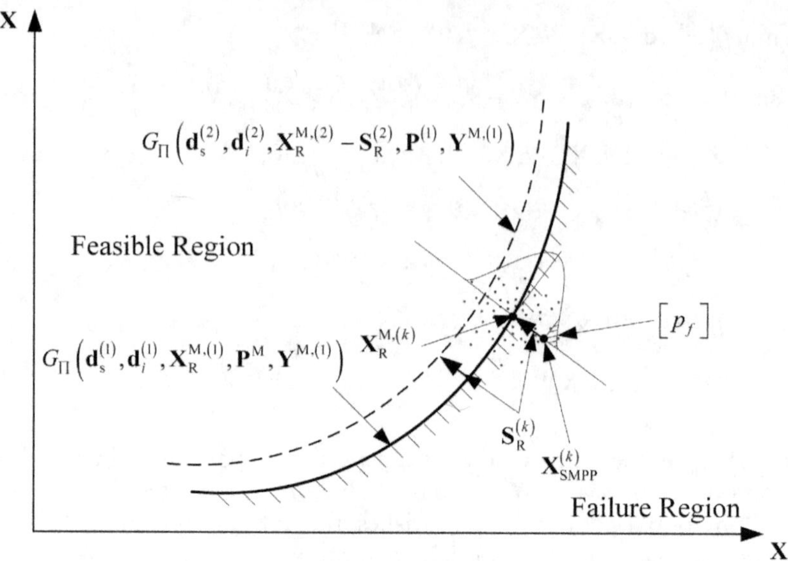

Figure 7.11 Schematic diagram of SSRA-MDO-SORA.

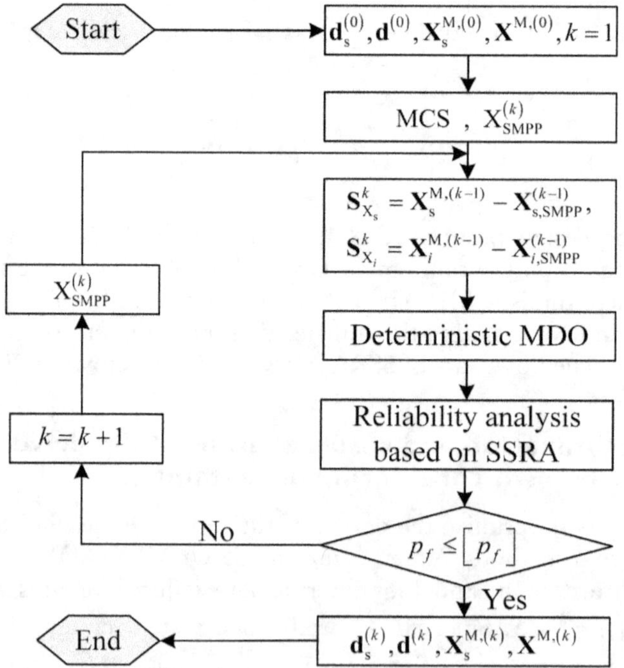

Figure 7.12 Flowchart of SSRA-MDO-SORA.

$$\begin{aligned}
\max\ & T_t = T_t\left(\hat{x}_{\mathrm{B}}^{\mathrm{M}}, \hat{y}_{\mathrm{B}}^{\mathrm{M}}, \hat{x}_{\mathrm{C}}^{\mathrm{M}}, \hat{y}_{\mathrm{C}}^{\mathrm{M}}, \hat{x}_{\mathrm{E}}^{\mathrm{M}}, \hat{y}_{\mathrm{E}}^{\mathrm{M}}, \hat{\beta}^{\mathrm{M}}, \hat{F}^{\mathrm{M}}\right) \\
\text{s.t.}\ & t_t = t_t\left(\hat{x}_{\mathrm{B}}^{\mathrm{M}}, \hat{y}_{\mathrm{B}}^{\mathrm{M}}, \hat{x}_{\mathrm{C}}^{\mathrm{M}}, \hat{y}_{\mathrm{C}}^{\mathrm{M}}, \hat{x}_{\mathrm{E}}^{\mathrm{M}}, \hat{y}_{\mathrm{E}}^{\mathrm{M}}, \hat{\beta}^{\mathrm{M}}, \hat{F}^{\mathrm{M}}\right) \le 35, \\
& M_t = M_t\left(\hat{x}_{\mathrm{B}}^{\mathrm{M}}, \hat{y}_{\mathrm{B}}^{\mathrm{M}}, \hat{x}_{\mathrm{C}}^{\mathrm{M}}, \hat{y}_{\mathrm{C}}^{\mathrm{M}}, \hat{x}_{\mathrm{E}}^{\mathrm{M}}, \hat{y}_{\mathrm{E}}^{\mathrm{M}}\right) \le 16.5, \\
& J_1 = \left\|x_{\mathrm{B}}^{\mathrm{M}} - \hat{x}_{\mathrm{B}}^{\mathrm{M}}\right\|_2^2 + \left\|y_{\mathrm{B}}^{\mathrm{M}} - \hat{y}_{\mathrm{B}}^{\mathrm{M}}\right\|_2^2 + \left\|F_1^{\mathrm{M}} - \hat{F}^{\mathrm{M}}\right\|_2^2 \\
& \quad + \left\|x_{\mathrm{C},1}^{\mathrm{M}} - \hat{x}_{\mathrm{C}}^{\mathrm{M}}\right\|_2^2 + \left\|y_{\mathrm{C},1}^{\mathrm{M}} - \hat{y}_{\mathrm{C}}^{\mathrm{M}}\right\|_2^2 = 0, \\
& J_2 = \left\|x_{\mathrm{E}}^{\mathrm{M}} - \hat{x}_{\mathrm{E}}^{\mathrm{M}}\right\|_2^2 + \left\|y_{\mathrm{E}}^{\mathrm{M}} - \hat{y}_{\mathrm{E}}^{\mathrm{M}}\right\|_2^2 + \left\|\beta^{\mathrm{M}} - \hat{\beta}^{\mathrm{M}}\right\|_2^2 \\
& \quad + \left\|F_2^{\mathrm{M}} - \hat{F}^{\mathrm{M}}\right\|_2^2 + \left\|x_{\mathrm{C},2}^{\mathrm{M}} - \hat{x}_{\mathrm{C}}^{\mathrm{M}}\right\|_2^2 + \left\|y_{\mathrm{C},2}^{\mathrm{M}} - \hat{y}_{\mathrm{C}}^{\mathrm{M}}\right\|_2^2 = 0, \\
& 0 \le \hat{x}_{\mathrm{B}}^{\mathrm{M}}, \hat{x}_{\mathrm{C}}^{\mathrm{M}}, \hat{x}_{\mathrm{E}}^{\mathrm{M}} \le 954, \\
& 208 \le \hat{y}_{\mathrm{B}}^{\mathrm{M}}, \hat{y}_{\mathrm{C}}^{\mathrm{M}}, \hat{y}_{\mathrm{E}}^{\mathrm{M}} \le 0, \\
& 4.5 \le \hat{\beta}^{\mathrm{M}} \le 6.5, \\
& 0 \le \hat{F}^{\mathrm{M}} \le 12.4 \times 10^5
\end{aligned} \tag{7.24}$$

where F_1^{M}, $x_{\mathrm{C},1}^{\mathrm{M}}$, and $y_{\mathrm{C},1}^{\mathrm{M}}$ are the mean value of the coupling variables of the power transmission discipline, F_2^{M}, $x_{\mathrm{C},2}^{\mathrm{M}}$, and $y_{\mathrm{C},2}^{\mathrm{M}}$ are the mean value of the coupling variables of the power input discipline.

The optimization model of power transmission discipline is shown in Eq. 7.25.

$$\begin{aligned}
\min\ & J_1 = \left\|x_{\mathrm{B}}^{\mathrm{M}} - \hat{x}_{\mathrm{B}}^{\mathrm{M}}\right\|_2^2 + \left\|y_{\mathrm{B}}^{\mathrm{M}} - \hat{y}_{\mathrm{B}}^{\mathrm{M}}\right\|_2^2 + \left\|F_1^{\mathrm{M}} - \hat{F}^{\mathrm{M}}\right\|_2^2 + \left\|x_{\mathrm{C},1}^{\mathrm{M}} - \hat{x}_{\mathrm{C}}^{\mathrm{M}}\right\|_2^2 + \left\|y_{\mathrm{C},1}^{\mathrm{M}} - \hat{y}_{\mathrm{C}}^{\mathrm{M}}\right\|_2^2 \\
\text{s.t.}\ & \sigma_c\left(x_{\mathrm{B}}^{\mathrm{M}} - \mathrm{S}_{x_{\mathrm{B}}}^{(k)}, y_{\mathrm{B}} - \mathrm{S}_{y_{\mathrm{B}}}^{(k)}, l - \mathrm{S}_l^{(k)}, \alpha - \mathrm{S}_\alpha^{(k)}, F_1 - \mathrm{S}_{F_1}^{(k)}\right) \le 505, \\
& x_{\mathrm{C}}^{\mathrm{M}} = l^{\mathrm{M}} \times \cos\alpha^{\mathrm{M}} + x_{\mathrm{B}}^{\mathrm{M}}, \\
& y_{\mathrm{C}}^{\mathrm{M}} = l^{\mathrm{M}} \times \sin\alpha^{\mathrm{M}} + y_{\mathrm{B}}^{\mathrm{M}}, \\
& 0 \le x_{\mathrm{B}}^{\mathrm{M}}, x_{\mathrm{C}}^{\mathrm{M}} \le 954, \\
& 208 \le y_{\mathrm{B}}^{\mathrm{M}}, y_{\mathrm{C}}^{\mathrm{M}} \le 0, \\
& 132 \le l^{\mathrm{M}} \le 160, \\
& 4.5 \le \alpha^{\mathrm{M}} \le 6.5, \\
& 0 \le F_1^{\mathrm{M}} \le 12.4 \times 10^5,
\end{aligned} \tag{7.25}$$

where $\mathrm{S}^{(k)}$ is the moving vector of the corresponding design variables in the kth ring deterministic design optimization model.

The optimization model of the power input discipline is shown in Eq. 7.26.

$$
\begin{aligned}
\min\ J_2 = & \left\| x_{\mathrm{E}}^{\mathrm{M}} - \hat{x}_{\mathrm{E}}^{\mathrm{M}} \right\|_2^2 + \left\| y_{\mathrm{E}}^{\mathrm{M}} - \hat{y}_{\mathrm{E}}^{\mathrm{M}} \right\|_2^2 + \left\| \beta^{\mathrm{M}} - \hat{\beta}^{\mathrm{M}} \right\|_2^2 \\
& + \left\| F_2^{\mathrm{M}} - \hat{F}^{\mathrm{M}} \right\|_2^2 + \left\| x_{\mathrm{C},2}^{\mathrm{M}} - \hat{x}_{\mathrm{C}}^{\mathrm{M}} \right\|_2^2 + \left\| y_{\mathrm{C},2}^{\mathrm{M}} - \hat{y}_{\mathrm{C}}^{\mathrm{M}} \right\|_2^2 \\
\text{s.t.}\ & \sigma_b \left(x_{\mathrm{C},2}^{\mathrm{M}} - S_{x_{\mathrm{C}}}^{(k)}, y_{\mathrm{C},2}^{\mathrm{M}} - S_{y_{\mathrm{C}}}^{(k)}, x_{\mathrm{E}}^{\mathrm{M}} - S_{x_{\mathrm{E}}}^{(k)}, y_{\mathrm{E}}^{\mathrm{M}} - S_{y_{\mathrm{E}}}^{(k)}, D^{\mathrm{M}} - S_{D}^{(k)} \right) \le 505, \\
& \sigma_s \left(\beta^{\mathrm{M}} - S_{\beta}^{(k)}, x_{\mathrm{C},2}^{\mathrm{M}} - S_{x_{\mathrm{C}}}^{(k)}, y_{\mathrm{C},2}^{\mathrm{M}} - S_{y_{\mathrm{C}}}^{(k)}, x_{\mathrm{E}}^{\mathrm{M}} - S_{x_{\mathrm{E}}}^{(k)}, y_{\mathrm{E}}^{\mathrm{M}} - S_{y_{\mathrm{E}}}^{(k)}, D^{\mathrm{M}} - S_{D}^{(k)} \right) \le 505, \\
& F_2^{\mathrm{M}} = P \times 10^6 \times \frac{\pi}{4} \times \left(D^{\mathrm{M}} \times 10^{-4} \right)^2, \\
& 0 \le x_{\mathrm{C}}^{\mathrm{M}}, x_{\mathrm{E}}^{\mathrm{M}} \le 954, \\
& 208 \le y_{\mathrm{C}}^{\mathrm{M}}, y_{\mathrm{E}}^{\mathrm{M}} \le 0, \\
& 4.5 \le \beta^{\mathrm{M}} \le 6.5, \\
& 80 \le D^{\mathrm{M}} \le 120, \\
& 0 \le F_2^{\mathrm{M}} \le 12.4 \times 10^5
\end{aligned}
\tag{7.26}
$$

In the CO method, the information exchange between the system layer and the discipline layer is shown in Figure 7.13.

The optimization process of the objective function T_t is shown in Figure 7.14.

The optimization results are shown in Table 7.3.

Table 7.3 shows that compared with the original design scheme, the optimized flip drive mechanism's maximum output torque is increased by 17.2%, and the weight is reduced by 9.43%. The structure of the flip drive mechanism has specifically changed.

a. Increase the length of the connecting rod, as shown in Figure 7.15

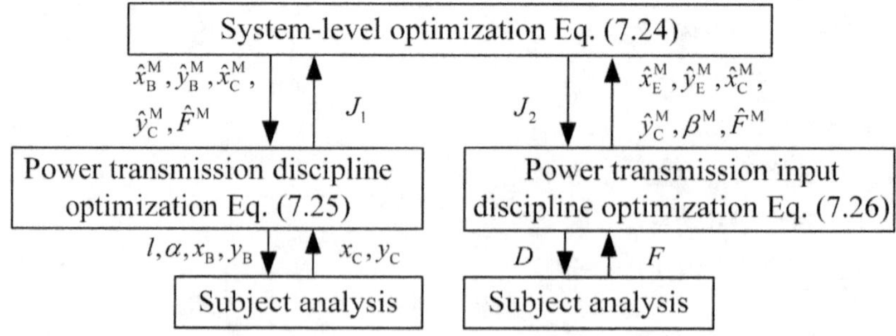

Figure 7.13 CO optimization strategy of the flip drive mechanism.

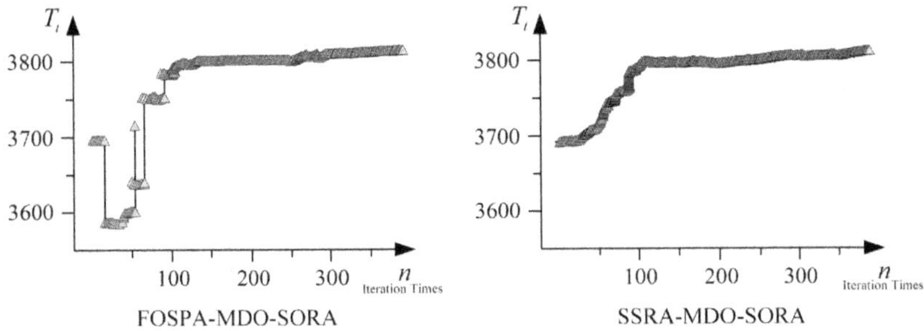

Figure 7.14 Turnover torque optimization process.

Table 7.3 The original value and optimization value of the design variables of the flip drive mechanism

Design variables	*Original value*	*FOSPA optimization value*	*SSRA optimization value*
x_B mm	125	126.86	126.56
y_B mm	-104	-109.01	-109.24
x_E mm	948	947.14	946.98
y_E mm	-114	-113.25	-113.41
D mm	100	109	110
l mm	146	148.53	148.42
α °	5.5	5.66	5.64
β °	5	3.80	3.72

b. As shown in Figure 7.16, reduce the position of the connecting rod and the hinge support point B of the triangular support
c. Increase the initial placement angle of the connecting rod (Figure 7.17)

These three changes can increase the vertical component of the hydraulic driving force used to promote the rotation of the connecting rod under the condition that the hydraulic driving force is constant to improve the output value of the turning torque. In order to further illustrate the influence of the above optimization scheme on the force of the connecting rod, the dynamic simulation of the original design scheme and the optimized design scheme is carried out, respectively.

Then, the maximum force value of the connecting rod at each moment during the flipping process of the flipping drive mechanism is compared. As shown in Figure 7.18 the connecting rod's optimized force value is greater

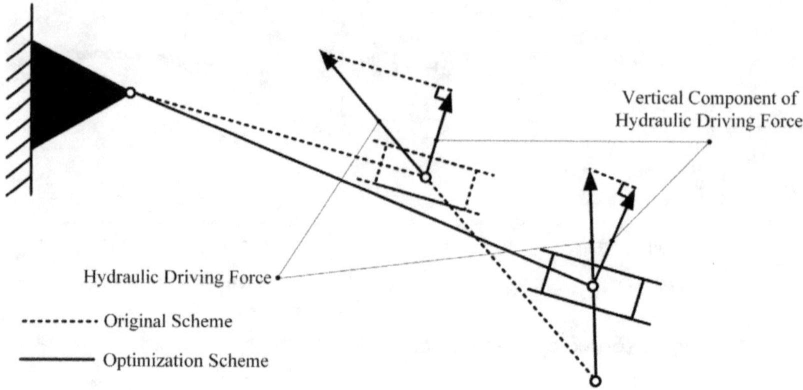

Figure 7.15 Comparative analysis of increasing connecting rod length.

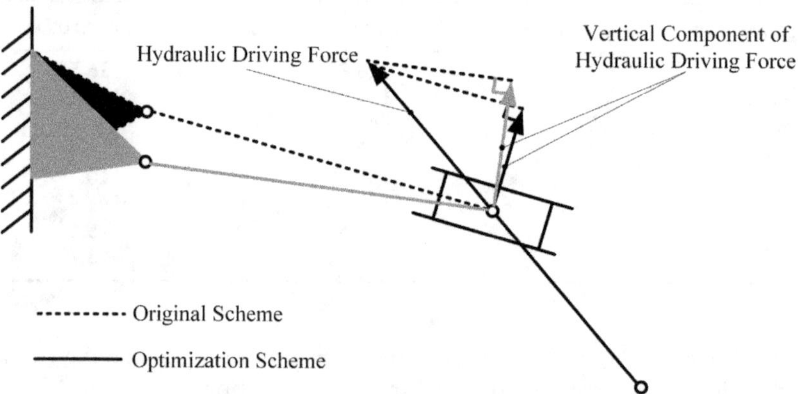

Figure 7.16 Comparative analysis of reducing the position of hinge fulcrum B of connecting rod and triangular support.

than the force value in the original design scheme to overcome its gravity's influence. Under the turning section, it completes the turning movement from the unfolding state to the folding state. This results from the increase of the vertical component of the hydraulic driving force that drives the rotation of the connecting rod.

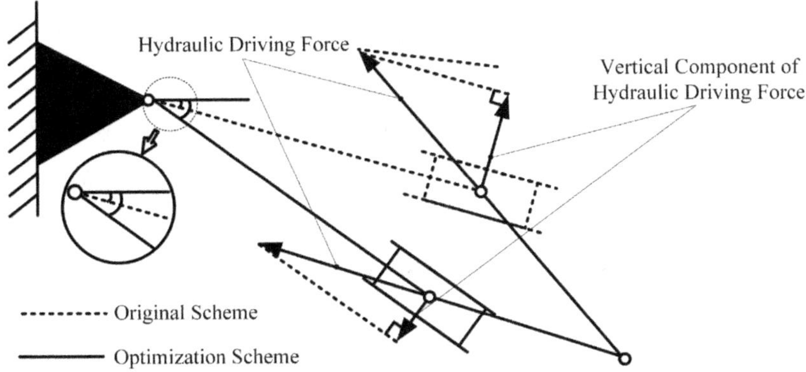

Figure 7.17 Comparative analysis of increasing the initial placement angle of the connecting rod.

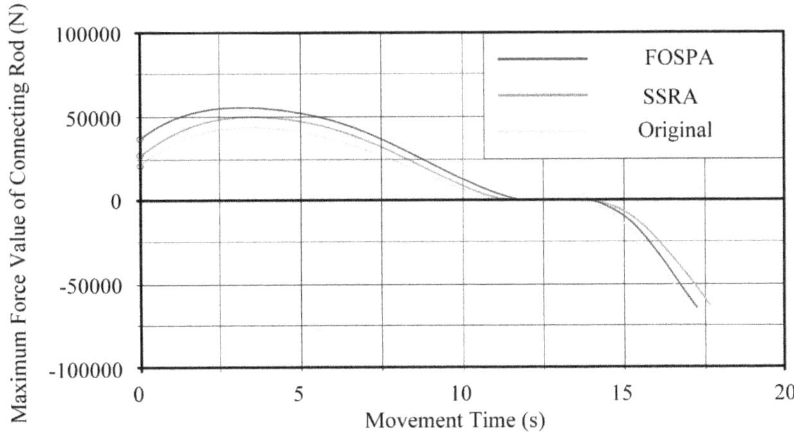

Figure 7.18 The comparison diagram of the force of the connecting rod during the turning process before and after optimization.

REFERENCES

[1] Cui E., Bai P., Yang J. (2007). The development path of intelligent morphing aircraft. Aviation Manufacturing Technology, (8): 38–41.

[2] Zhang Y. (2006). Future aircraft: morphing aircraft. Foreign Science and Technology Trends, (1): 43–48.

[3] Wlezien R. W., Horner G. C., McGowan, A. M. R., Padula, S. L., Scott, M. A., Silcox, R. J., Harrison, J. S. (1998). Aircraft morphing program. Smart

structures and materials 1998. Industrial and Commercial Applications of Smart Structures Technologies, 3326: 176–187.

[4] McGowan A. M. R., Washburn A. E., Horta, L. G., Bryant, R. G., Cox, D. E., Siochi, E. J., Padula S. L., Holloway, N. M. (2002). Recent results from NASA's morphing project. Smart structures and materials 2002. Industrial and Commercial Applications of Smart Structures Technologies, 4698: 97–111.

[5] Perkins D., Reed J., Havens E. (2004). Morphing wing structures for loitering air vehicles. 45th AIAA/ASME/ASCE/AHS/ASC Structures. Structural Dynamics & Materials Conference, 1888.

[6] Zhang J. (2012). Structural Optimization Design and Mechanical Analysis of Variable Swept-Back Wing. Master thesis of Harbin Institute of Technology.

[7] Bye D., McClure P. (2007). Design of a morphing vehicle. 48th AIAA/ASME/ASCE/AHS/ASC Structures. Structural Dynamics, and Materials Conference, 1728.

[8] Ivanco T., Scott R., Love M., Zink S., Weisshaar T. (2007). Validation of the Lockheed Martin morphing concept with wind tunnel testing. 48th AIAA/ASME/ASCE/AHS/ASC Structures. Structural Dynamics, and Materials Conference, 2235.

[9] Love M., Zink P., Stroud R., Bye D., Rizk S., White D. (2007). Demonstration of morphing technology through ground and wind tunnel tests. 48th AIAA/ASME/ASCE/AHS/ASC Structures. Structural Dynamics, and Materials Conference, 1729.

[10] Zhong Y., Chen B., Wang Z. (2006). Multidisciplinary Comprehensive Optimization Design Principles and Methods. Wuhan, China: Huazhong University of Science and Technology Press.

[11] Wang X. (2014). Research on structural dynamics and characteristics of construction machinery arm system. PhD thesis of Zhejiang University (in Chinese), Hangzhou, China.

[12] Du X. (2008). Saddlepoint approximation for sequential optimization and reliability analysis. Journal of Mechanical Design, 130(1): 011011.1–011011.11.

Chapter 8

Application of MDO considering random and interval uncertainty in energy equipment

People's awareness of environmental protection has gradually increased in recent years, and the demand for and attention to clean energy has become increasingly high. Compared with traditional fossil energy, clean energy has many advantages, such as environmental protection, renewable energy, and low carbon emissions. Maintaining clean energy equipment is becoming increasingly important with the popularization of clean energy. These equipment include solar panels, wind turbines, and hydroelectric generators, and their regular operation is significant for ensuring energy supply, reducing environmental pollution, and reducing energy consumption. The hydraulic turbine and rotor bracket are critical components in the hydraulic power generation system, which is a common clear energy equipment, and their performance directly affects the efficiency and stability of the whole system. In the design process of a rotor bracket, it is necessary to consider the knowledge of structural mechanics, fluid dynamics, thermodynamics, and other disciplines. At the same time, due to the influence of uncertain factors such as manufacturing process and material properties, the design result often has certain risks.

The traditional single-discipline design method makes it difficult to deal with these problems effectively, so the Multidisciplinary Design Optimization (MDO) method considering random and interval uncertainty is essential. Since the coupling relationship between the components in the hydraulic turbine rotor system is complex, integrated Multidisciplinary Analysis (MDA) and optimization can be employed. In this chapter, the MDA of the rotor mechanism is carried out through the coupling relationship between the components of the turbine rotor system. The Collaborative Optimization (CO) algorithm is applied to achieve MDO between the rotors. The influence of uncertainty on the reliability of the rotor mechanism is also quantified. A comprehensive MDO model for the turbine rotor mechanism, accounting for multi-source uncertainty, is established.

DOI: 10.1201/9781003464792-8

8.1 INTRODUCTION OF HYDRAULIC TURBINE AND ROTOR BRACKET

The hydraulic turbine is categorized as a type of new energy equipment. It is a power machine that can convert water flow energy into rotational mechanical energy and falls under the classification of turbomachinery in fluid machinery [1]. As early as around 100 B.C., the prototype of a hydraulic turbine, the water wheel, appeared in China, and it was used to lift irrigation and drive grain processing equipment. In contemporary settings, the majority of hydraulic turbines find application in hydroelectric power stations where they drive generators to produce electricity.

In a hydroelectric power station, water from the upstream reservoir is directed to the turbine through a diversion pipe. This process propels the turbine runner into rotation, subsequently powering the generator to generate electricity [2]. The finished water is then released to the downstream through the tailwater pipe. A greater head and increased flow contribute to a higher turbine output power. The classification of hydraulic turbines is shown in Figure 8.1.

Based on different water flow and structural principles, turbines are divided into two categories in Figure 8.1, namely the impulse turbine and the counterattack turbine the counterattack turbine harnesses both potential and water flow energy, whereas the impact turbine only uses water flow energy.

In addition to the various models listed above, reversible turbines have also been born with the development of modern engineering technology.

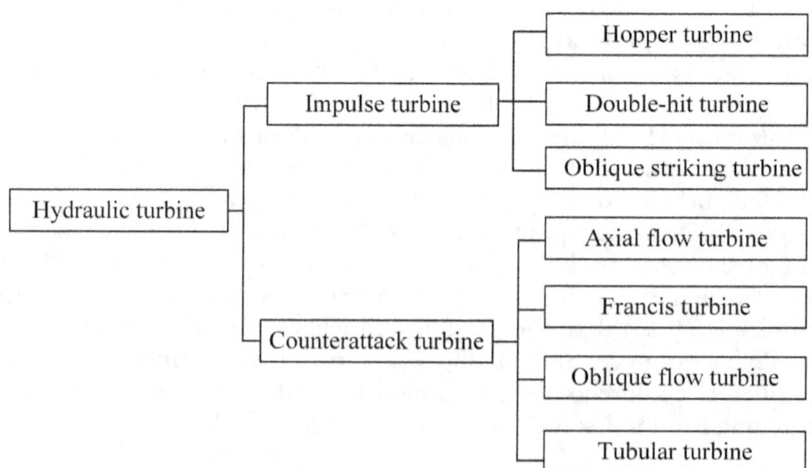

Figure 8.1 Classification of various types of hydraulic turbines.

Three common types of reversible turbines are: Francis flow, oblique flow, and axial flow. The Francis turbine is characterized by high efficiency, a straightforward structure, and broad adaptability. With these features, Francis turbines are extensively employed in countries worldwide. Their maximum power capacity exceeds 700,000 kilowatts, making them a staple in hydropower stations across the globe. Therefore, the rotor mechanism of this turbine type is chosen as the analysis example in this chapter.

The water flows radially from the four directions into the runner's center and then flows out from the lower outlet through the runner. When the water flows radially into the runner, the runner will rotate because the blade is passed and pushed by the water flow. At the same time, when the water flows out axially at the lower outlet, the runner's rotation is also caused by the blade's rotation. The above is the working principle of the Francis turbine runner. Additionally, the Francis turbine is sometimes called the amplitude axial flow turbine because of the above reasons. Figure 8.2 is a schematic diagram of the principle of the turbine.

The Francis turbine's main components encompass runner, main shaft, guide bearing, bottom ring, volute, seat ring, water guide mechanism, top cover, draft tube, rotor, and stator. The runner comprises a discharge cone, a lower ring, an upper crown, and several fixed blades. It is the core rotating component of the Francis turbine, which transmits torque and converts energy. Connected to the shaft, the runner's shape and dimensions often vary depending on different waterheads. The volute is a water diversion component, which is mainly made of metal with a circular cross-section. The seat ring is directly connected to the volute and positioned between

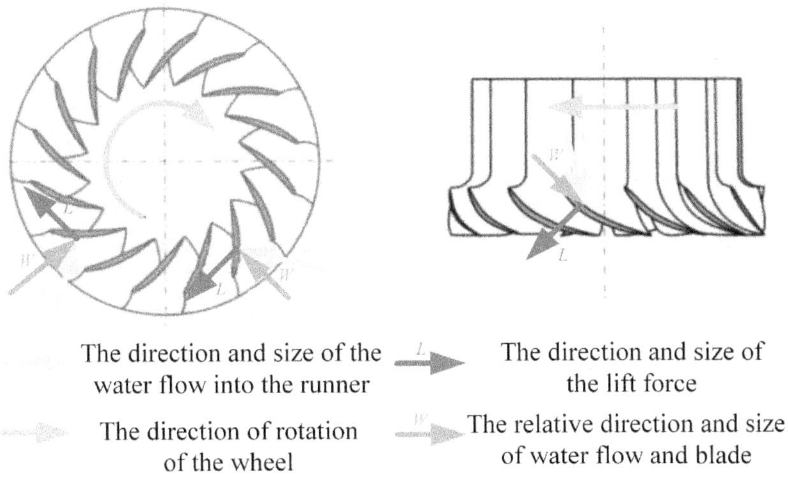

Figure 8.2 Principles of hydraulic turbine.

the volute and the guide vane. It consists of several columns and upper and lower rings. The fixed guide vane takes the form of a wing-shaped non-rotating column. The comprehensive water guide mechanism is a composite structure, including the crank arm, connecting rod, speed-regulating ring, movable guide vane, and other components. The elbow-shaped turbine discharge component is the draft tube, which guides the water flow from the runner outlet to the downstream [3].

The hydro-generator rotor transmits torque, transforms energy, and serves as a rotating component within the hydro-generator. Core components of the rotor include the rotor bracket, magnetic pole, yoke, and shaft. Among these, the rotor bracket stands out as the most complex and influential component. Therefore, examples in this chapter focus on the engineering research of the rotor support with a fixed-size shaft, yoke, and magnetic pole and conduct the design optimization studies. Given that the rotor bracket is the connecting link between the shaft and the yoke, it holds a pivotal position in the rotor's core functionality. Hence, when this chapter refers to the optimized object rotor, it predominantly denotes the rotor bracket.

Rotor brackets, combined rotor brackets, disc rotor brackets, and integral casting rotors are several ordinary structural rotor brackets. Among them, the disc rotor bracket contains a special structure. It has the advantage of good ventilation loss and large stiffness and is widely used both domestically and internationally. Therefore, examples in this chapter also focus on the engineering research of this type of rotor support, the disc rotor bracket.

The main components of the disc rotor bracket include the wheel, upper and lower discs, support plates, and vertical ribs. The detailed component structure is shown in Figure 8.3, where 1 is the lower disc, 2 is the wheel, 3 is the upper disc, 4 is the vertical rib, and 5 is the support plate.

8.2 DISCIPLINE DIVISION AND OPTIMIZATION PROBLEM DEFINITION OF HYDRAULIC TURBINE ROTOR BRACKET

As the core component of the hydro-generator unit, the dynamics of the turbine rotor mechanism are intricate. In addition to fulfilling the criteria of structural lightweight, it also needs to meet certain strength and stiffness requirements. Within the MDO problem of the turbine rotor, two crucial performance indicators, the rotor weight M_R and the stress σ of each rotor component, should be considered. When creating the optimization model, the objective function is aimed at minimizing the weight M_R of the rotor mechanism, and the constraint function is selected as the residual performance index.

According to relevant data, the design optimization problem incorporates four local variables, two coupling variables, five shared variables, and two parameters. These two parameters are the yoke section weakening coefficient

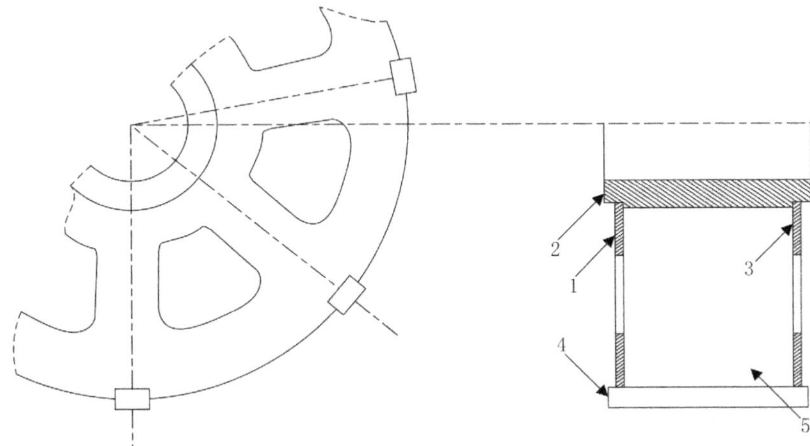

Figure 8.3 Rotor bracket structure composition.

Table 8.1 Variable information for rotor mechanism design optimization

	Design variable	*Lower bound*	*Upper bound*
The disc discipline local variables	*h* mm	30	45
	b mm	0	482
Tendon discipline local variables	*a* mm	60	80
	q mm	50	70
Coupling variables	G_{bm} kgf	23.5	44.1
	F_{cbm} kgf	315.8	642.1
Shared variables	z mm	0	20
	l_1 mm	140	200
	l_2 mm	0	780
	l_3 mm	0	780
	l_4 mm	0	780

β and correction coefficient ζ, respectively. Details of the variables are presented in Table 8.1 and Figure 8.4, where G_{bm} is the gravity of the vertical bar, F_{cbm} is the centrifugal force of the vertical bar, and the total number of vertical bars is denoted as eight.

In order to address the operational force requirements of the mechanism, the design optimization problem of the hydraulic turbine rotor mechanism needs to incorporate the effect of the magnetic yoke on the force of the rotor mechanism. In MDO, the definition and division of disciplines can depend on various factors, including domains, different phases of design, functions, and the components within a product [4]. Therefore, based on the different

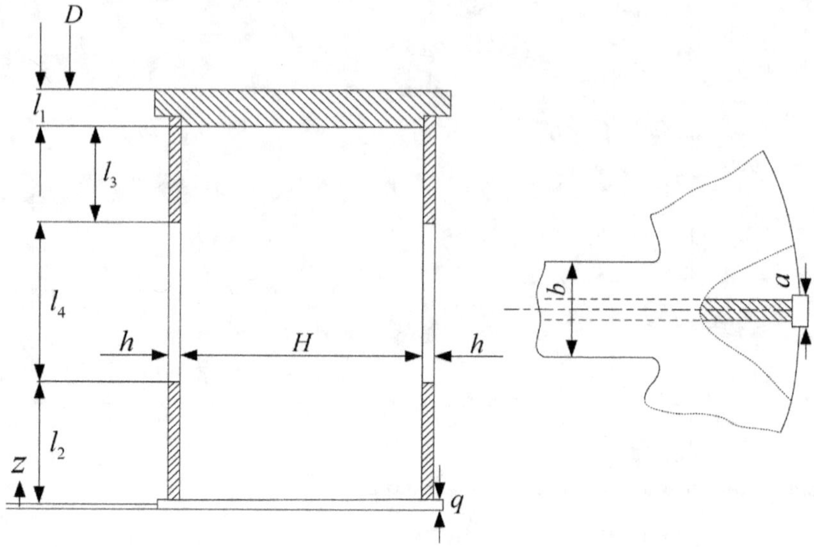

Figure 8.4 Practical implications of design variables.

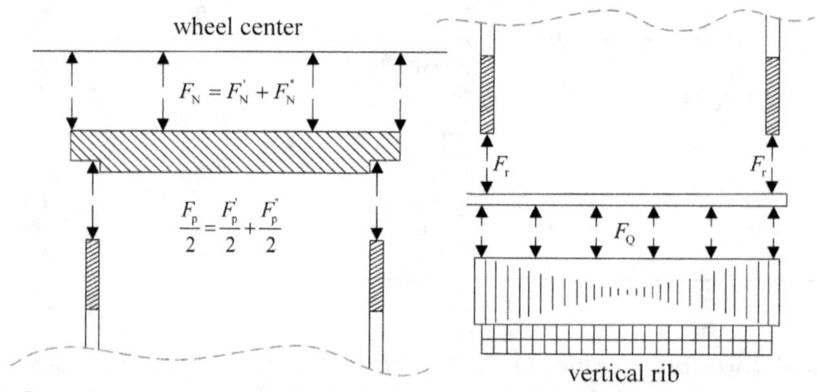

Figure 8.5 Force diagram of component.

components contained in the rotor mechanism, the hydraulic turbine rotor mechanism can be segmented into disc and standing bar disciplines. Forces on the components contained in the two disciplines are shown in Figure 8.5. The disciplinary relationships are shown in Figure 8.6. The purpose of the tendons is to enhance the disk's resistance to deformation and to increase its

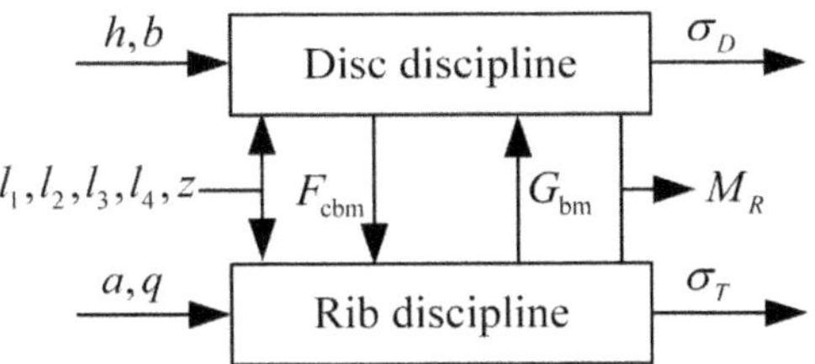

Figure 8.6 The coupling relationship of rotor mechanism disciplines.

rigidity. The magnitude of F_{cbm} acting on the disk can be limited by inputs to the disk discipline from the tendons discipline.

Based on the preceding analysis, minimizing M_R is taken as the objective function of rotor mechanism design optimization. Moreover, constraints include ensuring the strength of each mechanism component and adhering to a reasonable range for size variations.

The deterministic model for optimizing the design of the rotor mechanism is shown in Eq. 8.1.

$$\begin{cases} \min\ M_R = M_R\left(h,b,l_2,l_3,l_4,a,q,G_{bm},z,l_1\right) \\ \text{s.t.}\ \ x_R = x_R\left(h,b,l_2,l_3,l_4,a,q,G_{bm},z,l_1\right) \le 0.5, \\ \quad \sigma_D = \sigma_D\left(h,b,l_2,l_3,l_4,G_{bm},z,F_{cbm},l_1,\beta,\zeta\right) \le 235, \\ \quad \sigma_T = \sigma_T\left(a,q,l_2,l_3,l_4,G_{bm},z,F_{cbm},l_1,\beta,\zeta\right) \le 235, \\ \quad \sigma_H = \sigma_H\left(h,b,l_2,l_3,l_4,a,q,G_{bm},z,F_{cbm},l_1,\beta,\zeta\right) \le 500, \\ \quad 96 \le b \le 144,\ 30 \le h \le 45, \\ \quad 0 \le l_2,l_3,l_4 \le 780,\ 0 \le z \le 20, \\ \quad 1.05 \le \beta \le 1.08,\ 0 \le \zeta \le 0.96, \\ \quad 315.8 \le F_{cbm} \le 642.1, 140 \le l_1 \le 200, \\ \quad 60 \le a \le 80,\ 50 \le q \le 70,\ 23.5 \le G_{bm} \le 44.1, \end{cases} \tag{8.1}$$

where G_{bm} and F_{cbm} represent coupling variables, respectively.

$$G_{bm} = G_{bm}\left(l_2,l_3,l_4,a,q,z,l_1,F_{cbm}\right) \le \left[G_{bm}\right] \tag{8.2}$$

$$F_{cbm} = F_{cbm}\left(h,b,l_2,l_3,l_4,z,l_1,G_{bm}\right) \le \left[F_{cbm}\right] \tag{8.3}$$

where σ_D, σ_H, and σ_T are calculated by F_N, F_p, F_r, and F_Q.

$$F_{\mathrm{N}} = F'_{\mathrm{N}} + F''_{\mathrm{N}} \tag{8.4}$$

$$F'_{\mathrm{N}} = \frac{\Delta_z - \beta_{11}\Delta_{1\mathrm{g}}}{\left(\dfrac{1}{\beta_z} - \beta_{11}\right)\lambda_{\mathrm{g}}} \tag{8.5}$$

$$F''_{\mathrm{N}} = \beta_z\left(F''_{\mathrm{p}} - \frac{\Delta_{\mathrm{gz}}}{\lambda_{\mathrm{g}}}\right) \tag{8.6}$$

$$F_{\mathrm{p}} = F'_{\mathrm{p}} + F''_{\mathrm{p}} \tag{8.7}$$

$$F'_{\mathrm{p}} = \frac{\beta_z\Delta_z - \Delta_{1\mathrm{g}}}{\left(\dfrac{1}{\beta_{11}} - \beta_2\right)\lambda_{\mathrm{g}}} \tag{8.8}$$

$$F''_{\mathrm{p}} = \frac{\Delta_{1\mathrm{g}} + \beta_z\Delta_{\mathrm{gz}} + \beta_{\mathrm{e}}\Delta_{\mathrm{e}}}{\beta_{\mathrm{e}}\lambda_{12} - (1-\beta_z)\lambda_{\mathrm{g}} - \lambda_{11}} \tag{8.9}$$

$$F_{\mathrm{r}} = F_{\mathrm{Q}} - F_{\mathrm{cbm}}\left(\frac{n}{100}\right)^2 \tag{8.10}$$

$$F_{\mathrm{Q}} = \frac{1}{2}\beta_{\mathrm{e}}\left(F''_{\mathrm{p}} + \frac{\Delta_{\mathrm{e}}}{\lambda_{12}}\right) \tag{8.11}$$

where $\Delta_i, \beta_i, \lambda_i$ are the computation coefficients represented by design variables and design parameters [5].

The rotor is assembled through the welding of the wheel, the upper disc, the lower disc, the support plate, and the tendons. Since the wheel is made of steel castings, the strength constraint for the wheel is set at a permissible level of 500 MPa. The remaining components are crafted from Q235-A, with the allowable strength set at 235 Mpa for the disc and vertical bar.

8.3 UNCERTAINTY MODELING OF HYDRAULIC TURBINE ROTOR BRACKET

Before the uncertainty modeling, the virtual prototype and response surface modeling of the hydraulic turbine rotor bracket can be carried out. This facilitates a more intuitive comprehension for readers regarding the model of this hydraulic turbine rotor bracket and the specific basis for uncertainty modeling. The 3D model and mesh division are presented in Figures 8.7 and 8.8, respectively.

The example in this chapter closely parallels the one in the previous chapter, employing response surface modeling for model analysis. Eq. 8.12 is the objective function of the rotor mechanism.

$$M_R(\mathbf{D}) \approx \varphi_0 + \sum_{i=1}^{10} \varphi_i \mathrm{D}_i + \sum_{i=1}^{10} \varphi_{ii} \mathrm{D}_i^2 + \sum_{i} \sum_{j>i} \varphi_{ij} \mathrm{D}_i \mathrm{D}_j \tag{8.12}$$

where $\mathbf{D} = (\mathrm{D}_i, i = 1 \sim 10) = (h, b, l_2, l_3, l_4, a, q, G_{\mathrm{bm}}, z, l_1)$. It serves as the test factor for an orthogonal test when establishing the rotor mass response surface. Here, the coefficients φ in Eq. 8.12 can be determined by the least square method and the design points obtained in the orthogonal experimental

Figure 8.7 3D model of the rotor mechanism.

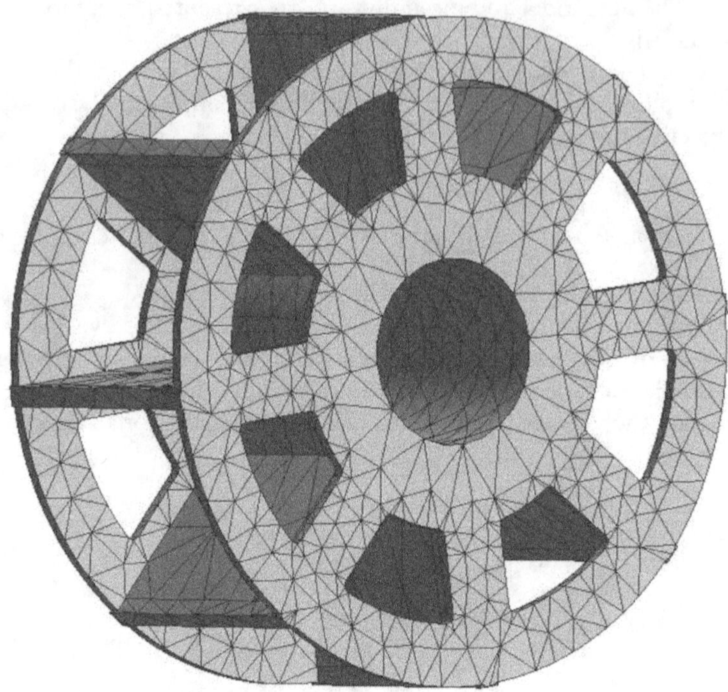

Figure 8.8 Mesh division of rotor mechanism.

design. The whole process of experimental design and response surface construction is illustrated in Figure 8.9.

Subsequently, the uncertainty analysis is carried out. When dealing with the design optimization problem of turbine rotor support, the sources of multiple uncertain factors should be considered, such as part manufacturing size uncertainty, component assembly uncertainty, material property uncertainty, load uncertainty, design uncertainty, and environmental work uncertainty. The example in this chapter will mainly consider the objective uncertainty factors (e.g., part manufacturing dimension uncertainty, assembly uncertainty), and the cognitive uncertainty factors (e.g., design uncertainty).

In this example, the design variables and parameters, involving cognitive and objective uncertainties, are characterized by interval and random variables, respectively. The fourth chapter mentions the mixed uncertainty analysis method, and various methods exist. The example in this chapter will employ the MDO method of mixed random and interval variables, to

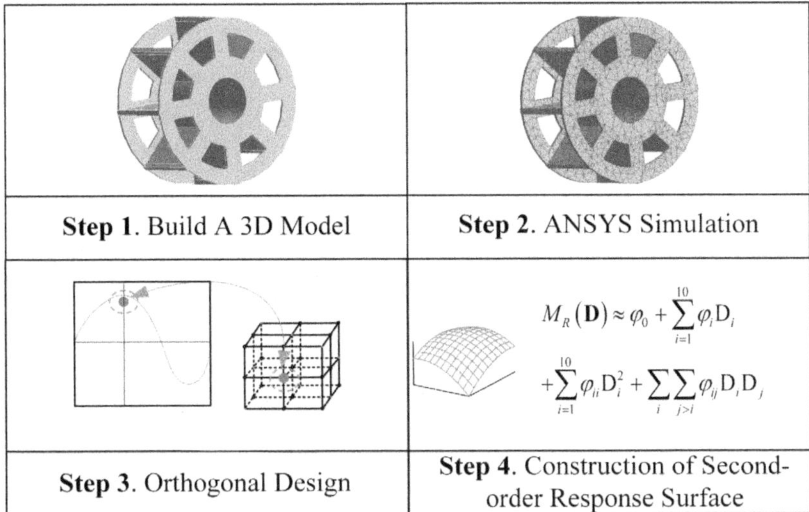

Figure 8.9 Orthogonal test process and response surface construction of rotor mechanism.

Table 8.2 The uncertainty description of each random variable in the optimization problem of rotor mechanism design

Variable	*Mean*	*Standard deviation*	*Distribution*
h	h^M	$0.01h^M$	Normal
b	b^M	$0.01b^M$	Normal
l_2	l_2^M	$0.01l_2^M$	Normal
l_3	l_3^M	$0.01l_3^M$	Normal
l_4	l_4^M	$0.01l_4^M$	Normal
a	a^M	$0.01a^M$	Normal
q	q^M	$0.01q^M$	Normal
G_{bm}	G_{bm}^M	$0.01G_{bm}^M$	Normal
z	z^M	$0.01z^M$	Normal
F_{cbm}	F_{cbm}^M	$0.01F_{cbm}^M$	Normal
l_1	l_1^M	$0.01l_1^M$	Normal

optimize the design of the rotor structure and finally generate the design scheme of the rotor mechanism under the interval and objective uncertainty. First, it is necessary to analyze and quantify the uncertainty associated with the rotor mechanism. With relevant technical data, the uncertainty information for design variables and design parameters is presented in Tables 8.2 and 8.3, respectively.

Table 8.3 The uncertainty description of each interval variable in the optimization problem of rotor mechanism design

Interval variable	*Upper bound*	*Lower bound*
β	1.05	1.08
ζ	0	0.96

The reliability design optimization model of the rotor mechanism concerning the uncertainty information of each random and interval variable can be expressed as Eq. 8.13.

$$\begin{cases} \min\ M_R = M_R\left(h^{\mathrm{M}}, b^{\mathrm{M}}, l_2^{\mathrm{M}}, l_3^{\mathrm{M}}, l_4^{\mathrm{M}}, a^{\mathrm{M}}, q^{\mathrm{M}}, G_{\mathrm{bm}}^{\mathrm{M}}, z^{\mathrm{M}}, l_1^{\mathrm{M}}\right) \\ \text{s.t.}\ \ \Pr\left[x_R = x_R\left(h, b, l_2, l_3, l_4, a, q, G_{\mathrm{bm}}, z^{\mathrm{M}}, l_1\right) \le 0.5\right] \ge [R], \\ \Pr\left[\sigma_D\left(h, b, l_2, l_3, l_4, G_{\mathrm{bm}}, z^{\mathrm{M}}, F_{\mathrm{cbm}}, l_1, \beta, \zeta\right) \le 235\right] \ge [R], \\ \Pr\left[\sigma_T\left(h, b, l_2, l_3, l_4, a, q, G_{\mathrm{bm}}, z^{\mathrm{M}}, F_{\mathrm{cbm}}, l_1, \beta, \zeta\right) \le 235\right] \ge [R], \\ \Pr\left[\sigma_H\left(h, b, l_2, l_3, l_4, a, q, G_{\mathrm{bm}}, z^{\mathrm{M}}, F_{\mathrm{cbm}}, l_1, \beta, \zeta\right) \le 500\right] \ge [R], \\ 96 \le b^{\mathrm{M}} \le 144,\ 0 \le z^{\mathrm{M}} \le 20, \\ 60 \le a^{\mathrm{M}} \le 80,\ 50 \le q^{\mathrm{M}} \le 70, \\ 315.8 \le F_{\mathrm{cbm}}^{\mathrm{M}} \le 642.1,\ 140 \le l_1^{\mathrm{M}} \le 200, \\ 0 \le l_2^{\mathrm{M}}, l_3^{\mathrm{M}}, l_4^{\mathrm{M}} \le 780,\ 23.5 \le G_{\mathrm{bm}}^{\mathrm{M}} \le 44.1, \\ 1.05 \le \overline{\beta} \le 1.08,\ 0 \le \overline{\zeta} \le 0.96,\ 30 \le h^{\mathrm{M}} \le 45 \end{cases} \tag{8.13}$$

8.4 UBMDO REALIZATION AND RESULT ANALYSIS OF HYDRAULIC TURBINE ROTOR BRACKET

As a high-end device in the field of energy application, the turbine rotor mechanism demands a high level of reliability. In this design optimization problem, the reliability threshold is set to 0.99. This chapter will use the MDO method, with random and interval uncertainty considered, to optimize the design. Then, this method will be introduced and elaborated upon.

8.4.1 Analysis method considering random and interval uncertainty

For a variable valued in a certain interval, if its distribution information is unavailable, its uncertainty should be described by interval theory. Compared with Reliability-Based Multidisciplinary Design Optimization (RBMDO), which considers only random variables, RBMDO considers interval variables, which will face more severe computational pressure in reliability analysis. In order to address this issue, a Multidisciplinary Reliability Analysis (MRA) method based on the worst reliability of interval variables within their value range is introduced here.

The RBMDO optimization model with random variables and interval mixed variables is

$$\begin{cases} \min\limits_{\mathrm{DV}} f\left(\boldsymbol{d}_i, \boldsymbol{d}_{\mathrm{S}}, \boldsymbol{X}_i^{\mathrm{R,M}}, \boldsymbol{X}_{\mathrm{S}}^{\mathrm{R,M}}, \boldsymbol{Y}_{ji}^{\mathrm{R,M}}, \bar{\boldsymbol{X}}_i^{\mathrm{I}}, \bar{\boldsymbol{X}}_{\mathrm{S}}^{\mathrm{I}}, \bar{\boldsymbol{Y}}_{ji}^{\mathrm{I}}\right) \\ \text{s.t.} \quad \Pr\left\{\boldsymbol{g}_i\left(\boldsymbol{d}_i, \boldsymbol{d}_{\mathrm{S}}, \boldsymbol{X}_i^{\mathrm{R}}, \boldsymbol{X}_{\mathrm{S}}^{\mathrm{R}}, \boldsymbol{Y}_{ji}^{\mathrm{R}}, \boldsymbol{X}_i^{\mathrm{I}}, \boldsymbol{X}_{\mathrm{S}}^{\mathrm{I}}, \boldsymbol{Y}_{ji}^{\mathrm{I}}\right) \geq 0\right\} \geq [R] = \Phi(\beta), \\ \qquad i, j = 1 \sim n,\ i \neq j, \\ \mathrm{DV} = \{\boldsymbol{d}, \boldsymbol{X}\} \end{cases} \tag{8.14}$$

where $\left(\boldsymbol{X}_i^{\mathrm{I}}, \boldsymbol{X}_{\mathrm{S}}^{\mathrm{I}}\right)$ is the interval variable, $\left[x_{\text{lower}}^{\mathrm{I}}, x_{\text{upper}}^{\mathrm{I}}\right]$ is the specific interval. $\bar{\bullet}$ denotes the numerical midpoint of the upper and lower limits of the range of the interval variable $\boldsymbol{X}^{\mathrm{I}}$.

In addition, the consistency analysis between different disciplines in the internal cycle can be formulated as Eq. 8.15.

$$\boldsymbol{Y}_{ij} = \boldsymbol{Y}_{ij}\left(\boldsymbol{d}_i, \boldsymbol{d}_{\mathrm{S}}, \boldsymbol{X}_i^{\mathrm{R}}, \boldsymbol{X}_{\mathrm{S}}^{\mathrm{R}}, \boldsymbol{Y}_{ji}^{\mathrm{R}}, \boldsymbol{X}_i^{\mathrm{I}}, \boldsymbol{X}_{\mathrm{S}}^{\mathrm{I}}, \boldsymbol{Y}_{ji}^{\mathrm{I}}\right) \tag{8.15}$$

where $\boldsymbol{Y}_{ij} = \left\{\boldsymbol{Y}_{ij}^{\mathrm{R}}, \boldsymbol{Y}_{ij}^{\mathrm{I}}\right\}$.

Due to random and interval uncertainties, the performance function and interval uncertainty should be evaluated in solving RBMDO for the uncertainty constraints outlined in Eq. 8.13. This chapter will also introduce double-loop and three-loop optimization strategies for the mixed random and interval uncertainty analysis in RBMDO.

The optimization model of the outer ring is to identify the worst reliability caused by interval uncertainty, as depicted in Eq. 8.16.

$$\begin{cases} \min\limits_{\mathrm{DV}} \ \beta = \left\|\boldsymbol{U}^*\left(\boldsymbol{X}_i^{\mathrm{I}}, \boldsymbol{X}_{\mathrm{S}}^{\mathrm{I}}, \boldsymbol{Y}_{ji}^{\mathrm{I}}\right)\right\| \\ \text{s.t.} \quad x_{\text{lower}}^{\mathrm{I}} \leq \boldsymbol{X}_i^{\mathrm{I}}, \boldsymbol{X}_{\mathrm{S}}^{\mathrm{I}} \leq x_{\text{upper}}^{\mathrm{I}}, \\ \qquad i, j = 1 \sim n,\ i \neq j, \\ \mathrm{DV} = \left\{\boldsymbol{X}_i^{\mathrm{I}}, \boldsymbol{X}_{\mathrm{S}}^{\mathrm{I}}\right\} \end{cases} \tag{8.16}$$

The optimization model of the middle and inner rings is to identify the Most Probable Point (MPP) when maintaining the multidisciplinary balance, as depicted in Eq. 8.17.

$$\begin{cases} \min\limits_{\mathrm{DV}} \ \beta = \left\| \boldsymbol{U}^{\mathrm{R}} \right\| \\ \text{s.t.} \ \ \boldsymbol{g}_i\left(\boldsymbol{d}_i, \boldsymbol{d}_{\mathrm{S}}, \boldsymbol{U}_i^{\mathrm{R}}, \boldsymbol{U}_{\mathrm{S}}^{\mathrm{R}}, \boldsymbol{Y}_{ji}^{\mathrm{R}}, \boldsymbol{X}_i^{\mathrm{I}}, \boldsymbol{X}_{\mathrm{S}}^{\mathrm{I}}, \boldsymbol{Y}_{ji}^{\mathrm{I}}\right) = 0, \\ \qquad \boldsymbol{Y}_{ij} = \boldsymbol{Y}_{ij}\left(\boldsymbol{d}_i, \boldsymbol{d}_{\mathrm{S}}, \boldsymbol{U}_i^{\mathrm{R}}, \boldsymbol{U}_{\mathrm{S}}^{\mathrm{R}}, \boldsymbol{Y}_{ji}^{\mathrm{R}}, \boldsymbol{X}_i^{\mathrm{I}}, \boldsymbol{X}_{\mathrm{S}}^{\mathrm{I}}, \boldsymbol{Y}_{ji}^{\mathrm{I}}\right), \\ \mathrm{DV} = \left\{\boldsymbol{U}_i^{\mathrm{R}}, \boldsymbol{U}_{\mathrm{S}}^{\mathrm{R}}\right\}, \ i, j = 1 \sim n, \ i \neq j \end{cases} \tag{8.17}$$

It is worth mentioning that the optimization model of Eq. 8.17 is incorporated into the optimization model of Eq. 8.16. The solution in Eq. 8.16 represents the worst-case scenario combination $\boldsymbol{X}_{\mathrm{worst}}^{\mathrm{I}} = \left(\boldsymbol{X}_i^{\mathrm{I}}, \boldsymbol{X}_{\mathrm{S}}^{\mathrm{I}}\right)$, while the solution in Eq. 8.17 represents the worst-case scenario MPP $\boldsymbol{U}_{\mathrm{worst}}^{\mathrm{R},*} = \left(\boldsymbol{U}_i^{\mathrm{R},*}, \boldsymbol{U}_{\mathrm{S}}^{\mathrm{R},*}\right)$. Given this situation, PMA-based RBMDO is usually more efficient than the original RBMDO. Therefore, the three-loop optimization strategy using PMA is described as follows.

For the outer ring, the model in Eq. 8.16 can be transformed into

$$\begin{cases} \min\limits_{\mathrm{DV}} \ g\left(\boldsymbol{d}_i, \boldsymbol{d}_{\mathrm{S}}, \boldsymbol{U}_i^{\mathrm{R},*}, \boldsymbol{U}_{\mathrm{S}}^{\mathrm{R},*}, \boldsymbol{Y}_{ji}^{\mathrm{R}}, \boldsymbol{X}_i^{\mathrm{I}}, \boldsymbol{X}_{\mathrm{S}}^{\mathrm{I}}, \boldsymbol{Y}_{ji}^{\mathrm{I}}\right) \\ \text{s.t.} \ \ \boldsymbol{x}_{\mathrm{lower}}^{\mathrm{I}} \leq \boldsymbol{X}_i^{\mathrm{I}}, \boldsymbol{X}_{\mathrm{S}}^{\mathrm{I}} \leq \boldsymbol{x}_{\mathrm{upper}}^{\mathrm{I}}, \\ \qquad i, j = 1 \sim n, \ i \neq j, \\ \mathrm{DV} = \left\{\boldsymbol{X}_i^{\mathrm{I}}, \boldsymbol{X}_{\mathrm{S}}^{\mathrm{I}}\right\} \end{cases} \tag{8.18}$$

For the middle and inner rings, the model in Eq. 8.17 can be transformed into

$$\begin{cases} \min\limits_{\mathrm{DV}} \ g\left(\boldsymbol{d}_i, \boldsymbol{d}_{\mathrm{S}}, \boldsymbol{U}_i^{\mathrm{R}}, \boldsymbol{U}_{\mathrm{S}}^{\mathrm{R}}, \boldsymbol{Y}_{ji}^{\mathrm{R}}, \boldsymbol{X}_i^{\mathrm{I}}, \boldsymbol{X}_{\mathrm{S}}^{\mathrm{I}}, \boldsymbol{Y}_{ji}^{\mathrm{I}}\right) \\ \text{s.t.} \ \ \left\| \boldsymbol{U} \right\| = \boldsymbol{\Phi}^{-1}\left([R]\right), \\ \qquad \boldsymbol{Y}_{ij} = \boldsymbol{Y}_{ij}\left(\boldsymbol{d}_i, \boldsymbol{d}_{\mathrm{S}}, \boldsymbol{U}_i^{\mathrm{R}}, \boldsymbol{U}_{\mathrm{S}}^{\mathrm{R}}, \boldsymbol{Y}_{ji}^{\mathrm{R}}, \boldsymbol{X}_i^{\mathrm{I}}, \boldsymbol{X}_{\mathrm{S}}^{\mathrm{I}}, \boldsymbol{Y}_{ji}^{\mathrm{I}}\right), \\ \mathrm{DV} = \left\{\boldsymbol{U}_i^{\mathrm{R}}, \boldsymbol{U}_{\mathrm{S}}^{\mathrm{R}}\right\}, \ i, j = 1 \sim n, \ i \neq j \end{cases} \tag{8.19}$$

where $[R]$ denotes the required reliability.

The worst-case performance metric $G_{\mathrm{MPP,worst}}^{[R]}$ can be calculated as

$$G_{\mathrm{MPP,worst}}^{[R]} = g\left(\boldsymbol{d}_i, \boldsymbol{d}_{\mathrm{S}}, \boldsymbol{U}_{\mathrm{worst}}^{\mathrm{R},*,[R]}, \boldsymbol{Y}_{ji}^{\mathrm{R}}, \boldsymbol{X}_{\mathrm{worst}}^{\mathrm{I}}, \boldsymbol{Y}_{ji}^{\mathrm{I}}\right) \tag{8.20}$$

It can be seen that the uncertainty analysis in RBMDO includes a three-loop optimization process. When accounting for system optimization, the whole RBMDO process would extend to a four-loop optimization process. In order to simplify the RBMDO process as much as possible and reduce the amount of calculation, the three-loop optimization process of uncertainty analysis can be condensed into a double-loop optimization process by combining Eqs. 8.16 and 8.17

$$\begin{cases} \min\limits_{\mathrm{DV}} \ \beta = \left\| \boldsymbol{U}^* \left(\boldsymbol{X}_i^{\mathrm{I}}, \boldsymbol{X}_{\mathrm{S}}^{\mathrm{I}}, \boldsymbol{Y}_{ji}^{\mathrm{I}} \right) \right\| \\ \text{s.t.} \ \ \boldsymbol{g}_i \left(\boldsymbol{d}_i, \boldsymbol{d}_{\mathrm{S}}, \boldsymbol{U}_i^{\mathrm{R}}, \boldsymbol{U}_{\mathrm{S}}^{\mathrm{R}}, \boldsymbol{Y}_{ji}^{\mathrm{R}}, \boldsymbol{X}_i^{\mathrm{I}}, \boldsymbol{X}_{\mathrm{S}}^{\mathrm{I}}, \boldsymbol{Y}_{ji}^{\mathrm{I}} \right) = 0, \\ \qquad \boldsymbol{Y}_{ij} = \boldsymbol{Y}_{ij} \left(\boldsymbol{d}_i, \boldsymbol{d}_{\mathrm{S}}, \boldsymbol{U}_i^{\mathrm{R}}, \boldsymbol{U}_{\mathrm{S}}^{\mathrm{R}}, \boldsymbol{Y}_{ji}^{\mathrm{R}}, \boldsymbol{X}_i^{\mathrm{I}}, \boldsymbol{X}_{\mathrm{S}}^{\mathrm{I}}, \boldsymbol{Y}_{ji}^{\mathrm{I}} \right), \\ \qquad x_{\mathrm{lower}}^{\mathrm{I}} \le \boldsymbol{X}_i^{\mathrm{I}}, \boldsymbol{X}_{\mathrm{S}}^{\mathrm{I}} \le x_{\mathrm{upper}}^{\mathrm{I}}, \\ \qquad i, j = 1 \sim n, \ i \neq j, \\ \mathrm{DV} = \left\{ \boldsymbol{U}_i^{\mathrm{R}}, \boldsymbol{U}_{\mathrm{S}}^{\mathrm{R}}, \boldsymbol{X}_i^{\mathrm{I}}, \boldsymbol{X}_{\mathrm{S}}^{\mathrm{I}} \right\} \end{cases} \tag{8.21}$$

Subsequently, based on the same strategy, Eqs. 8.18 and 8.19 can be merged to derive the worst-case performance measure, as shown below:

$$\begin{cases} \min\limits_{\mathrm{DV}} \ g \left(\boldsymbol{d}_i, \boldsymbol{d}_{\mathrm{S}}, \boldsymbol{U}_i^{\mathrm{R}}, \boldsymbol{U}_{\mathrm{S}}^{\mathrm{R}}, \boldsymbol{Y}_{ji}^{\mathrm{R}}, \boldsymbol{X}_i^{\mathrm{I}}, \boldsymbol{X}_{\mathrm{S}}^{\mathrm{I}}, \boldsymbol{Y}_{ji}^{\mathrm{I}} \right) \\ \text{s.t.} \ \ \left\| \boldsymbol{U} \right\| = \boldsymbol{\Phi}^{-1} \left([R] \right), \\ \qquad x_{\mathrm{lower}}^{\mathrm{I}} \le \boldsymbol{X}_i^{\mathrm{I}}, \boldsymbol{X}_{\mathrm{S}}^{\mathrm{I}} \le x_{\mathrm{upper}}^{\mathrm{I}}, \\ \qquad \boldsymbol{Y}_{ij} = \boldsymbol{Y}_{ij} \left(\boldsymbol{d}_i, \boldsymbol{d}_{\mathrm{S}}, \boldsymbol{U}_i^{\mathrm{R}}, \boldsymbol{U}_{\mathrm{S}}^{\mathrm{R}}, \boldsymbol{Y}_{ji}^{\mathrm{R}}, \boldsymbol{X}_i^{\mathrm{I}}, \boldsymbol{X}_{\mathrm{S}}^{\mathrm{I}}, \boldsymbol{Y}_{ji}^{\mathrm{I}} \right), \\ \qquad i, j = 1 \sim n, \ i \neq j, \\ \mathrm{DV} = \left\{ \boldsymbol{U}_i^{\mathrm{R}}, \boldsymbol{U}_{\mathrm{S}}^{\mathrm{R}}, \boldsymbol{X}_i^{\mathrm{I}}, \boldsymbol{X}_{\mathrm{S}}^{\mathrm{I}} \right\} \end{cases} \tag{8.22}$$

8.4.2 MDO method considering random and interval uncertainty

Next, the multidisciplinary uncertainty analysis method is introduced, taking into account both random and interval uncertainties. The fundamental concept of this approach involves decoupling the reliability analysis loop and the deterministic optimization design loop through the Sequential Optimization and Reliability Assessment (SORA) decoupling strategy. Within the deterministic multidisciplinary optimization loop, the CO algorithm is employed. In contrast, the multidisciplinary uncertainty analysis method under random and interval uncertainty, introduced in the previous section, is applied in the reliability analysis loop.

Concerning the worst case of the pickup combination, the RBMDO problem in Eq. 8.14 can be reformulated as Eq. 8.23.

$$\begin{cases}\min\limits_{\mathrm{DV}} f\left(\boldsymbol{d}_i,\boldsymbol{d}_\mathrm{S},\boldsymbol{U}_i^{\mathrm{R,M}},\boldsymbol{U}_\mathrm{S}^{\mathrm{R,M}},\boldsymbol{Y}_{ji}^{\mathrm{R,M}},\bar{\boldsymbol{X}}_i^\mathrm{I},\bar{\boldsymbol{X}}_\mathrm{S}^\mathrm{I},\bar{\boldsymbol{Y}}_{ji}^\mathrm{I}\right)\\ \text{s.t.}\quad \Pr\left\{\boldsymbol{g}_i\left(\boldsymbol{d}_i,\boldsymbol{d}_\mathrm{S},\boldsymbol{U}_i^\mathrm{R},\boldsymbol{U}_\mathrm{S}^\mathrm{R},\boldsymbol{Y}_{ji}^\mathrm{R},\boldsymbol{X}_i^\mathrm{I},\boldsymbol{X}_\mathrm{S}^\mathrm{I},\boldsymbol{Y}_{ji}^\mathrm{I}\right)\ge 0\right\}=\boldsymbol{\Phi}\left(\left\|\boldsymbol{U}_\mathrm{worst}^{\mathrm{R},*,[R]}\right\|\right)\ge[R],\\ \qquad \boldsymbol{Y}_{ij}=\boldsymbol{Y}_{ij}\left(\boldsymbol{d}_i,\boldsymbol{d}_\mathrm{S},\boldsymbol{U}_i^\mathrm{R},\boldsymbol{U}_\mathrm{S}^\mathrm{R},\boldsymbol{Y}_{ji}^\mathrm{R},\boldsymbol{X}_i^\mathrm{I},\boldsymbol{X}_\mathrm{S}^\mathrm{I},\boldsymbol{Y}_{ji}^\mathrm{I}\right),\\ \mathrm{DV}=\left\{\boldsymbol{d}_i,\boldsymbol{d}_\mathrm{S},\boldsymbol{U}_i^{\mathrm{R,M}},\boldsymbol{U}_\mathrm{S}^{\mathrm{R,M}},\bar{\boldsymbol{X}}_i^\mathrm{I},\bar{\boldsymbol{X}}_\mathrm{S}^\mathrm{I}\right\},\ i,j=1\sim n,\ i\ne j\end{cases}\tag{8.23}$$

The Performance Measure Approach (PMA) is added here to improve the efficiency of RBMDO. The PMA-based RBMDO model can be expressed as Eq. 8.24.

$$\begin{cases}\min\limits_{\mathrm{DV}} f\left(\boldsymbol{d}_i,\boldsymbol{d}_\mathrm{S},\boldsymbol{U}_i^{\mathrm{R,M}},\boldsymbol{U}_\mathrm{S}^{\mathrm{R,M}},\boldsymbol{Y}_{ji}^{\mathrm{R,M}},\bar{\boldsymbol{X}}_i^\mathrm{I},\bar{\boldsymbol{X}}_\mathrm{S}^\mathrm{I},\bar{\boldsymbol{Y}}_{ji}^\mathrm{I}\right)\\ \text{s.t.}\quad G_{\mathrm{MPP,worst}}^{[R]}=g\left(\boldsymbol{d}_i,\boldsymbol{d}_\mathrm{S},\boldsymbol{U}_\mathrm{worst}^{\mathrm{R},*,[R]},\boldsymbol{Y}_{ji}^\mathrm{R},\boldsymbol{X}_\mathrm{worst}^\mathrm{I},\boldsymbol{Y}_{ji}^\mathrm{I}\right)\ge 0,\\ \qquad \boldsymbol{Y}_{ij}=\boldsymbol{Y}_{ij}\left(\boldsymbol{d}_i,\boldsymbol{d}_\mathrm{S},\boldsymbol{U}_i^\mathrm{R},\boldsymbol{U}_\mathrm{S}^\mathrm{R},\boldsymbol{Y}_{ji}^\mathrm{R},\boldsymbol{X}_i^\mathrm{I},\boldsymbol{X}_\mathrm{S}^\mathrm{I},\boldsymbol{Y}_{ji}^\mathrm{I}\right),\\ \mathrm{DV}=\left\{\boldsymbol{d}_i,\boldsymbol{d}_\mathrm{S},\boldsymbol{U}_i^{\mathrm{R,M}},\boldsymbol{U}_\mathrm{S}^{\mathrm{R,M}},\bar{\boldsymbol{X}}_i^\mathrm{I},\bar{\boldsymbol{X}}_\mathrm{S}^\mathrm{I}\right\},\ i,j=1\sim n,\ i\ne j\end{cases}\tag{8.24}$$

As uncertainty analysis has not been conducted yet, the first cycle of SORA comprises only deterministic design optimization problems, as shown in Eq. 8.25.

$$\begin{cases}\min\limits_{\mathrm{DV}} f\left(\boldsymbol{d}_i,\boldsymbol{d}_\mathrm{S},\boldsymbol{X}_i^{\mathrm{R,M}},\boldsymbol{X}_\mathrm{S}^{\mathrm{R,M}},\boldsymbol{Y}_{ji}^{\mathrm{R,M}},\bar{\boldsymbol{X}}_i^\mathrm{I},\bar{\boldsymbol{X}}_\mathrm{S}^\mathrm{I},\bar{\boldsymbol{Y}}_{ji}^\mathrm{I}\right)\\ \text{s.t.}\quad g\left(\boldsymbol{d}_i,\boldsymbol{d}_\mathrm{S},\boldsymbol{X}_i^{\mathrm{R,M}},\boldsymbol{X}_\mathrm{S}^{\mathrm{R,M}},\boldsymbol{Y}_{ji}^{\mathrm{R,M}},\bar{\boldsymbol{X}}_i^\mathrm{I},\bar{\boldsymbol{X}}_\mathrm{S}^\mathrm{I},\bar{\boldsymbol{Y}}_{ji}^\mathrm{I}\right)\ge 0,\\ \qquad \boldsymbol{Y}_{ij}^\mathrm{M}=\boldsymbol{Y}_{ij}\left(\boldsymbol{d}_i,\boldsymbol{d}_\mathrm{S},\boldsymbol{X}_i^{\mathrm{R,M}},\boldsymbol{X}_\mathrm{S}^{\mathrm{R,M}},\boldsymbol{Y}_{ji}^{\mathrm{R,M}},\bar{\boldsymbol{X}}_i^\mathrm{I},\bar{\boldsymbol{X}}_\mathrm{S}^\mathrm{I},\bar{\boldsymbol{Y}}_{ji}^\mathrm{I}\right),\\ \mathrm{DV}=\left\{\boldsymbol{d}_i,\boldsymbol{d}_\mathrm{S},\boldsymbol{X}_i^{\mathrm{R,M}},\boldsymbol{X}_\mathrm{S}^{\mathrm{R,M}},\bar{\boldsymbol{X}}_i^\mathrm{I},\bar{\boldsymbol{X}}_\mathrm{S}^\mathrm{I}\right\},\ i,j=1\sim n,\ i\ne j\end{cases}\tag{8.25}$$

In the deterministic optimization design of sequence 1, both $\boldsymbol{X}_\mathrm{worst}^\mathrm{I}=\left(\boldsymbol{X}_{i,\mathrm{worst}}^\mathrm{I},\boldsymbol{X}_{\mathrm{S,worst}}^\mathrm{I}\right)$ and $\boldsymbol{U}_\mathrm{worst}^{\mathrm{R},*,[R]}=\left(\boldsymbol{U}_i^{\mathrm{R},*,[R]},\boldsymbol{U}_\mathrm{S}^{\mathrm{R},*,[R]}\right)$ can be set as the mean values of interval variables and random variables, respectively. Following deterministic optimization and reliability analysis of sequence 1, the optimal design $\left[\boldsymbol{d}_i^{(1)}, \boldsymbol{d}_{\mathrm{S}}^{(1)}, \boldsymbol{X}_i^{\mathrm{R,M},(1)}, \boldsymbol{X}_{\mathrm{S}}^{\mathrm{R,M},(1)}, \overline{\boldsymbol{X}}_i^{\mathrm{I},(1)}, \overline{\boldsymbol{X}}_{\mathrm{S}}^{\mathrm{I},(1)}\right]$ can be obtained. Due to a lack of consideration for uncertainty in the first cycle, the reliability of the design scheme is not high. Based on PMA, sequence 1 interval variable $\boldsymbol{X}_{\mathrm{worst}}^{\mathrm{I},(1)} = \left(\boldsymbol{X}_{i,\mathrm{worst}}^{\mathrm{I},(1)}, \boldsymbol{X}_{\mathrm{S,worst}}^{\mathrm{I},(1)}\right)$ and MPP $\boldsymbol{U}_{\mathrm{worst}}^{\mathrm{R},*,[R],(1)} = \left(\boldsymbol{U}_i^{\mathrm{R},*,[R],(1)}, \boldsymbol{U}_{\mathrm{S}}^{\mathrm{R},*,[R],(1)}\right)$ can be identified. If the performance function $G_{\mathrm{MPP,worst}}^{[R]} = g\left(\boldsymbol{d}_i^{(1)}, \boldsymbol{d}_{\mathrm{S}}^{(1)}, \boldsymbol{X}_i^{\mathrm{R,M},(1)}, \boldsymbol{X}_{\mathrm{S}}^{\mathrm{R,M},(1)}, \overline{\boldsymbol{X}}_i^{\mathrm{I},(1)}, \overline{\boldsymbol{X}}_{\mathrm{S}}^{\mathrm{I},(1)}\right) \geq 0$ is under the corresponding conditions, $\boldsymbol{X}_{\mathrm{worst}}^{\mathrm{I},(1)} = \left(\boldsymbol{X}_{i,\mathrm{worst}}^{\mathrm{I},(1)}, \boldsymbol{X}_{\mathrm{S,worst}}^{\mathrm{I},(1)}\right)$ and $\boldsymbol{U}_{\mathrm{worst}}^{\mathrm{R},*,[R],(1)} = \left(\boldsymbol{U}_i^{\mathrm{R},*,[R],(1)}, \boldsymbol{U}_{\mathrm{S}}^{\mathrm{R},*,[R],(1)}\right)$ are then taken as the initial points, and the optimization model for the next sequence can be constructed.

Then, the deterministic optimization model in sequence 2 can be expressed as Eq. 8.26.

$$\begin{cases} \min\limits_{\mathrm{DV}} f\left(\boldsymbol{d}_i, \boldsymbol{d}_{\mathrm{S}}, \boldsymbol{U}_i^{\mathrm{R,M}}, \boldsymbol{U}_{\mathrm{S}}^{\mathrm{R,M}}, \boldsymbol{Y}_{ji}^{\mathrm{R,M}}, \overline{\boldsymbol{X}}_i^{\mathrm{I}}, \overline{\boldsymbol{X}}_{\mathrm{S}}^{\mathrm{I}}, \overline{\boldsymbol{Y}}_{ji}^{\mathrm{I}}\right) \\ \text{s.t.} \quad G_{\mathrm{MPP,worst}}^{[R]} = g\left(\boldsymbol{d}_i, \boldsymbol{d}_{\mathrm{S}}, \boldsymbol{U}_{\mathrm{worst}}^{\mathrm{R},*,[R],(1)}, \boldsymbol{Y}_{ji}^{\mathrm{R}}, \boldsymbol{X}_{\mathrm{worst}}^{\mathrm{I},(1)}, \boldsymbol{Y}_{ji}^{\mathrm{I}}\right) \geq 0, \\ \qquad \boldsymbol{Y}_{ij}^{\mathrm{M}} = \boldsymbol{Y}_{ij}\left(\boldsymbol{d}_i, \boldsymbol{d}_{\mathrm{S}}, \boldsymbol{X}_i^{\mathrm{R,M}}, \boldsymbol{X}_{\mathrm{S}}^{\mathrm{R,M}}, \boldsymbol{Y}_{ji}^{\mathrm{R,M}}, \overline{\boldsymbol{X}}_i^{\mathrm{I}}, \overline{\boldsymbol{X}}_{\mathrm{S}}^{\mathrm{I}}, \overline{\boldsymbol{Y}}_{ji}^{\mathrm{I}}\right), \\ \mathrm{DV} = \left\{\boldsymbol{d}_i, \boldsymbol{d}_{\mathrm{S}}, \boldsymbol{U}_i^{\mathrm{R,M}}, \boldsymbol{U}_{\mathrm{S}}^{\mathrm{R,M}}, \overline{\boldsymbol{X}}_i^{\mathrm{I}}, \overline{\boldsymbol{X}}_{\mathrm{S}}^{\mathrm{I}}\right\}, \quad i, j = 1 \sim n, \ i \neq j \end{cases} \tag{8.26}$$

If the optimal design scheme $\left[\boldsymbol{d}_i^{(2)}, \boldsymbol{d}_{\mathrm{S}}^{(2)}, \boldsymbol{X}_i^{\mathrm{R,M},(2)}, \boldsymbol{X}_{\mathrm{S}}^{\mathrm{R,M},(2)}, \overline{\boldsymbol{X}}_i^{\mathrm{I},(2)}, \overline{\boldsymbol{X}}_{\mathrm{S}}^{\mathrm{I},(2)}\right]$ yielded by sequence 2 still violates some probability constraints, then $\boldsymbol{X}_{\mathrm{worst}}^{\mathrm{I}} = \left(\boldsymbol{X}_{i,\mathrm{worst}}^{\mathrm{I}}, \boldsymbol{X}_{\mathrm{S,worst}}^{\mathrm{I}}\right)$ and $\boldsymbol{U}_{\mathrm{worst}}^{\mathrm{R},*,[R]} = \left(\boldsymbol{U}_i^{\mathrm{R},*,[R]}, \boldsymbol{U}_{\mathrm{S}}^{\mathrm{R},*,[R]}\right)$ yielded by the reliability analysis loop in sequence 2 are taken as the new initial points of the next sequence. A new deterministic optimization model is constructed in turn. The whole operation flowchart is shown in Figure 8.10.

8.4.3 Optimization and result analysis of turbine rotor mechanism considering uncertainty

According to the above analysis and MDO method, Figure 8.11 shows the flowchart of RBMDO for the turbine rotor mechanism.

Since the rotor optimization problem has only a single system objective function of the rotor mass, the deterministic MDO can be carried out by the CO method under the SORA strategy. The optimization model at the system level for the rotor self-optimization problem is represented as Eq. 8.27.

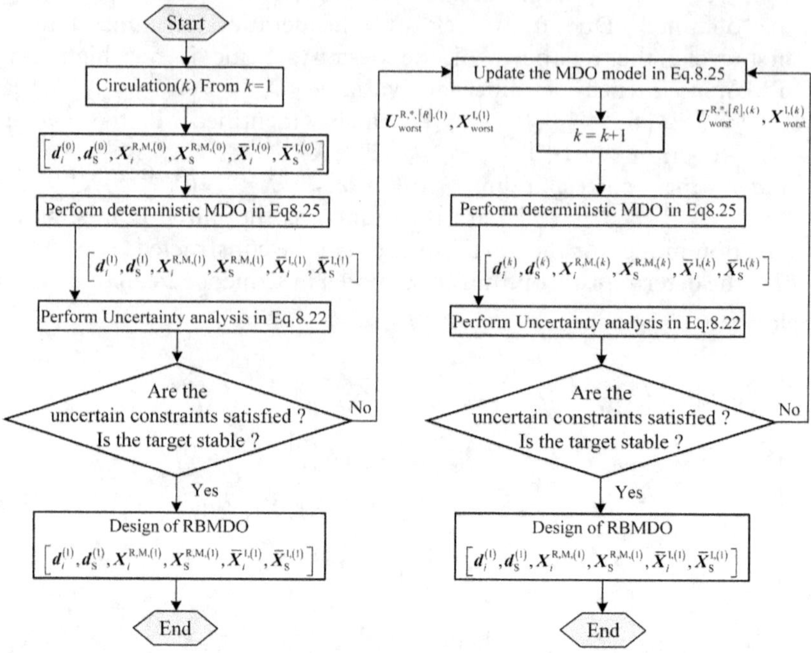

Figure 8.10 MDO solution flowchart of mixed random and interval uncertainty.

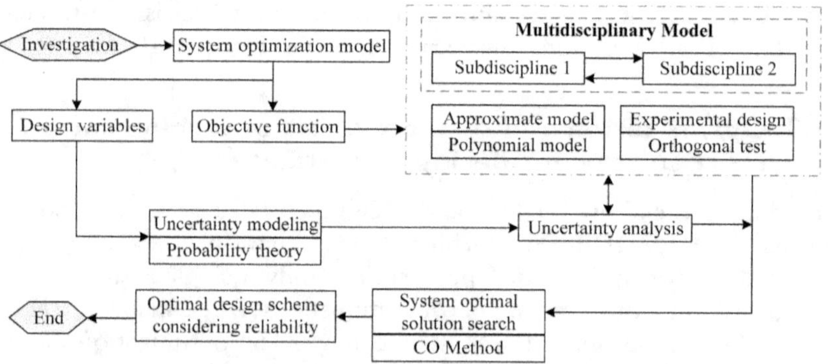

Figure 8.11 Rotor mechanism considering mixed uncertainty optimization flowchart.

$$
\begin{cases}
\min \ M_R = M_R\left(\hat{h}^{\mathrm{M}},\hat{b}^{\mathrm{M}},\hat{l}_2^{\mathrm{M}},\hat{l}_3^{\mathrm{M}},\hat{l}_4^{\mathrm{M}},\hat{a}^{\mathrm{M}},\hat{q}^{\mathrm{M}},\hat{G}_{\mathrm{bm}}^{\mathrm{M}},\hat{z}^{\mathrm{M}},\hat{l}_1^{\mathrm{M}}\right) \\
\text{s.t.} \ \ x_R = x_R\left(\hat{h}^{\mathrm{M}},\hat{b}^{\mathrm{M}},\hat{l}_2^{\mathrm{M}},\hat{l}_3^{\mathrm{M}},\hat{l}_4^{\mathrm{M}},\hat{a}^{\mathrm{M}},\hat{q}^{\mathrm{M}},\hat{G}_{\mathrm{bm}}^{\mathrm{M}},\hat{z}^{\mathrm{M}},\hat{l}_1^{\mathrm{M}}\right) \le 0.5, \\
\quad \sigma_H = \sigma_H\left(\hat{h}^{\mathrm{M}},\hat{b}^{\mathrm{M}},\hat{l}_2^{\mathrm{M}},\hat{l}_3^{\mathrm{M}},\hat{l}_4^{\mathrm{M}},\hat{a}^{\mathrm{M}},\hat{q}^{\mathrm{M}},\hat{G}_{\mathrm{bm}}^{\mathrm{M}},\hat{z}^{\mathrm{M}},\hat{l}_1^{\mathrm{M}},\hat{F}_{\mathrm{cbm}}^{\mathrm{M}},\hat{\bar{\beta}},\hat{\bar{\zeta}}\right) \le 500, \\
\quad J_1 = \left\|h^{\mathrm{M}} - \hat{h}^{\mathrm{M}}\right\|_2^2 + \left\|b^{\mathrm{M}} - \hat{b}^{\mathrm{M}}\right\|_2^2 + \left\|l_2^{\mathrm{M}} - \hat{l}_2^{\mathrm{M}}\right\|_2^2 + \left\|l_3^{\mathrm{M}} - \hat{l}_3^{\mathrm{M}}\right\|_2^2 + \left\|l_4^{\mathrm{M}} - \hat{l}_4^{\mathrm{M}}\right\|_2^2 \\
\qquad + \left\|G_{\mathrm{bm},1}^{\mathrm{M}} - \hat{G}_{\mathrm{bm}}^{\mathrm{M}}\right\|_2^2 + \left\|z_1^{\mathrm{M}} - \hat{z}^{\mathrm{M}}\right\|_2^2 + \left\|F_{\mathrm{cbm},1}^{\mathrm{M}} - \hat{F}_{\mathrm{cbm}}^{\mathrm{M}}\right\|_2^2 + \left\|l_1^{\mathrm{M}} - \hat{l}_1^{\mathrm{M}}\right\|_2^2 = 0, \\
\quad J_2 = \left\|a^{\mathrm{M}} - \hat{a}^{\mathrm{M}}\right\|_2^2 + \left\|q^{M} - \hat{q}^{\mathrm{M}}\right\|_2^2 + \left\|l_2^{\mathrm{M}} - \hat{l}_2^{\mathrm{M}}\right\|_2^2 + \left\|l_3^{\mathrm{M}} - \hat{l}_3^{\mathrm{M}}\right\|_2^2 + \left\|l_4^{\mathrm{M}} - \hat{l}_4^{\mathrm{M}}\right\|_2^2 \\
\qquad + \left\|G_{\mathrm{bm},2}^{\mathrm{M}} - \hat{G}_{\mathrm{bm}}^{\mathrm{M}}\right\|_2^2 + \left\|z_1^{\mathrm{M}} - \hat{z}^{\mathrm{M}}\right\|_2^2 + \left\|F_{\mathrm{cbm},2}^{\mathrm{M}} - \hat{F}_{\mathrm{cbm}}^{\mathrm{M}}\right\|_2^2 + \left\|l_1^{\mathrm{M}} - \hat{l}_1^{\mathrm{M}}\right\|_2^2 = 0, \\
\quad 96 \le \hat{b}^{\mathrm{M}} \le 144, \ 0 \le \hat{z}^{\mathrm{M}} \le 20, \\
\quad 60 \le \hat{a}^{\mathrm{M}} \le 80, \ 50 \le \hat{q}^{\mathrm{M}} \le 70, \\
\quad 315.8 \le \hat{F}_{\mathrm{cbm}}^{\mathrm{M}} \le 642.1, \ 140 \le \hat{l}_1^{\mathrm{M}} \le 200, \\
\quad 0 \le \hat{l}_2^{\mathrm{M}},\hat{l}_3^{\mathrm{M}},\hat{l}_4^{\mathrm{M}} \le 780, \ 23.5 \le \hat{G}_{\mathrm{bm}}^{\mathrm{M}} \le 44.1, \\
\quad 1.05 \le \hat{\bar{\beta}} \le 1.08, \ 0 \le \hat{\bar{\zeta}} \le 0.96, \ 30 \le \hat{h}^{\mathrm{M}} \le 45
\end{cases}
\tag{8.27}
$$

where $G_{\mathrm{bm},1}^{\mathrm{M}}$, z_1^{M}, and $F_{\mathrm{cbm},1}^{\mathrm{M}}$ are the corresponding mean values of the coupling variables of the disc disciplines $G_{\mathrm{bm},2}^{\mathrm{M}}$, z_2^{M}, and $F_{\mathrm{cbm},2}^{\mathrm{M}}$ are the corresponding mean values of the coupling variables of the vertical reinforcement discipline.

The optimization model of the disc discipline is shown in Eq. 8.28.

$$
\begin{cases}
\min \ J_1 = \left\|h^{\mathrm{M}} - \hat{h}^{\mathrm{M}}\right\|_2^2 + \left\|b^{\mathrm{M}} - \hat{b}^{\mathrm{M}}\right\|_2^2 + \left\|l_2^{\mathrm{M}} - \hat{l}_2^{\mathrm{M}}\right\|_2^2 + \left\|l_3^{\mathrm{M}} - \hat{l}_3^{\mathrm{M}}\right\|_2^2 + \left\|l_4^{\mathrm{M}} - \hat{l}_4^{\mathrm{M}}\right\|_2^2 \\
\qquad + \left\|G_{\mathrm{bm},1}^{\mathrm{M}} - \hat{G}_{\mathrm{bm}}^{\mathrm{M}}\right\|_2^2 + \left\|z_1^{\mathrm{M}} - \hat{z}^{\mathrm{M}}\right\|_2^2 + \left\|F_{\mathrm{cbm},1}^{\mathrm{M}} - \hat{F}_{\mathrm{cbm}}^{\mathrm{M}}\right\|_2^2 + \left\|l_1^{\mathrm{M}} - \hat{l}_1^{\mathrm{M}}\right\|_2^2 = 0 \\
\text{s.t.} \ \ \sigma_D\begin{pmatrix} h^{\mathrm{M}},b^{\mathrm{M}},l_2^{\mathrm{M}},l_3^{\mathrm{M}},l_4^{\mathrm{M}},G_{\mathrm{bm},1}^{\mathrm{M}}, \\ z_1^{\mathrm{M}},F_{\mathrm{cbm},1}^{\mathrm{M}},l_1^{\mathrm{M}},\bar{\beta},\bar{\zeta} \end{pmatrix} \le 235, \\
\quad 0 \le \hat{z}^{\mathrm{M}} \le 20, \\
\quad 96 \le \hat{b}^{\mathrm{M}} \le 144, \\
\quad 315.8 \le \hat{F}_{\mathrm{cbm}}^{\mathrm{M}} \le 642.1, \ 140 \le \hat{l}_1^{\mathrm{M}} \le 200, \\
\quad 0 \le \hat{l}_2^{\mathrm{M}},\hat{l}_3^{\mathrm{M}},\hat{l}_4^{\mathrm{M}} \le 780, \ 23.5 \le \hat{G}_{\mathrm{bm}}^{\mathrm{M}} \le 44.1, \\
\quad 1.05 \le \hat{\bar{\beta}} \le 1.08, \ 0 \le \hat{\bar{\zeta}} \le 0.96, \ 30 \le \hat{h}^{\mathrm{M}} \le 45
\end{cases}
\tag{8.28}
$$

The optimization model of the tendons discipline is shown in Eq. 8.29.

$$\begin{cases} \min\ J_2 = \left\|a^{\mathrm{M}} - \hat{a}^{\mathrm{M}}\right\|_2^2 + \left\|q^{\mathrm{M}} - \hat{q}^{\mathrm{M}}\right\|_2^2 + \left\|l_2^{\mathrm{M}} - \hat{l}_2^{\mathrm{M}}\right\|_2^2 + \left\|l_3^{\mathrm{M}} - \hat{l}_3^{\mathrm{M}}\right\|_2^2 + \left\|l_4^{\mathrm{M}} - \hat{l}_4^{\mathrm{M}}\right\|_2^2 \\ \quad + \left\|G_{\mathrm{bm},2}^{\mathrm{M}} - \hat{G}_{\mathrm{bm}}^{\mathrm{M}}\right\|_2^2 + \left\|z_2^{\mathrm{M}} - \hat{z}^{\mathrm{M}}\right\|_2^2 + \left\|F_{\mathrm{cbm},2}^{\mathrm{M}} - \hat{F}_{\mathrm{cbm}}^{\mathrm{M}}\right\|_2^2 + \left\|l_1^{\mathrm{M}} - \hat{l}_1^{\mathrm{M}}\right\|_2^2 = 0 \\ \text{s.t.}\ \ \sigma_T\begin{pmatrix} a^{\mathrm{M}}, q^{\mathrm{M}}, l_2^{\mathrm{M}}, l_3^{\mathrm{M}}, l_4^{\mathrm{M}}, G_{\mathrm{bm},2}^{\mathrm{M}}, \\ z_2^{\mathrm{M}}, F_{\mathrm{cbm},2}^{\mathrm{M}}, l_1^{\mathrm{M}}, \bar{\beta}, \bar{\zeta} \end{pmatrix} \le 235, \\ \quad 0 \le \hat{z}^{\mathrm{M}} \le 20, \\ \quad 60 \le \hat{a}^{\mathrm{M}} \le 80, \\ \quad 315.8 \le \hat{F}_{\mathrm{cbm}}^{\mathrm{M}} \le 642.1,\ 140 \le \hat{l}_1^{\mathrm{M}} \le 200, \\ \quad 0 \le \hat{l}_2^{\mathrm{M}}, \hat{l}_3^{\mathrm{M}}, \hat{l}_4^{\mathrm{M}} \le 780,\ 23.5 \le \hat{G}_{\mathrm{bm}}^{\mathrm{M}} \le 44.1, \\ \quad 1.05 \le \hat{\bar{\beta}} \le 1.08,\ 0 \le \hat{\bar{\zeta}} \le 0.96,\ 50 \le \hat{q}^{\mathrm{M}} \le 70 \end{cases} \tag{8.29}$$

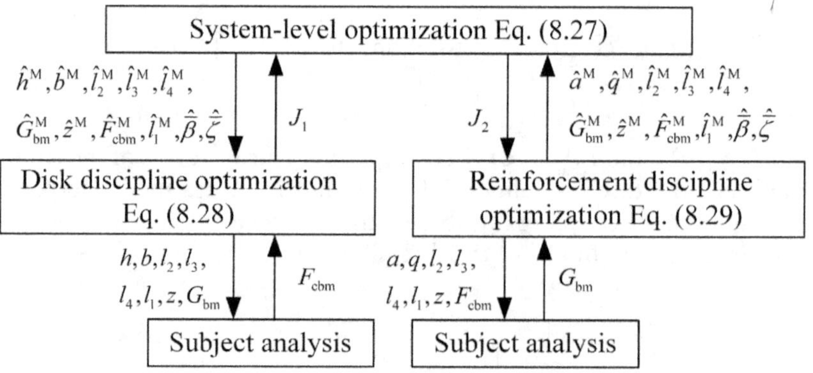

Figure 8.12 CO optimization strategy of the rotor mechanism.

Figure 8.12 illustrates the interaction between the system layer and the discipline layer in the CO model. The optimization results are shown in Table 8.4.

According to Table 8.4, the optimized rotor mass is 7640 kg, a 2.6% reduction compared to the original data. Although the maximum stress and strain are increased, they are still within the scope of the design requirements. This optimization saves materials and reduces the manufacturing cost of products, which would be more conducive to promoting and popularizing hydropower.

Table 8.4 Optimization value of design variables of the rotor mechanism

Design variables	*Optimized value*	*Design variables*	*Optimized value*
h (mm)	41	l_4 (mm)	331
b (mm)	240	a (mm)	71
l_2 (mm)	221	q (mm)	65
l_3 (mm)	138	l_1 (mm)	158
z (mm)	18		

In order to improve the feasibility and authenticity of the optimization scheme, a static analysis of the optimized rotor structure can be conducted. Under the condition of rated operation, the rotor is subjected to a fixed displacement constraint. This constraint is loaded on the end face of the connection between the center body and the spindle. The working principle of the turbine rotor mechanism is to realize the energy conversion from mechanical energy to electric energy by overcoming the moment of electromagnetic resistance. Under ideal design conditions, the stator and the rotor have a fixed distance gap. However, in actual working conditions, due to processing errors, one side of the installed rotor is close to the stator, while the other side deviates from it. This unequal gap distance triggers a unilateral magnetic pull. In this example, the concept of extreme values is employed to superimpose the unilateral magnetic pull and the gravity. Furthermore, the worst force condition of the system is simulated. At the same time, it is also necessary to specify the speed and torque of the rotor mechanism under rated conditions.

Once the loading is completed, the simulation of stress and strain is performed. Figures 8.13 and 8.14 show the analysis results of stress and strain under the rated working condition of the rotor.

According to the analysis of the results in the figures, the stress value of the rotor under rated conditions reaches a maximum of 140.7 MPa, which satisfies the strength constraint. The maximum deformation is 0.499 mm, less than the set maximum deformation of 0.5 mm. In general, with the lightweight optimization of the structure, the volume size of the target mechanism decreases in the original working state, and the stress and strain will increase. The simulation outcomes of the optimized design scheme adhere to the strength requirements. The strain is close to the extreme value of the constraint. It shows that the optimization result is the optimal solution under the condition of considering the influence of random and interval uncertainty.

In order to further verify the feasibility of this design scheme, the working conditions of the rotor shutdown and runaway operation can also be simulated and analyzed. Under the working condition of the rotor shutdown, the spindle rotation stops. At this time, there are only magnetic poles

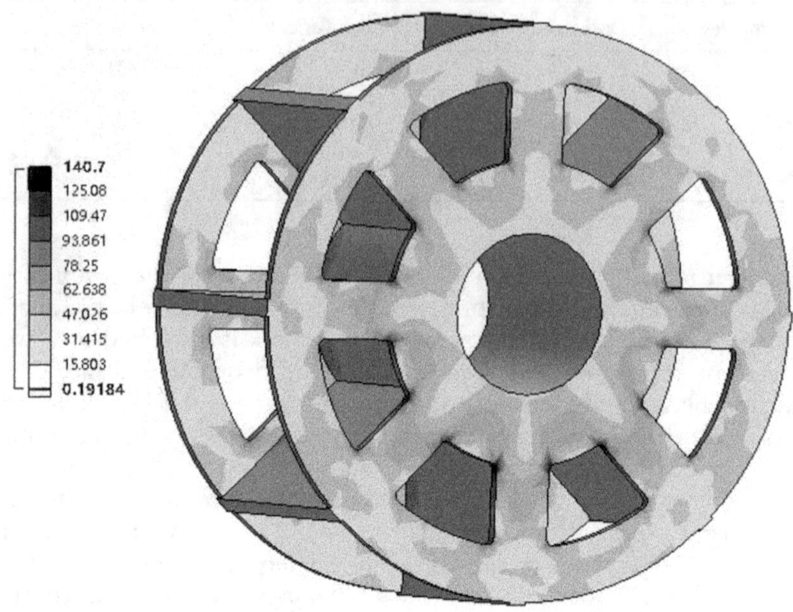

Figure 8.13 Rated stress distribution.

and gravity. The results obtained after loading according to this are shown in Figures 8.15 and 8.16.

The diagram shows that the maximum stress of the rotor during shutdown is 0.47 MPa, and the maximum deformation is 0.005 mm, both of which meet the constraint requirements. If the turbine suddenly loses its load during operation and cannot close the water guide, the motor's output power will become zero, causing the turbine speed to rapidly increase. When the mechanical energy generated during the increase of the rotational speed can be balanced with the water flow, the rotational speed reaches an extreme and stable value. This speed is generally referred to as the runaway speed of the turbine. The secondary working state is called runaway when the turbine rotor is at this speed. For small and medium-sized units, the runaway state must be considered in the design process, and it is significant to ensure that the turbine does not fail under this condition. Therefore, it is imperative to analyze the design scheme in the runaway state.

Under the runaway condition, the water flow reaches its maximum, the speed regulation protection system malfunctions, and the speed attains its extreme value. Since no load is put on the system under this working condition, there is no unilateral magnetic pull but only the gravitational force

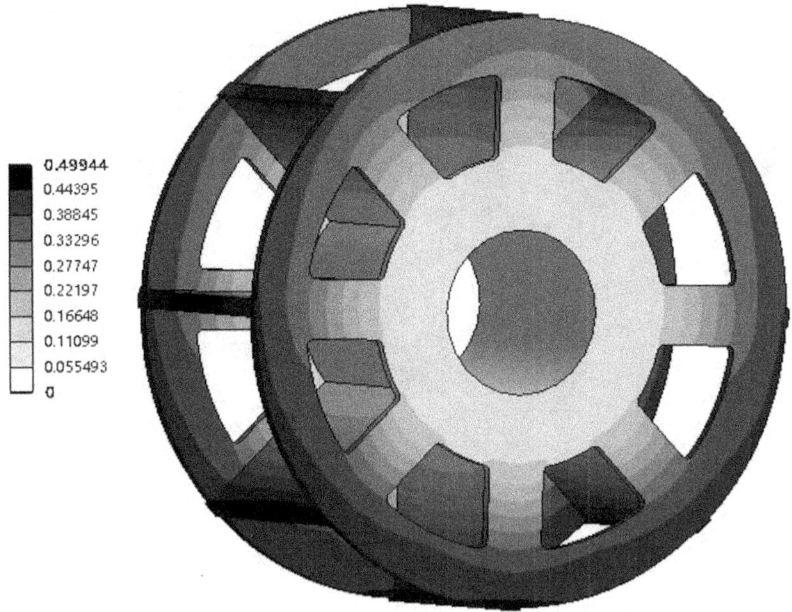

Figure 8.14 Rated strain distribution.

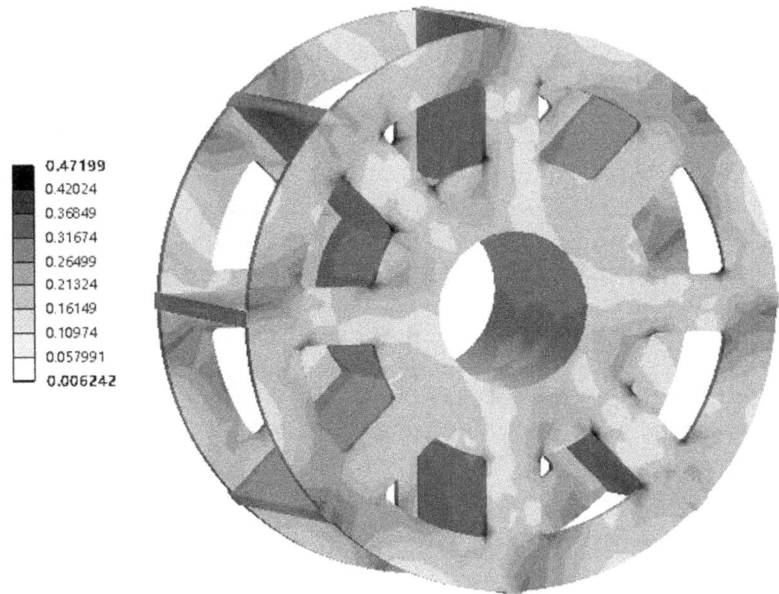

Figure 8.15 Shutdown stress distribution.

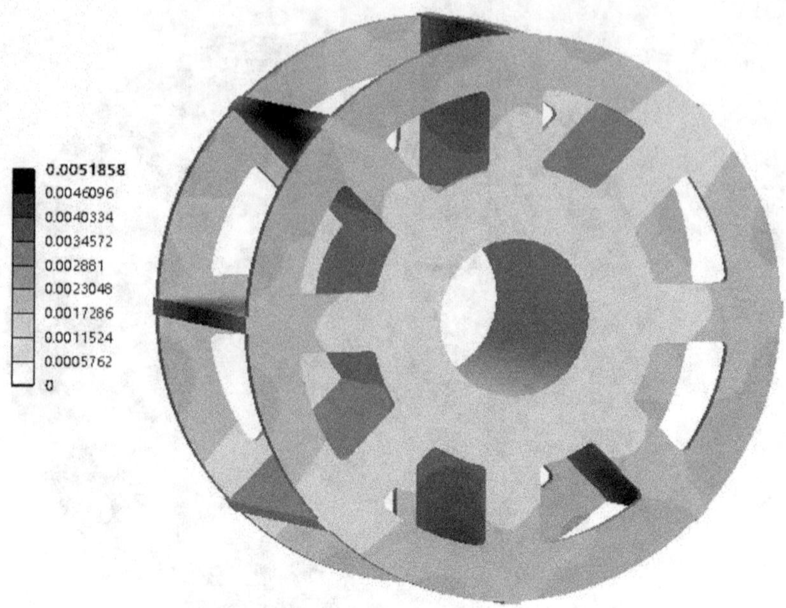

Figure 8.16 Shutdown strain distribution.

acting on the mechanism itself. The rotor displacement constraint under this condition aligns with the conditions mentioned above. After the displacement constraint and force loading are completed, Figures 8.17 and 8.18 show the analysis results of stress and strain under the condition of rotor runaway.

The above diagram shows that the maximum stress exhibited by the rotor under the runaway condition is 170.94 MPa, and the maximum deformation is 0.255 mm. These values align with the constraints set forth in the model. Therefore, through the above simulation results, the design scheme fulfills the operational requirements across all operating conditions.

This optimization example combines the actual engineering requirements and considers the influence of multi-source uncertainty to establish the RBMDO model of the turbine rotor mechanism. The resulting optimized design scheme successfully lowers the weight of the rotor mechanism while ensuring strength performance and achieving the goal of structural lightweight.

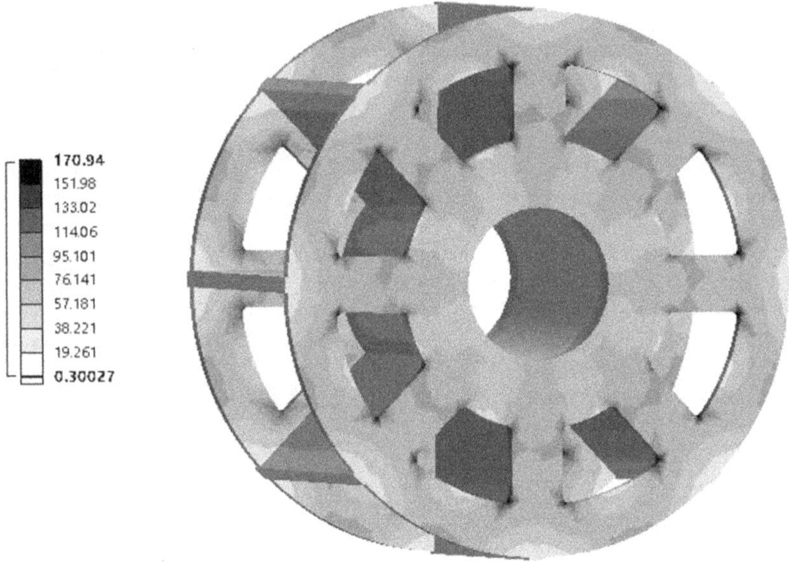

Figure 8.17 Runaway stress distribution.

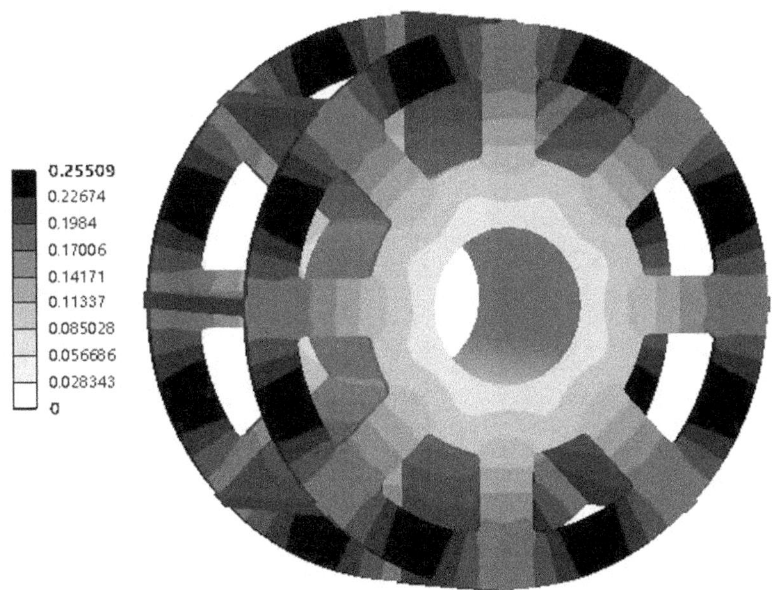

Figure 8.18 Runaway train distribution.

REFERENCES

[1] Patryniak K., Collu M., Coraddu A. (2022). Multidisciplinary design analysis and optimisation frameworks for floating offshore wind turbines: State of the art. Ocean Engineering, 251: 111002.

[2] Yuan R., Li H., Xie T., Lv Z., Meng D., Yang W. (2022). Multidisciplinary collaborative design and optimization of turbine rotors considering aleatory and interval mixed uncertainty under a SORA framework. Machines, 10(6): 445.

[3] Wang L. (2012). Installation and Maintenance of Hydro-generator Unit. Beijing, China: China Water Resources and Hydropower Publishing House.

[4] Zhong Y., Chen B., Wang Z. (2006). Multidisciplinary Comprehensive Optimization Design Principles and Methods. Wuhan, China: Huazhong University of Science and Technology Press.

[5] Bai Y. (1982). Design and Calculation of Hydro-generator. Beijing, China: China Machine Press.

Chapter 9

Application of adaptive surrogate model-assisted UBMDO in propulsion equipment

Promoting the performance optimization and design of equipment has gradually become a crucial research field for the continuous improvement of equipment performance requirements. In this context, adaptive surrogate model-assisted Uncertainty-Based Multidisciplinary Design Optimization (UBMDO) provides an innovative approach to advancing device design. This chapter will discuss the application and advantages of this advanced method through a specific case – the design optimization of the ducted fan.

As an efficient propulsion device, the ducted fan is widely used in Unmanned Aerial Vehicle (UAV), Vertical Take-off and Landing (VTOL), and other fields. Its design involves many disciplines, such as aerodynamic performance, structural strength, and thermodynamics. At the same time, it faces the influence of uncertain factors such as manufacturing tolerance and material properties. Traditional design methods often make finding the optimal design scheme challenging in a complex multidisciplinary environment.

The adaptive surrogate model-assisted UBMDO method combines the efficient computing power of the surrogate model with the global optimization capability of Multidisciplinary Design Optimization (MDO). First, the computational efficiency is greatly improved by constructing a surrogate model to approximate a complex multidisciplinary analysis model. Then, the uncertainty quantization technique is used to deal with various uncertainty factors to ensure the robustness of the design scheme. Finally, the optimization algorithm finds the optimal design scheme to meet the multidisciplinary performance requirements.

9.1 INTRODUCTION OF DUCTED FAN

9.1.1 Concept and application of ducted fan

The ducted fan is a mechanical structure with several rotatable blades surrounded by an annular duct. The ducted fan propulsion system is an electric drive power device composed of a ducted fan, drive motor, and

DOI: 10.1201/9781003464792-9

Table 9.1 Comparison of helicopter configuration and development time

	Ducted tail rotor type	*Conventional tail rotor type*	*Proportion of duct tail propeller type*
Overall amount	14	128	9.86%
After 1990	10	25	28.57%

controller. By inputting the appropriate voltage and electric power to drive the high-speed rotation of the blade, continuous and controllable thrust can be generated.

The ducted fan first appeared in the 1840s. Its application can be divided into two categories, as a propulsion component and as a lift component. In the role of a thrust component, such as the propeller used in ships, submarines, and propeller aircraft, it is used for large inflows. As a lift component, it is divided into two application directions. It can be employed as the helicopter's tail rotor to provide the torque against the main rotor torque for the helicopter or applied to VTOL equipment as the main source of lift.

The tail rotor of the helicopter was first used in 1968. The SA341 'Little Antelope' helicopter developed by Aerospace France adopted the configuration of the ducted tail rotor for the first time [1]. The French Aerospace has always favored the configuration of the ducted tail rotor. Subsequently, the dolphin AS365 and EC135 developed by the European Helicopter Company adopted the configuration of the ducted tail rotor [2].

After 1990, the ducted tail rotor helicopters experienced a blowout growth. The ducted tail rotor helicopters accounted for 28.57% of the newly developed helicopters, mainly used in light and small helicopters of 6 tons and below, as shown in Table 9.1.

As a lift component of VTOL equipment, the United States Navy successfully tested the Hiller VZ-1 Pawnee individual VTOL aircraft as early as 1955, which controls the movement of the aircraft by moving the center of gravity.

In recent years, with the rise of UAV, ducted fan has become a vital configuration scheme for Unmanned VTOL (UVTOL) vehicle. Currently, ducted fans are commonly used in VTOL four-rotor, personal flight devices, and other small aircraft.

9.1.2 Advantages and disadvantages of ducted fan

9.1.2.1 Advantages

Compared with the conventional rotor, the ducted fan has many advantages, which are mainly reflected in the following aspects.

First, the ducted fan has higher aerodynamic efficiency. Compared with the open rotor with the same rotor size, the ducted fan can obtain higher thrust at the same power or lower power at the same thrust. This is mainly due to the three aerodynamic characteristics of the ducted fan.

When the airflow enters the culvert, it will be gathered. In the process of gathering, a negative pressure will be generated on the surface of the culvert inlet. That is, the culvert will also produce a part of the tension [3]. As the airflow is gathered, the velocity of the inflow above the rotor will increase, thereby reducing the load of the propeller disk and the required power.

The distance between the duct and the blade tip is very close, which can effectively prevent the backflow of the rotor blade tip, reduce the blade tip loss, and improve the aerodynamic efficiency of the rotor.

The airflow will produce a wake after passing through the fan, the wake diameter will shrink in the open rotor, and the airflow speed will be accelerated. In the ducted fan, the wake will continue to expand with the duct outlet section, or the diameter will remain unchanged, causing the wake velocity to be decreased, the kinetic energy loss reduced, and the hovering efficiency improved.

Second, the ducted fan is extremely safe. For users, high-speed rotating blades are very dangerous; for the rotor itself, if the rotor is disturbed by obstacles, it will lose a stable flight attitude and even cause fatal damage to the mechanical structure. Therefore, it is difficult for aircraft with rotors to work in a narrow space or an environment with dense obstacles. The existence of the culvert can protect the rotating parts, prevent encountering obstacles or hurting people, facilitate cooperation with people, and enable them to perform tasks in the case of dense obstacles. The security of the structure is an important reason why many helicopter tail rotors and VTOL adopt ducted fan configurations.

The third benefit refers to a lower noise level. In the application of UAV, low noise level is a major advantage. The rectification effect of the duct can suppress the tip reflux, thereby reducing the tip Mach number and effectively reducing the noise level.

9.1.2.2 Disadvantages

Compared with an open rotor, the disadvantages of a ducted fan are mainly reflected in the following two aspects.

The first is the larger structural weight. The increase in weight mainly comes from the weight of the culvert, and the weight of the culvert will offset the increase in tension caused by the increase in efficiency. In recent years, with the popularization of composite materials, culverts can have lighter weight while ensuring stiffness and strength; shorter culverts can also reduce structural weight.

The second is the resistance and torque under the lateral flow. Compared to the open rotor, in the case of lateral flow, the ducted fan will produce greater resistance due to the larger ducted windward area. The lateral flow will also generate torque simultaneously: The head-up torque for VTOL and the yaw torque for the helicopter's ducted tail rotor. This is because the pressure distribution at the culvert inlet will change under the action of the incoming flow. The negative pressure on the side near the incoming flow will become larger, and the negative pressure on the other side will become smaller, resulting in pitching or yawing torque.

9.1.3 The research object of this chapter

It can be seen from the above that the ducted fan is widely used as a practical mechanical device. Its structure can be composed of two or more blades and hubs. When rotating in the medium, multiple blade movements push the air backward, and the force and reaction can generate thrust, which converts the rotating power of the engine into a driving force or a force that promotes airflow. The model diagram of a ducted fan is shown in Figure 9.1, which is drawn by SolidWorks software.

In this chapter, the ducted fan shown in the above figure is selected as the subject of our research, with the goal of optimizing its design for lightweight. The initial step involves defining the design variables, as shown in Table 9.2. The distribution of the fan design variables is illustrated in Figure 9.2. The

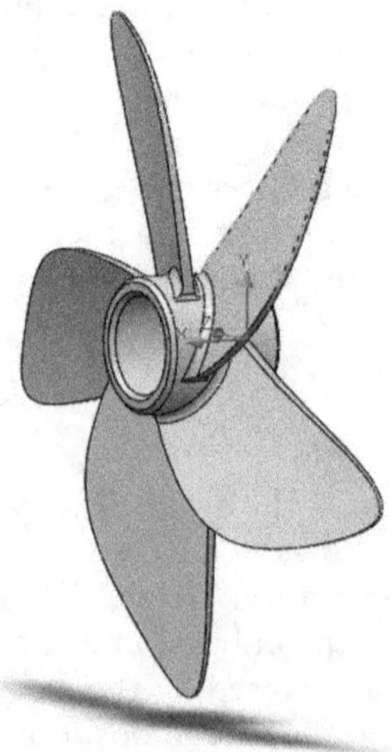

Figure 9.1 3D model of ducted fan.

Table 9.2 Information of design variables in the fan example

Variable	*Name*	*Unit*	*Distribution*	*Lower bound*	*Upper bound*
x_1	Blade inlet angle	Degree (°)	–	70	75
x_2	Blade outlet angle	Degree (°)	–	25	30
x_3	Chord length	mm	Normal distribution	80	90
x_4	The thickest diameter of the blade	mm	Normal distribution	5	10
x_5	Blade incidence	Degree (°)	–	50	55

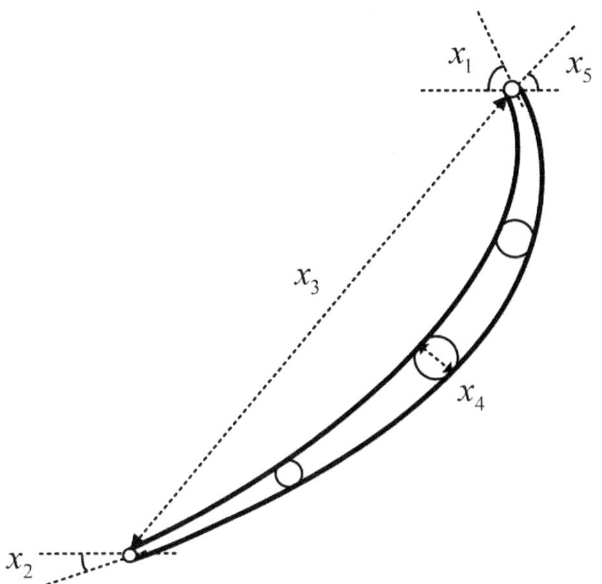

Figure 9.2 The distribution of design variables in the fan.

material chosen for the ducted fan is copper alloy. By incorporating other metal elements into copper, this alloy can change its material properties and improve its mechanical characteristics. The material properties, such as density and mechanical properties of copper alloys in this example, are shown in Table 9.3.

According to the initial size data of the ducted fan, the fan's finite element simulation analysis is carried out under extreme working conditions. First, the model is meshed, where the size and density of the mesh will affect the simulation results. Therefore, on the premise of combining the design scheme, the outcomes of meshing the model are presented in Figure 9.3, which indicates a total of 52,996 nodes and 22,637 units in the grid.

Table 9.3 Material performance parameters of copper alloy

Name	*Unit*	*Numerical value*
Density	kg/m^3	8300
Shear modulus	MPa	41045
Tensile yield strength	MPa	280
Compressive yield strength	MPa	280
Ultimate tension	MPa	430

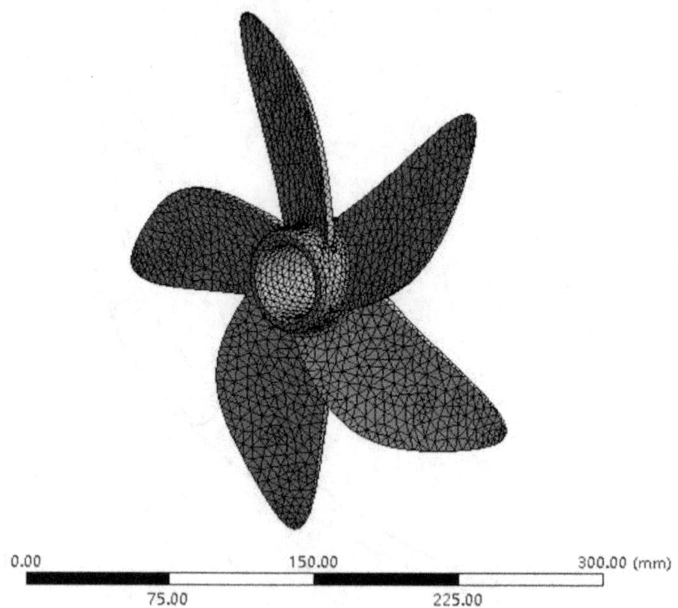

Figure 9.3 The grid division diagram of the fan.

9.2 DISCIPLINARY DIVISION AND OPTIMIZATION PROBLEM DEFINITION OF DUCTED FAN BLADE

9.2.1 Multidisciplinary optimization methods

The overall MDO solution process can be represented in Figure 9.4. Among them, the discipline update refers to the necessary information in the calculation process of optimization design that needs to be updated in time after the decoupling analysis of multiple disciplines. The pre-processing of the model should be carried out before the MDO solution, and the purpose is to search for loopholes in the subject model for pre-processing.

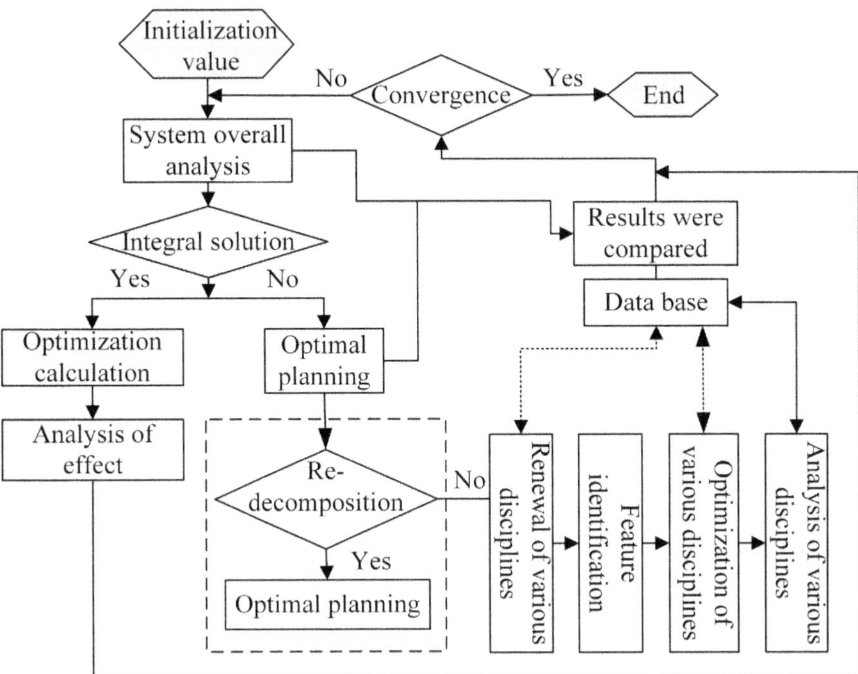

Figure 9.4 MDO overall solution flowchart.

The Collaborative Optimization (CO) method is extensively utilized in MDO at the present stage. The algorithm framework of the CO method presents a two-layer distribution comprising the system layer and the discipline layer. The system layer encompasses the design objectives of the optimization problem, all design variables, and consistency constraints. The discipline layer includes consistency constraints and uses them as the performance function of the discipline layer and the constraints of different sub-disciplines. The CO method has unique characteristics, such as a high degree of discipline autonomy, multi-level optimization, and distributed computing. It is suitable for design optimization problems in engineering fields such as large and complex products and has reasonable practicability. The algorithm flow is illustrated in Figure 9.5.

The CO method is an MDO hierarchical design optimization strategy. The complex MDO issue is divided into a system-level optimization issue and a sub-discipline-level optimization issue. The system-level optimization model is

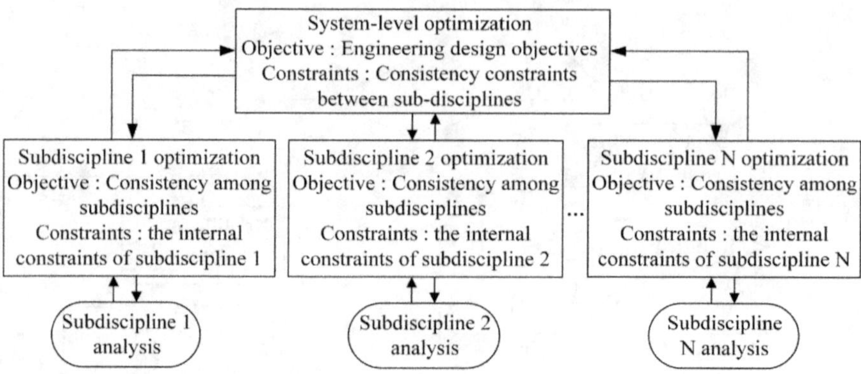

Figure 9.5 Flowchart of the collaborative optimization algorithm.

$$\begin{aligned}
&\min_{DV} f(\mathbf{d}_c,\mathbf{d}_d,\mathbf{Z}_j^c,\mathbf{Z}_j^d,\mathbf{P}_c,\mathbf{P}_d) \\
&s.t.\ g_0(\mathbf{d}_c,\mathbf{d}_d,\mathbf{Z}_j^c,\mathbf{Z}_j^d,\mathbf{P}_c,\mathbf{P}_d) \le 0 \\
&\qquad h_0\left(\mathbf{d}_c,\mathbf{d}_d,\mathbf{Z}_j^c,\mathbf{Z}_j^d,\mathbf{P}_c,\mathbf{P}_d\right) = 0 \\
&\qquad J_i\left(\mathbf{d}_c,\mathbf{d}_d,\mathbf{X}_{ij}^c,\mathbf{X}_{ij}^d,\mathbf{Z}_j^c,\mathbf{Z}_j^d,\mathbf{P}_c,\mathbf{P}_d\right) = 0 \\
&DV = \{\mathbf{d}_c,\mathbf{d}_d,\mathbf{X}_{ij}^c,\mathbf{X}_{ij}^d,\mathbf{Z}_j^c,\mathbf{Z}_j^d\}
\end{aligned} \tag{9.1}$$

where **d** signifies the deterministic design variable; Z signifies the system layer design variables; X signifies the random design variables at the discipline level; **P** signifies a parameter with uncertainty; $g_0(\bullet) \le 0$ refers to the inequality constraint condition of the system layer; $h_0(\bullet) = 0$ refers to the equality constraint condition of the system layer; the superscripts or subscripts c and d represent continuous type and discrete type; i denotes the category of disciplines; j denotes the number of target design variables; J_i represents the consistency constraint of the ith discipline, expressed by the under formula:

$$J_i = \left\|\mathbf{X}_{ij}^c - \mathbf{Z}_j^c\right\|_2^2 + \left\|\mathbf{X}_{ij}^d - \mathbf{Z}_j^d\right\|_2^2 + \left\|\mathbf{Y}_i - \hat{\mathbf{Y}}_i\right\|_2^2 = 0 \tag{9.2}$$

where **Y** denotes the set of deterministic conditions, including deterministic variables and parameters.

The subject optimization model of the first sub-discipline in the i sub-discipline level of the system as follows:

$$
\begin{aligned}
&\min_{\mathrm{DV}} J_i = J_i\left(\mathbf{d}_c, \mathbf{d}_d, \mathbf{X}_{ij}^c, \mathbf{X}_{ij}^d, \mathbf{Z}_j^c, \mathbf{Z}_j^d, \mathbf{P}_c, \mathbf{P}_d\right) \\
&\text{s.t.} \;\; g_i\left(\mathbf{d}_c, \mathbf{d}_d, \mathbf{X}_{ij}^c, \mathbf{X}_{ij}^d, \mathbf{P}_c, \mathbf{P}_d\right) \le 0, \\
&\qquad h_i\left(\mathbf{d}_c, \mathbf{d}_d, \mathbf{X}_{ij}^c, \mathbf{X}_{ij}^d, \mathbf{P}_c, \mathbf{P}_d\right) = 0, \\
&DV = \left\{\mathbf{d}_c, \mathbf{d}_d, \mathbf{X}_{ij}^c, \mathbf{X}_{ij}^d\right\}
\end{aligned}
\tag{9.3}
$$

This paper employs the CO method as the MDO approach. Since its algorithm structure is similar to the division of labor in modern engineering design, it is adopted as the solution strategy for the MDO problem in this paper. Furthermore, due to the structural characteristics of its hierarchical optimization, each discipline's analysis procedures and strategies can be applied unchanged, facilitating a reduction in the time cost required for design. In addition, the collaboration effect between sub-disciplines in the discipline layer of the CO method is represented by the consistency constraint in the system layer, which mitigates the computational difficulty caused by the complex discipline coupling.

9.2.2 Collaborative optimization method based on the adaptive surrogate model

Based on the adaptive surrogate model, the CO method is used as a multidisciplinary optimization design algorithm to approximate the system layer's objective function and the sub-disciplinary layer's constraints in the design optimization process. It is of great significance and value. The construction flowchart is shown in Figure 9.6.

9.2.2.1 Adaptive surrogate model of system layer

In engineering optimization problems, it is highly meaningful to accurately fit the mathematical model for the objective function of the system layer. Taking the CO method as the MDO method, and combined with the actual engineering situation, the mathematical model of the system layer optimization can be approximated by the adaptive surrogate model to obtain a relatively accurate mathematical model. First, the system layer optimization model of the CO method is simplified based on a mathematical logic solution, resulting in the following expression:

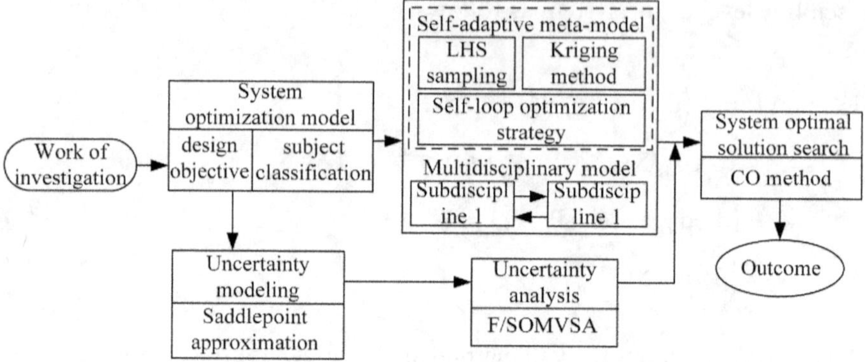

Figure 9.6 Flowchart of multidisciplinary adaptive surrogate model.

$$\text{System:}\begin{cases}\text{Find: } x_{ij}, z_j \ (i,j = 1,2,3,\ldots,n) \\ \text{Min: } F = f\left(z_j\right) \\ s.t.\ J_i^*\left(x_{ij}^*, z_j\right) = 0\end{cases} \tag{9.4}$$

where x_{ij} refers to the jth objective design variable in the ith sub-discipline; z_j refers to the jth objective design variable; $F(\bullet)$ represents the objective performance function of system layer optimization; x_{ij}^* represents the optimization result of x_{ij} after the analysis of sub-disciplines at the discipline level; $J_i^*(\bullet)$ represents the system-level consistency constraint, which can be solved by the following equation:

$$J_i^*(z) = \sum_{j=1}^{s_i}\left(z_j - x_{ij}^*\right)^2, i = 1,2,3,\ldots,n \tag{9.5}$$

where s_i represents the number of target design variables in sub-discipline i; n represents the number of sub-disciplines at the discipline level.

The initial surrogate model is used to approximate the system layer of the CO method, which is expressed as follows:

$$\text{System:}\begin{cases}\text{Min: } F = \hat{f}(z) \\ s.t.\ J_i^*\left(x_{ij}^*, z_j\right) = 0\end{cases} \tag{9.6}$$

where $\hat{f}(\bullet)$ represents the approximate fitting objective function, and can be defined using the following formula:

$$\hat{f}(z) = q(z)^{\mathrm{T}} \beta + r^{\mathrm{T}}(z) R^{-1}(F - Q\beta) \tag{9.7}$$

where $q(z)$ represents the polynomial vector set of the system layer; β represents the polynomial parameter matrix of the system layer; $r^{\mathrm{T}}(z)$ represents the correlation vector between the sample set and the unknown design variable; $R(\bullet)$ is shown in the following Eq. 9.8; F denotes the response matrix; and Q denotes the regression polynomial matrix. Specifically, the Eq. 9.9

$$R_{ij}\left(\theta, x_i, x_j\right) = \exp\left(-\sum_{k=1}^{m} \theta_k \left|x_k^i - x_k^j\right|^2\right) \tag{9.8}$$

$$\begin{cases} q(z) = \left[q_1(z), q_2(z), \ldots, q_k(z)\right]^{\mathrm{T}} \\ \beta = \left[\beta_1, \beta_2, \ldots, \beta_k\right]^{\mathrm{T}} \\ r^{\mathrm{T}}(z) = \left[R(z, z_1), R(z, z_2), \ldots, R(z, z_n)\right] \\ R\left(z_i, z_j\right) = \exp\left(-\sum_{k=1}^{n} \theta_k \left|z_k^i - z_k^j\right|^2\right) \\ \mathrm{Q} = \left[\mathrm{q}^{\mathrm{T}}(\mathrm{z}_1), \mathrm{q}^{\mathrm{T}}(\mathrm{z}_2), \ldots, \mathrm{q}^{\mathrm{T}}(\mathrm{z}_n)\right]^{\mathrm{T}} \\ \mathrm{F} = \left[f_1, f_2, \ldots, f_n\right]^{\mathrm{T}} \end{cases} \tag{9.9}$$

$$\max\left\{-\left[n \ln(\alpha^2) + \ln|R|\right]/2\right\} \tag{9.10}$$

where θ_k is a random parameter greater than 0. Different values of parameter θ will lead to changes in the correlation of the correlation function. Its computation can be expressed by Eq. 9.10, and when the formula takes the maximum value the θ value is the obtained; $f_i\,(i = 1, 2, \ldots, n)$ represents the response of design variable z_i. The effect of θ_k on matrix correlation is shown in Figure 9.7.

On the basis of the initial surrogate model, the system-level approximate fitting objective function of the CO method is modified by the self-circulation optimization strategy based on the EI function.

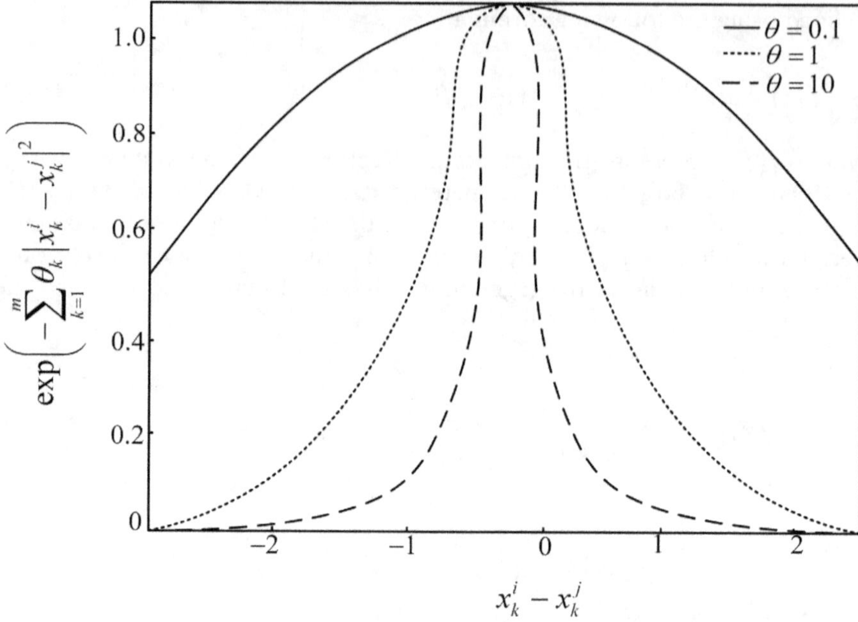

Figure 9.7 θ_k value on the influence of the matrix correlation diagram.

$$\text{System:}\begin{cases}\text{Min: } F = \left\{\hat{f}(\text{z}),\ E\left[I(\text{z})\right]\right\} \\ s.t.\ J_i^*\left(x_{ij}^*, \text{z}_j\right) = 0\end{cases} \tag{9.11}$$

where $E\left[I(\text{z})\right]$ represents the EI function of the system-level design variables.

An approximate fitting mathematical model for the system-level objective function of the CO method based on the adaptive surrogate model can be constructed through the above methods and processes. This approach allows for the preliminary solution of the problem that the objective function is difficult to express in mathematical form due to the intricate system structure and the diverse subject category in practical engineering design optimization problems. It further provides a certain reference value for developing high-confidence numerical simulation design methods and enables analysts to explore other possible solutions beyond the design sample set within the scope of existing computing resources.

9.2.2.2 *Adaptive surrogate model of discipline level*

In the MDO problem, the discipline layer contains some target design variables and the analysis tasks of the discipline itself. The discipline analysis includes the self-constraints of the sub-disciplines. If the constraints of the sub-disciplines cannot be well satisfied, the design optimization results cannot coordinate and solve the complex coupling relationship between different disciplines, which brings difficulties to the search for the global optimal solution. Taking the CO method as the MDO method, and combined with the actual engineering situation, the approximate fitting of the constraint conditions of the subject layer based on the adaptive surrogate model can facilitate design optimization to a certain extent. The subject layer optimization model of the CO method is simplified based on a mathematical logic solution, and the following expression can be obtained:

$$\text{Sub-discipline-}i\text{:}\begin{cases}\text{Find: } x_i, z_j^* \\ \text{Min: } J_i(x_i) \\ s.t.\ G_i(x_i) \le 0\end{cases} \tag{9.12}$$

where x_i signifies part of the design variables belonging to the ith sub-discipline; z_j^* signifies the expected value of the system level for the jth objective design variable; $G_i(\bullet)$ signifies the constraint condition of the ith sub-discipline; $J_i(\bullet)$ signifies the objective function of sub-discipline i, which can be expressed as follows

$$J_i(x_i) = \sum_{j=1}^{s_i}\left(x_{ij} - z_j^*\right)^2, i = 1,2,3,\ldots,n \tag{9.13}$$

Through the approximate fitting of the optimization model of the sub-discipline by the adaptive surrogate model, the following can be derived

$$\text{Sub-discipline-}i\text{:}\begin{cases}\text{Min: } J_i(x_i) \\ s.t.\ \widehat{G}_i(x_i) \le 0\end{cases} \tag{9.14}$$

where $\hat{G}_i(x_i)$ represents the fitting constraint condition of the ith sub-discipline, which can be specified by the following formula:

$$\widehat{G}_i(x_i) = q(x_i)^{\mathrm{T}}\chi + r^{\mathrm{T}}(x_i)\mathrm{R}^{-1}(\mathrm{H} - \mathrm{P}\chi) \tag{9.15}$$

where $q(x_i)$ represents the set of polynomial vectors of sub-disciplines; χ represents the polynomial parameters of sub-disciplines; $r^{\mathrm{T}}(x_i)$ represents the correlation vector between the unknown design variables and the sample set; $\mathrm{R}(\bullet)$ represents the correlation matrix, and its calculation is shown in the previous Eq. 9.8; H represents the matrix composed of the response of the sample points; P represents the matrix of vector $q(x_i)$.

On the basis of the initial surrogate model, the approximate fitting constraints of the discipline level of the CO method are modified by the self-circulation optimization strategy based on the EI function.

$$\text{Sub-discipline-}i\text{:}\begin{cases}\text{Min: } J_i(x_i) \\ s.t.\ \left\{\widehat{G}_i(x_i),\ \mathrm{E}\left[I(x_i)\right]\right\} \le 0\end{cases} \tag{9.16}$$

where the EI function of some design variables of sub-discipline i of discipline $\mathrm{E}\left[I(x_i)\right]$.

Through the above methods and processes, an approximate fitting mathematical model for the constraints of the discipline layer of the CO method based on the adaptive surrogate model can be constructed. The number of disciplines involved in modern engineering design optimization problems is relatively large, and there may be strong coupling between different disciplines, resulting in the constraints of sub-disciplines being generally more complex and difficult to express through mathematical models. The adaptive surrogate model can solve the problem to a certain extent, and the mathematical expression of its approximate fitting can be derived. The design optimization problem of multidisciplinary coupling provides a method to solve the problem of the sub-disciplines that cannot meet the requirements of various disciplines due to the complex analysis of disciplines and the unclear relationship between disciplines.

9.2.3 Subject division and optimization of ducted fan blades

9.2.3.1 Division of disciplines

The disciplinary division is the process of dividing the overall optimization of complex engineering systems into smaller and easier-to-handle disciplines, which is one of the essential foundations for the research and application of MDO [4]. The division of disciplines generally follows the following principles: Realizing the overall integrated optimization of the system, reflecting the coupling relationship between disciplines, and minimizing information transmission between disciplines to improve optimization efficiency.

In general, the discipline division structure can be established from the perspective of system composition and analysis methods [5]. These two division methods have distinct starting points, each with its own advantages and disadvantages. The division method of product composition is intuitive

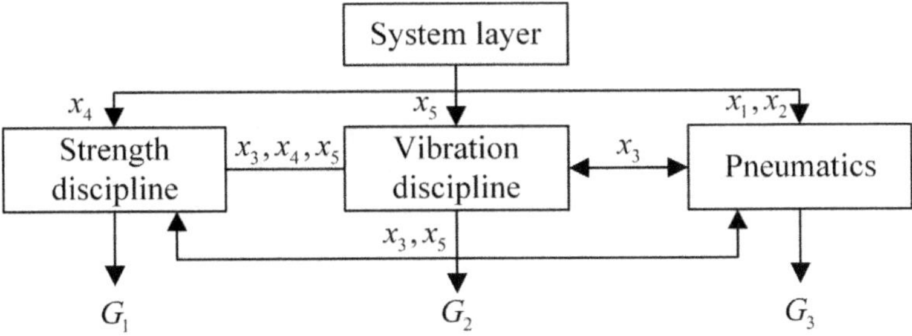

Figure 9.8 MDO problem of fan optimization.

and easy to understand. The division structure is closest to the industrial organization system in marine engineering, which is conducive to the parallel design of each subsystem. However, a disadvantage is that it is difficult to consider the coupling relationship between subsystems, hindering the achievement of comprehensive integration optimization. On the other hand, the division method oriented to the analysis method can fully consider the coupling relationship between disciplines, which is conducive to the overall planning and coordination between disciplines and realizes the overall integrated optimization. However, its shortcoming is that the difficulty of subject division is increased.

A certain type of ducted fan blade used in this paper will resonate due to the air's influence during the fan's operation in the working environment while ensuring the fan's strength conditions. Therefore, aerodynamics, vibration, and strength should be considered in the division of disciplines. The aerodynamics, vibration, and strength disciplines should be taken as sub-disciplines to optimize the lightweight design of the structure. Under the framework of the CO method, Figure 9.8 depicts the multidisciplinary relationship diagram in the ducted fan optimization design problem.

In Figure 9.8, the design variable between the strength discipline and the vibration discipline is shared, and it can be known as x_3, x_4, x_5 according to the discipline analysis. The coupling between other disciplines is also shown in the figure.

9.2.3.2 Definition of MDO problem

MDO has been developed for nearly 40 years. However, there is no unified understanding of the definition of MDO globally. The MDO Technical Committee of the American Aerospace Society gives the following three definitions [6].

Definition 1: MDO is a methodology for designing complex systems and subsystems by fully exploring and utilizing the synergistic mechanisms of interactions in the system.

Definition 2: MDO refers to the design optimization method that must analyze the interaction between disciplines (or subsystems) in the design of complex engineering systems and make full use of these interactions to optimize the system.

Definition 3: When each factor in the design affects all the other factors, a design method is used to determine which factor to change and to what extent.

Although the above definitions have different emphases, it can be concluded that MDO has the following characteristics: ① From the perspective of a theoretical system, MDO is a design methodology; ② Its research object is the complex product design process, which directly serves the complex product design; ③ Applicable to the entire product design life cycle; ④ Its basic idea is to use the synergy between multiple disciplines to achieve the optimal overall performance of the product [7].

In short, MDO is a comprehensive method for addressing the design challenges of complex engineering systems. By thoroughly investigating and utilizing the synergistic interaction mechanism in engineering systems, and considering the interaction between various disciplines, complex engineering systems are optimized and designed systematically, ultimately enhancing product performance, reducing costs, and expediting the design cycle [8].

For the ducted fan, the main optimization goal is lightweight. Lightweight can reduce the consumables and production cycle of the fan and can increase the flight time of the flight machine using the ducted fan, which is of great significance for improving flight quality and flight reliability.

9.3 UNCERTAINTY MODELING OF DUCTED FAN BLADES

9.3.1 Uncertainty design optimization modeling based on adaptive surrogate model

Uncertainty exists in every stage of practical engineering design, which is mainly divided into random and cognitive uncertainty. In the disciplinary analysis of engineering design optimization problems, uncertainty propagates step by step in the system [8, 9]. Taking the MDO problem as an example, the propagation process of uncertainty is shown in Figure 9.9. In Figure 9.9, X_s denotes the cognitive uncertainty variable; $X_i\ (i = 1,2,3)$ denotes the random uncertainty variables from three disciplines; $y_{ij}\ (i, j = 1,2,3)$ denotes the variable of coupling between disciplines.

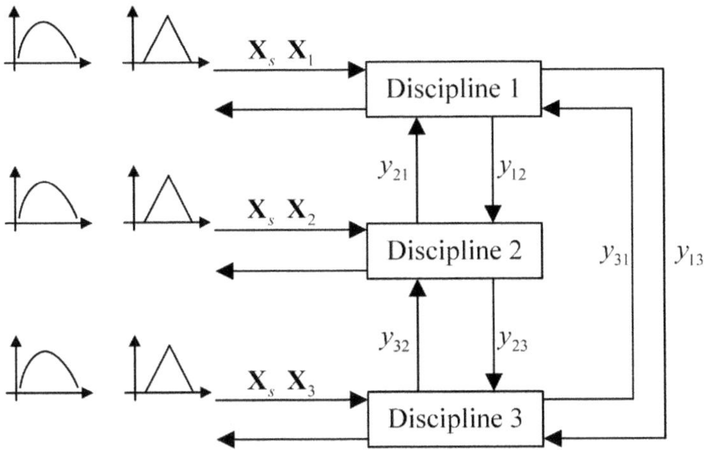

Figure 9.9 Propagation of stochastic and epistemic uncertainty in MDO.

The logical structure of this chapter is progressive. In the design optimization problem, the role and significance of the adaptive surrogate model in the uncertainty-based single-disciplinary design optimization method are first considered. The mathematical model of single-disciplinary design optimization is then given, as shown in Eq. 9.17.

$$
\begin{aligned}
&\min_{DV} f(\mathbf{d}_c,\mathbf{d}_d,\mathbf{X}_c^M,\mathbf{X}_d^M,\mathbf{P}_c^M,\mathbf{P}_d^M) \\
&s.t.\ \ g(\mathbf{d}_c,\mathbf{d}_d,\mathbf{X}_c^M,\mathbf{X}_d^M,\mathbf{P}_c^M,\mathbf{P}_d^M) \le 0 \\
&\qquad h\left(\mathbf{d}_c,\mathbf{d}_d,\mathbf{X}_c^M,\mathbf{X}_d^M,\mathbf{P}_c^M,\mathbf{P}_d^M\right) = 0 \\
&\qquad \mathbf{d}_c^L \le \mathbf{d}_c \le \mathbf{d}_c^U,\ \ \mathbf{d}_d^L \le \mathbf{d}_d \le \mathbf{d}_d^U \\
&\qquad \mathbf{X}_c^{M,L} \le \mathbf{X}_c^M \le \mathbf{X}_c^{M,U},\ \mathbf{X}_d^{M,L} \le \mathbf{X}_d^M \le \mathbf{X}_d^{M,U} \\
&DV = \{\mathbf{d}_c,\mathbf{d}_d,\mathbf{X}_c^M,\mathbf{X}_d^M\}
\end{aligned}
\tag{9.17}
$$

where **d** denotes the deterministic design variable; X denotes a random design variable; P denotes the parameter with uncertainty; $h(\bullet) = 0$ denotes equality design constraint; $g(\bullet) \le 0$ denotes the inequality design constraint; the subscripts c and d denote continuous and discrete, respectively. The superscript M represents the mean of the variables and parameters; the superscripts U and L represent the upper and lower bounds of the value interval. The process of single-discipline analysis is relatively simple, and there is only one discipline constraint condition. An adaptive surrogate model approximates the single-disciplinary design optimization model, and the following mathematical model can be obtained:

$$\begin{aligned}
&\min_{DV} \hat{f}(\mathbf{d}_c,\mathbf{d}_d,\mathbf{X}_c,\mathbf{X}_d,\mathbf{P}_c,\mathbf{P}_d)\\
&s.t.\ \hat{g}(\mathbf{d}_c,\mathbf{d}_d,\mathbf{X}_c,\mathbf{X}_d,\mathbf{P}_c,\mathbf{P}_d)\le 0\\
&\qquad \hat{h}\left(\mathbf{d}_c,\mathbf{d}_d,\mathbf{X}_c,\mathbf{X}_d,\mathbf{P}_c,\mathbf{P}_d\right)=0\\
&DV=\{\mathbf{d}_c,\mathbf{d}_d,\mathbf{X}_c,\mathbf{X}_d\}
\end{aligned} \tag{9.18}$$

where $\hat{f}(\bullet)$ represents the approximate objective function; $\hat{g}(\bullet)\le 0$ denotes the approximate inequality constraint condition; $\hat{h}(\bullet)=0$represents the approximate equality constraint condition; $\mathrm{X}=\left(X_1,X_2,\cdots,X_n\right)$ denotes the sample set of random design variables; $\mathbf{d}=\left(d_1,d_2,\cdots,d_n\right)$ denotes the sample set of deterministic design variables; **P** denotes a parameter with uncertainty, and the value of the parameter in the surrogate model is generally deterministic.

The construction process of the adaptive surrogate model under the consideration of only single-disciplinary analysis is shown in Figure 9.10.

At present, the majority of engineering design optimization problems involve MDO, while the situation where only a single discipline is considered is relatively rare. Based on the single-disciplinary analysis conditions, this section focuses on the single-disciplinary design optimization under uncertainty through the adaptive surrogate model. It provides its approximate mathematical model, which lays a foundation for the subsequent case analysis considering single-disciplinary.

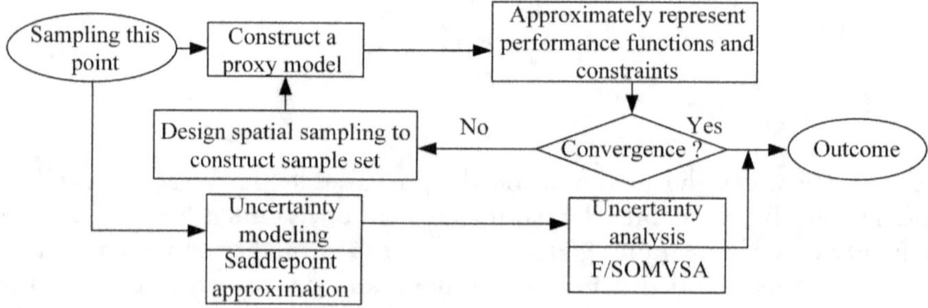

Figure 9.10 Flowchart of single-disciplinary adaptive surrogate model construction.

9.3.2 Uncertainty design optimization model of ducted fan blade

According to the collaborative uncertainty MDO method based on an adaptive surrogate model proposed in this chapter, the mathematical model of the fan design optimization problem is as follows:

$$\text{system}:\begin{cases}\min F_V=\left\{\hat{f}\left(x_1,x_2,x_3,x_4,x_5\right),E\left[I\left(x_1,x_2,x_3,x_4,x_5\right)\right]\right\}\\ s.t.\ J_i^*=\sum_{j=1}^{5}\left(\overline{x}_i-x_{ij}^*\right)^2\\ P_{failure}\left(F_P-G_{fource}\geq 0\right)\leq\sigma\end{cases}\tag{9.19}$$

$$\text{sub-discipline}\,i:\begin{cases}\min J_i^*=\sum_{j=1}^{5}\left(\overline{x}_i-x_{ij}^*\right)^2\\ s.t. G_1=\left\{\hat{G}_1\left(x_3,x_4,x_5\right),E_1\left[I\left(x_3,x_4,x_5\right)\right]\right\}\\ \quad G_2=\left\{\hat{G}_2\left(x_3,x_4,x_5\right),E_2\left[I\left(x_3,x_4,x_5\right)\right]\right\}\\ \quad G_3=\left\{\hat{G}_2\left(x_1,x_2,x_3,x_5\right),E_2\left[I\left(x_1,x_2,x_3,x_5\right)\right]\right\}\\ \quad 70\leq x_1\leq 75\quad 25\leq x_2\leq 30\\ \quad 80\leq x_3\leq 90\quad 5\leq x_4\leq 10\quad 50\leq x_5\leq 55\end{cases}\tag{9.20}$$

where F_V represents the objective function of design optimization, which represents the lightweight performance of the structure, and is approximately fitted by the adaptive surrogate model. F_P represents the strength performance of the structure; G_1, G_2, and G_3 represent the discipline objectives and constraints of strength discipline, vibration discipline and aerodynamic discipline, and are approximately fitted by adaptive surrogate model. J_i^* indicates that the consistency constraint condition of the system layer is also the performance function of the discipline layer. $\overline{x}_i$ denotes the expected value of the system layer for the ith design variable. x_{ij}^* represents the optimization result of the jth sub-discipline level for the ith design variable.

In particular, for the vibration discipline, the natural frequency of the structure should not be greater than 10% of the known excitation frequency of the structure, that is

$$\begin{aligned}&F_{excitation}=(R/60)\times n\times l\\ &\eta=F_{intrinsic}/F_{excitation}\leq 10\%\end{aligned}\tag{9.21}$$

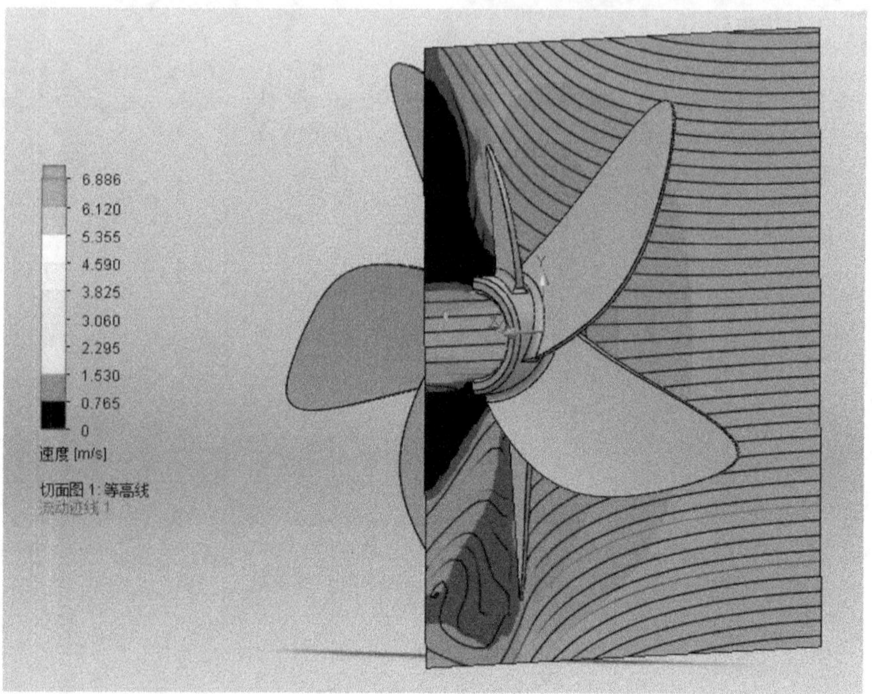

Figure 9.11 The speed of the fan at different positions.

where R signifies the rotation speed of the fan, measured in r/min; n signifies the number of blades of the fan; l signifies an integer, which is used to detect the order of the natural frequency of the structure. The speed of the fan also affects the performance of the blade. Figure 9.11 shows the speed of different positions when the fan rotates.

The performance analysis of the aerodynamics in the sub-discipline of this example optimization problem is mainly characterized by the pressure ratio of the front and rear parts of the gas flow on the blade surface. The specific analysis method is as follows. The total pressure ratio can be defined as [10]

$$PR = \frac{P_{0in}}{P_{0out}} \tag{9.22}$$

where p_0 denotes the pressure on the blade, and the lower entry and exit indicate the inlet and outlet positions of the compressor.

9.4 UBMDO IMPLEMENTATION AND RESULT ANALYSIS OF DUCTED FAN BLADES

According to the above conditions, MDO is carried out for the design problem of the ducted fan. The adaptive surrogate model is used to approximate the complex performance functions and constraints within the optimization model. The optimized natural frequency is determined by modal analysis for the vibration discipline, and its compliance with the discipline constraints is evaluated. For the strength discipline, static stress is employed to analyze the most equivalent stress of the fan surface under limited conditions. Then, whether it falls within the yield limit of the material is examined. For aerodynamics, fluid analysis is utilized to assess whether the pressure ratio of the inlet and outlet ends is enhanced after optimization. The optimized data are shown in Table 9.4.

This example uses the Flow Simulation plug-in in SolidWorks software for fluid simulation. The air condition is set to ideal air, and the flow direction of the air is one-way flow; without considering the turbulence situation, the fluid simulation of the ducted fan under the airflow condition can be obtained. Figure 9.12 displays the outcomes of the simulation.

The results show that the design variables of the ducted fan have been optimized to a certain extent. From the perspective of the overall structure, the volume of the fan will be relatively reduced, resulting in an overall improvement in fan weight. At the same time, in the pneumatic sub-discipline, it can be seen that there is an increase in the pressure ratio between the inlet and outlet ends. This suggests that the gas flow rate at both ends of the ducted fan is faster and more efficient. Moreover, in the strength sub-discipline, the equivalent stress distribution diagram reveals a decrease in the maximum equivalent stress of the optimized ducted fan blade surface under extreme working conditions, indicating higher working safety of the fan.

According to the above simulation results and data, it is evident that the MDO method considering the coupling of multiple disciplines aligns

Table 9.4 Optimization results and data comparison

Name	x_1	x_2	x_3	x_4	x_5	*PR*	*Maximum equivalent stress*	*Reliability*
Initialization value	71.31	28.73	82.10	7.56	52.44	0.9875	417.53	0.84
Optimal values	71.04	28.70	81.39	6.85	51.27	0.9912	409.90	0.86

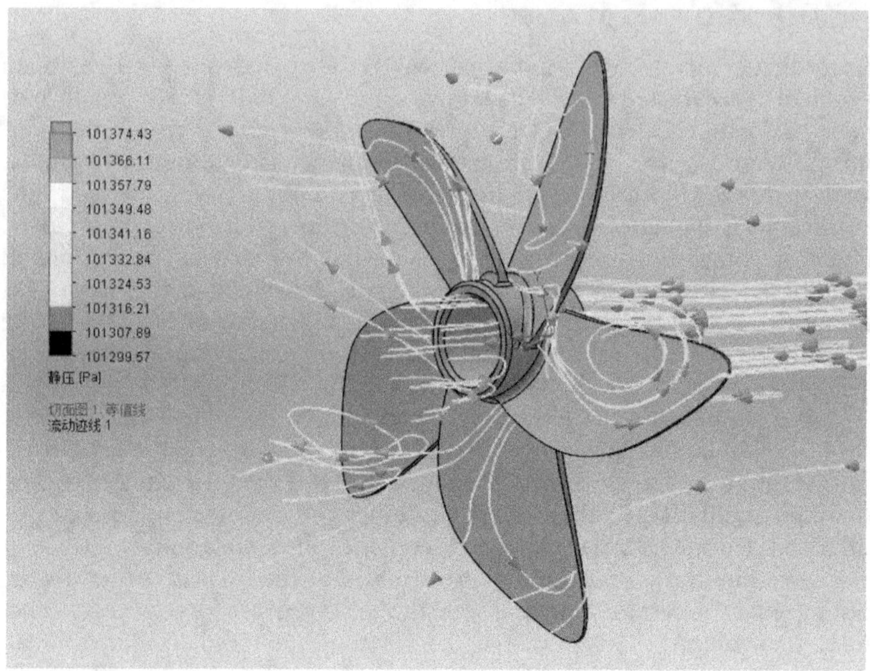

Figure 9.12 The pressure distribution of the inlet and outlet position of the fan.

more closely with the actual engineering situation compared to the single-disciplinary design optimization method considering only a single discipline. With structural lightweight as the design goal, this method successfully achieves a certain degree of lightweight while adhering to the discipline constraints of each sub-discipline.

In accordance with the optimization results of the ducted fan's design variables, modal analysis of the model is carried out to ascertain its six-order natural frequency. Taking the sixth order as an example, as shown in Figure 9.13, the excitation frequency of the mechanism under the existing variable data can be determined by combining the natural frequency of each order with Eq. 9.21. Based on this, the ratio of each order modal frequency to excitation frequency can be obtained, which is within the range of design optimization requirements.

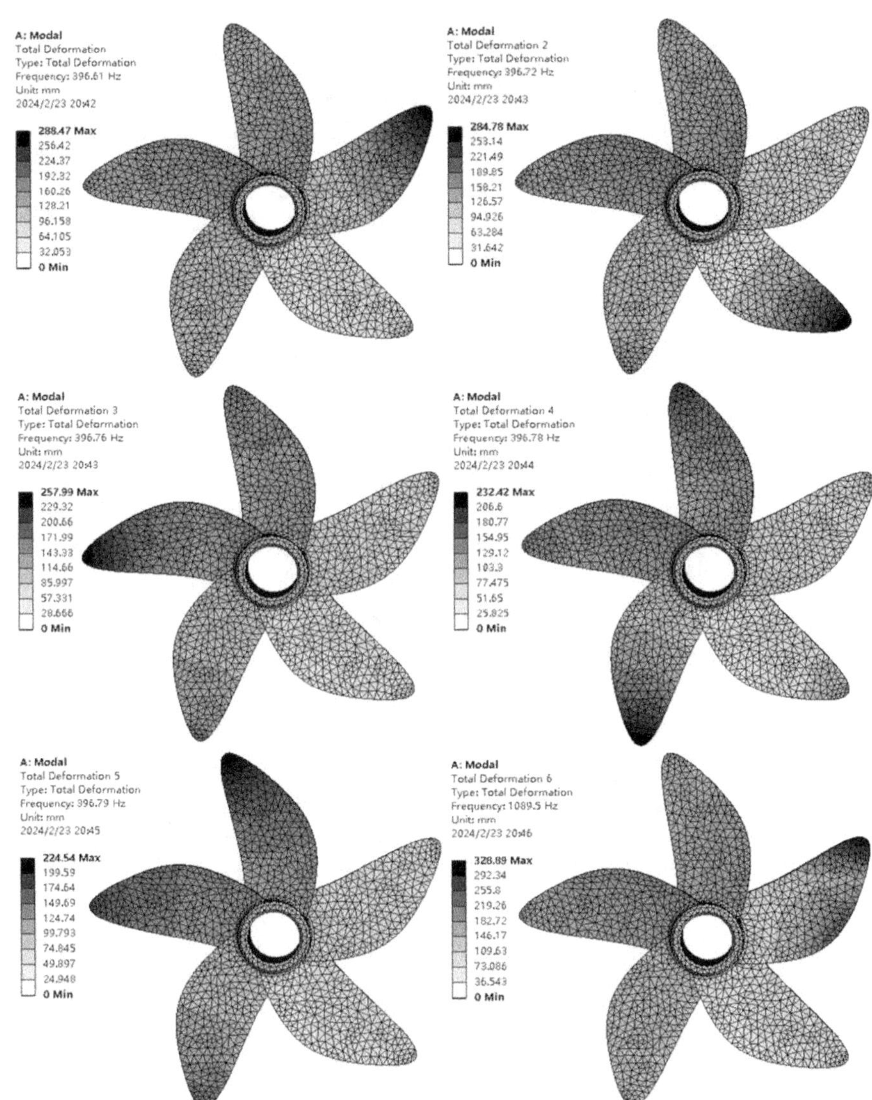

Figure 9.13 Fan six-order modal analysis simulation diagram.

REFERENCES

[1] Mouille R. (1970). The "Fenestron," shrouded tail rotor of the SA. 341 Gazelle. Journal of the American Helicopter Society, 15(4): 31–37.
[2] Rajagopalan R. G., Keys C. N. (1997). Detailed Aerodynamic Analysis of the RAH-66 Fantailtm (TM) Using CFD [J]. Journal of the American Helicopter Society, 42(4): 310–320.
[3] Lakshminarayan V. K., Baeder J. D. (2011). Computational investigation of microscale shrouded rotor aerodynamics in Hover [J]. Journal of the American Helicopter Society, 56(4): 1–15.
[4] Zhang D. Y. (2017). Multidisciplinary optimization algorithm and its application in underwater vehicle. PhD thesis of Northwestern Polytechnical University (in Chinese), Xi'an, China.
[5] Wang J. (2017). The application of MDO method in conceptual design of surface ship is studied. PhD thesis of China Shipbuilding Research Institute (in Chinese), Beijing, China.
[6] Sobieszczanski-Sobieski J., Haftka R. T. (1997). Multidisciplinary aerospace design optimization: Survey of recent developments. Structural Optimization, 14: 1–23.
[7] Alexandrov N. M., Hussaini M. Y. (Eds.). (1997). MDO: State of the Art. Philadelphia, PA: Society for Industrial and Applied Mathematics, Philadelphia.
[8] Zhang X. D. (2011) Research on MDO under uncertainty. PhD thesis of University of Electronic Science and Technology of China (in Chinese), Chengdu, China.
[9] Zhang X., Huang H. Z., Xu H. (2010). Multidisciplinary design optimization with discrete and continuous variables of various uncertainties. Structural and Multidisciplinary Optimization, 42: 605–618.
[10] Shen J., Huang M., Liu X. (2017). Reliability-based design optimization of a centrifugal compressor considering manufacturing uncertainties. 2017 Second International Conference on Reliability Systems Engineering (ICRSE), 1–7.

Chapter 10

The development prospect of UBMDO

In this book, we explain the basic knowledge of Uncertainty-based Multidisciplinary Design Optimization (UBMDO), including approximate models, uncertainty analysis, multidisciplinary optimization, etc. At the same time, the UBMDO method analysis and problem-solving are explained in detail through the application examples of flip drive mechanism, turbine rotor mechanism, and ducted fan blade. The above is the main content of this book.

With the development and extension of the Multidisciplinary Design Optimization (MDO) theory, the UBMDO theory has also made significant progress in the long research time [1]. Considerable academic achievements have been made in uncertainty quantification and modeling, reliability evaluation systems, and multidisciplinary uncertainty analysis. At the same time, it has also been widely used in engineering. However, as a discipline, the theoretical system of UBMDO is not perfect. Although the basic terms, methods, and general procedures have been established, detailed computational algorithms, and organizational procedures are still needed to solve the three main problems of modeling complexity, computational complexity, and organizational complexity. This requires more extensive and in-depth research by scholars in future development.

This chapter will summarize and prospect the future development directions of UBMDO to a certain extent, and we hope to make some contributions to future research direction.

10.1 UBMDO TECHNOLOGY

UBMDO can improve the optimization design process and enhance the reliability and robustness of optimization design results by using the synergistic effect of coupled discipline collaborative optimization [2]. This section will analyze and prospect from how to construct an accurate UBMDO model and explore an efficient UBMDO solution method.

DOI: 10.1201/9781003464792-10

10.1.1 Accurate UBMDO model

The construction of the UBMDO model is the basis of design optimization, and the accuracy of the model has an essential influence on the reliability of the optimization results. The degree of refinement of the model determines the computational efficiency of the entire UBMDO. The aging characteristics of the model reflect the external environment's influence on the product's overall performance.

10.1.1.1 Considering model uncertainty and human decision uncertainty comprehensively

In the current academic field, uncertainties of multiple sources are considered in uncertainty analysis and reliability analysis. The uncertainty analysis of stochastic uncertainty and cognitive uncertainty has also been significantly developed. However, human cognition is always limited, and there are still many uncertainties that designers need to explore, discover, and solve. For example, the study of model and decision uncertainty is a field that needs to be developed.

Model uncertainty refers to the inconsistency or uncertainty between the predicted and actual results of a model. In transforming the physical model of the actual engineering system into the mathematical model, not all the physical nonlinear characteristics can be accurately transformed into mathematical equations. Many models with different fidelity can be used for mathematical analysis when transforming mathematical models into computer models. These mathematical equations can be solved by various technical means, resulting in differences in accuracy between the models. There are two main sources of this uncertainty: Model error and parameter uncertainty.

a. The uncertainty of model error comes from the model itself. If the structure of the model or theoretical assumptions cannot fully reflect reality, it will produce model error. For example, a common situation is overfitting. One of the models performs well on training data but poorly on new or unseen data
b. Parameter uncertainty stems from the selection or use of model parameters. For example, parameters may be estimated based on sample data, which is subject to certain random errors

Model uncertainty will affect the accuracy of model predictions and the researchers' confidence in their results. Researchers and data scientists typically use statistical methods and techniques, such as confidence intervals, hypothesis tests, or Bayesian analysis, to quantify and address this uncertainty.

Decision uncertainty is a situation faced in the decision-making process in which the result or influence of the decision is unpredictable. This may be because the relevant information is incomplete or poor quality. Decision uncertainty can usually be divided into the following three types.

a. Results uncertainty. This refers to the uncertainty of the specific outcome caused by the decision. For example, launching a product in a new market may succeed. The uncertainty of the results is often caused by the fact that we cannot fully predict the future
b. Information uncertainty. This is because the decision-makers do not have enough information, or the quality of information is not high, leading to uncertainty. For example, if we lack sufficient data about a market, we may be uncertain about the success of new products in that market
c. Structural uncertainty. This means that understanding the relationship between the decision-making problem and its related factors is uncertain

One way to manage uncertainty of this type is to collect more information, conduct more analysis, or use a variety of decision-making tools, such as risk assessment or decision trees. However, it is worth noting that all decisions involve a certain degree of uncertainty to some extent. Furthermore, in many cases, we cannot eliminate uncertainty. Therefore, decision-makers need to accept and learn to make the best decisions in uncertainty.

10.1.1.2 UBMDO under mixed uncertainty considering variable correlation and failure mode correlation

So far, the MDO method considering multi-source uncertainty is generally based on the independence of various uncertain variables. However, in practical engineering, there may be correlations between the variables involved in the structure and the system. At the same time, in the long-life cycle of a complex product, multiple failure modes may also have a certain correlation due to shared variables. These correlations have an essential influence on the reliability of the structure. Ignoring the correlation in some design cases may lead to large errors. Therefore, it is imperative to consider the correlation between variables and the correlation between multiple failure modes. Two main problems in reliability analysis considering correlation are: one is how to construct a mathematical model for the correlation between different types of uncertain variables, and the other is how to improve the efficiency of uncertainty analysis and reliability analysis by considering correlation. Reliability-Based Multidisciplinary Design Optimization (RBMDO) is an integral part of UBMDO, and improving analysis efficiency is a problem that must be solved.

10.1.1.3 UBMDO considering the time factor

In practical engineering, the parameters of uncertain design variables that affect the function of complex products may also change with time. Products may also have different failure modes in different life cycles. Therefore, in the uncertainty analysis optimization and reliability modeling of the product, the product's reliability at the moment should be evaluated, and the product's reliability in different life cycles should be predicted. This is of great significance for improving product life cycle and quality. The time-varying reliability analysis of products is still in the process of exploration and discovery, which needs to be further studied by relevant scholars [3, 4]. Next, we introduce the factors that may affect the time in RBMDO analysis.

a. The time-varying of the MDO evaluation system. The focus of MDO will change at different stages of design. Introducing the whole life cycle concept into the design process of complex products is meaningful. In the different design stages of complex product solutions, it is necessary to consider the needs of manufacturing, assembly, logistics, and other aspects throughout the product's life cycle. At the same time, the evaluation indexes and feasibility methods of multidisciplinary optimization design considering time are formulated, and the evaluation results are fed back to the product's design cycle in a computer-recognizable way. This can enhance the credibility and accuracy of the design results, which is necessary to promote the application of MDO
b. Time-varying of parameter uncertainty. Many uncertain factors in mechanical products will change over time. For example, material properties will degrade over time, and the degree of wear and the completeness of uncertain information will increase over time. Dynamic uncertainty is the study of uncertain change considering the time factor, mainly considering the following two cases. One is that the same uncertainty needs to be reconstructed and calibrated in a changing environment; the other is that different types of uncertainty may transform with data completeness and the designer's cognition. It is necessary to analyze and discuss the influence of time on uncertain factors. This is of great significance for further research on MDO, considering dynamic uncertainty
c. Time variability of the reliability analysis model. In practical engineering, the load acting on the product is mostly time-varying, so the product's reliability is also affected by time. The existing RBMDO method mainly focuses on the theoretical exploration and engineering application research of mechanical system reliability design at a certain time with random and cognitive uncertainty. The performance

degradation and external load of mechanical systems in engineering are changing with time. In order to ensure the reliability of mechanical systems in each period, it is necessary to introduce the time-varying reliability analysis method into the reliability optimization method. How to apply the corresponding time-varying reliability analysis method and mixed uncertainty analysis method in RBMDO to establish an effective RBMDO model based on time-varying reliability constraints under mixed uncertainty is a research topic with crucial practical significance

d. The reliability analysis method is based on field and experimental data. The following problems in physics-based reliability analysis. The first is that the failure mode mechanism is unknown, and the failure model cannot be established, so it cannot include all the failure cases. Second, not all the calculation models are feasible. Therefore, the reliability optimization design method based on field and experimental data can be used as a supplement. For some products, the performance function cannot fully represent the failure mechanism of the product. A large error may occur if only the reliability obtained by the simulation model method is used for RBMDO. When the product design is carried out to a certain stage, some field or test data can be used to reduce the error caused by reliability evaluation. This method establishes a reliability optimization design model based on field data and physical models. The statistical-based reliability analysis method and the physical-based reliability analysis method can be effectively connected.

10.1.2 Efficient UBMDO solution method

10.1.2.1 Multidisciplinary uncertainty analysis

Multidisciplinary uncertainty analysis is usually used for complex products or systems involving multiple disciplines or fields. It is challenging to analyze problems clearly in a single discipline, which requires an interdisciplinary approach. In such analyses, researchers will pay attention to the input parameters of all relevant models and try to understand the influence and correlation of these parameters on the results. This includes establishing a detailed model, understanding how uncertainty from different sources affects the model, and how it affects the overall prediction results. The analysis also considers the uncertainty caused by the interrelationship or dependence between parameters. This analysis can help researchers understand which parameters or disciplines of uncertainty impact the overall results so they can focus on reducing these uncertainties. It can also provide a comprehensive approach to better understanding and optimizing complex systems' behavior.

10.1.2.2 An efficient deterministic MDO method

MDO has been developed for many years, and researchers have proposed many related methods. However, different MDO methods have advantages and disadvantages, and the degree of adaptation to different situations is also different. So far, there is still no common solution for the MDO method, which cannot perform well in all evaluation indicators. The large-scale, multi-coupled complex engineering system model contains continuous design variables and may contain discrete design variables. However, due to its various defects, the traditional MDO method has problems with low computational efficiency and difficult convergence. Therefore, it is urgent to integrate artificial intelligence technology, computational complexity theory, and traditional MDO methods and propose efficient MDO methods with fast convergence speed, no derivative information, and less disciplinary analysis, for example, in the probability-based optimization methods such as Genetic Algorithms, Particle Swarm Optimization, Simulated Annealing, etc. These effectively solve the global optimal solution and deal with non-linear and multimodal problems. This random search strategy can avoid falling into local optimal solutions.

10.1.2.3 Decoupling of multidisciplinary reliability multi-layer nested optimization

The multidisciplinary reliability design optimization of complex engineering systems represented by aerospace products is a typical multi-layer nested loop optimization problem. To this end, relevant research can be carried out from the following two aspects.

a. Efficient overall process solution strategy for RBMDO. When the MDO problem involves discrete, continuous, and multi-source uncertainty, it will be a serious multi-layer nested loop optimization process. In order to improve computational efficiency and engineering practicability, the integrated decoupling theory, the approximation technology KKT equivalent condition, and the related computational complexity theory will be important research directions. Researchers have proposed efficient multidisciplinary reliability design optimization strategies for multidisciplinary, multi-objective, multi-variable, multi-constraint, multi-coupling, multi-uncertainty, and highly nonlinear RBMDO problems
b. Distributed architecture decoupling. For multidisciplinary issues, a distributed architecture can increase design autonomy. Therefore, using the MDO of the current distributed architecture is necessary. In this way, the distributed architecture decoupling of multidisciplinary reliability design optimization under various uncertainties is established. Therefore, the discipline autonomy of multidisciplinary reliability

design optimization is realized, and the engineering applicability of multidisciplinary reliability design optimization is further improved

10.1.3 Better handle complex and large-scale problems

Processing complex and large-scale problems is often accompanied by high computational complexity and complex uncertainty relations, which is a major problem faced by the UBMDO method. Accurate models will bring higher accuracy and a sharp increase in computational complexity, which poses a serious challenge to the UBMDO method. The execution time and engineering resources are significantly squeezed. When dealing with complex and large-scale UBMDO problems, the following strategies can be used to reduce the difficulty of solving and improve the efficiency of solving.

a. Decomposition of the problem. By appropriately decomposing large-scale problems into several relatively independent and smaller problems, each sub-problem can be easier to handle and solve. Through decoupling and decomposition strategies, the interaction between subsystems can be minimized to deal with each sub-problem more independently in the optimization process
b. High-Performance Computing (HPC). This type of computing technology can provide greater processing power for higher levels of computing. Usually, word processing requires a lot of computing or data processing. HPC usually relies on parallel computing methods, such as multi-processor systems, multi-core CPUs, and GPUs for accelerated computing. These parallel computing technologies are based on performing multiple computing tasks simultaneously to improve the calculation speed and efficiency. One of the main types of high-performance computing is distributed computing, which distributes the task load to multiple connected computers to form a computer cluster or grid. The other is supercomputing, which usually involves a single computer with extremely high processing power. This has a positive effect on the computational problem of UBMDO, especially after the decomposition of the problem. It can better cope with the difficulties

10.2 THE COMBINATION OF UBMDO AND ADVANCED TECHNOLOGY

With the continuous development of today's scientific research environment, the UBMDO method also needs to keep pace with the times. The traditional UBMDO method has inevitable limitations. Therefore, combining the traditional UBMDO method with advanced technology is the mainstream development trend.

10.2.1 The development trend of UBMDO working environment

As product development complexity increases, it includes multiple disciplines and design stages. Uncertainty factors have also increased dramatically. The whole UBMDO process is time-consuming, computationally intensive, and iterative. It involves many disciplines. Under a modern scientific research background, organizing and integrating design resources, and sharing data has become an important research direction. It mainly has the following development trends.

a. Integration. This broad sense of integration includes data integration, tool and process integration. The MDO process consists of multidisciplinary design, multidisciplinary analysis, multidisciplinary optimization, and performance analysis. It has the characteristics of multi-process and frequent iteration, so seamless integration between design processes is of great significance to the realization of MDO
b. Intelligent. Because MDO has a complex design process and coupling relationship, the intelligent requirements of the MDO framework can be divided into the following two categories. One is an intelligent design flowchart for various engineering designs to achieve workflow-driven design; the other is the intelligent design technology library composed of experimental design, approximate method, and intelligent optimization algorithm. Both experimental design and approximation methods can replace high-precision models by establishing surrogate models. Intelligent optimization algorithms are used to achieve more complex optimization functions and improve computational efficiency without losing computational accuracy. Of course, the ideal MDO framework can select the most suitable algorithm for users according to the actual MDO problem
c. Distributed working environment. Complex engineering system design is a multidisciplinary design composed of different design groups. These design groups are distributed in different geographical locations and use heterogeneous platforms. For example, the aerodynamic design of an aircraft can be performed by a high-performance computing workstation, while the structural design is performed on a personal computer. Therefore, cross-platform and heterogeneous operation is an essential requirement of the MDO framework. In addition, an MDO framework should also provide a collaborative environment to allow designers, managers, and decision-makers to work together. At the same time, the data and information transmission between different groups should be guaranteed, and the timely update of each workflow and tree node in the design stage and different models

d. Customization. The MDO process of complex engineering design includes multidisciplinary collaborative design and multidisciplinary design team collaboration. Because of the differences between disciplines, each discipline also needs a working environment. Therefore, the MDO framework should provide a customized working environment according to the special design requirements of different disciplines
e. Data management. The MDO process includes many design and process data, such as design objectives, constraints, and design variables of systems and disciplines. In addition, the analysis and optimization of iterative process data, the experimental design data used to construct the surrogate model, and the relationship data management requirements of coupled disciplines hinder the application of MDO in practical engineering design. Therefore, the storage, description, and transmission of data is the key to realize the engineering application of MDO. Describing, managing, integrating, and sharing these data in the network environment is particularly important for the collaboration of participants
f. Visualization and Monitoring (VM). VM of MDO plays an increasingly important role in the development of complex products. Visualization of the complexity, diversity, design variables, constraints, and uncertainties of the MDO process and dynamic monitoring of the design process are essential. It can encourage designers to intervene in the MDO process or modify the new optimization's set parameters or mathematical models. The visualization of MDO mainly includes the following three aspects. The first one is the visualization of the approximate model, which can help the designer select the appropriate approximation technique to ensure the required accuracy. The second is the visualization of the search algorithm, which can guide the designer to determine the changing trend of key design variables. The third is the visualization of previous design and optimization results

10.2.2 UBMDO solution strategy based on data mining

Data mining is searching for information hidden in a large amount of data through algorithms. The MDO process is a numerical iterative calculation, so a large amount of data without obvious rules will be generated during optimization. The data has a loose or close relationship before. How to find the relationship between these data is significant for design optimization. Data mining provides such a function. The application of data mining in UBMDO is mainly in three aspects. They are optimization modeling, optimization calculation, and multi-objective optimization scheme decision-making processes. Next, we mainly introduce data mining in these three aspects.

10.2.2.1 Optimize data mining in the modeling process

The main optimization modeling process in UBMDO is to convert the physical model of the system to be optimized into a mathematical model that can be numerically calculated. The modeling process is shown in Figure 10.1.

The simulation model determines the design parameters and their ranges. This process mainly counts the existing or experimental design data and then uses expert knowledge to determine the design parameters. This process will inevitably involve a large amount of data, so using data mining technology to determine the design parameters will be effective. The technology of converting physical models into numerical models is called model validation and calibration technology. Model validation and model calibration belong to the category of test methods. It is used to calibrate and determine the numerical model of the system by using the data generated by the test. The adequate information provided by data mining technology is vital for complex models. In recent years, big data technology, which has been studied hotly and developed rapidly, will also help for data mining in MDO.

10.2.2.2 Data mining in the process of optimization calculation

Multidisciplinary optimization is the process of numerical iteration to obtain the convergence solution. Due to the complexity of the model and the limitations of the existing iterative methods, the optimization process is often very tortuous. There may be more iterations or even no convergence. Using data mining to find the optimal optimization path, judging the approximate range of the optimal solution, and combining expert experience to solve the optimization problem in UBMDO is a new and effective technical means. The optimal path selection based on data mining is shown in Figure 10.2.

10.2.2.3 Data mining in the decision-making process of a multi-objective optimization scheme

The results of multi-objective optimization are often a set of Pareto solutions rather than a single solution. How to obtain the optimal solution

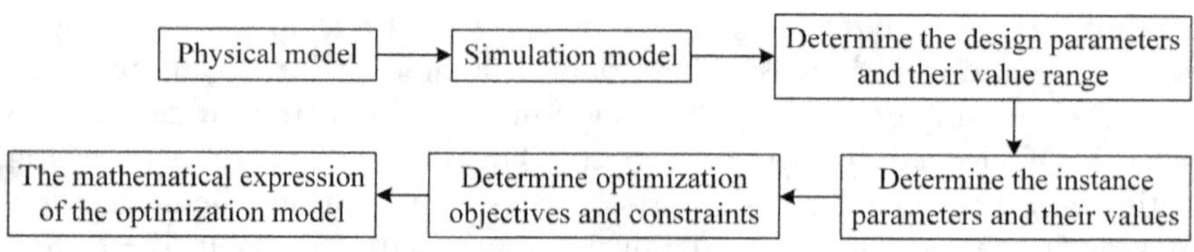

Figure 10.1 The process of optimization modeling.

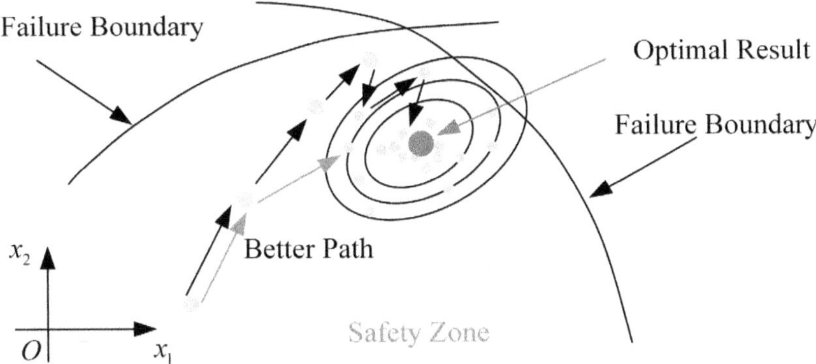

Figure 10.2 Optimized path selection based on data mining.

from this set of solutions is an important research content of multi-objective optimization. The traditional method assumes the weight value of each objective, transforming the multi-objective optimization problem into a single-objective optimization problem. However, this does not align with the actual situation, nor can it adapt to different design scenarios. Data mining provides the possibility for multi-objective optimization scheme decision-making and can also solve the final result of multi-objective optimization between previous data and expert knowledge.

10.2.3 3D printing design technology based on UBMDO

3D printing technology is also known as additive manufacturing technology. It is widely known as the representative technology of the Third Industrial Revolution. One of the advantages of 3D printing technology is that it can print products of any complex shape, which makes it possible to obtain products that are difficult to manufacture or cannot be manufactured by traditional processing methods. This advantage of 3D printing will also greatly liberate product designers. Designers do not need to consider the constraints of traditional manufacturing when designing and have a greater design space. 3D printing has brought many changes to the design. It makes the birth of new functional products possible and provides an important means to achieve lightweight and high-performance products through optimization design.

This section introduces a technique that uses the UBMDO method to optimize the design of 3D-printed objects. This technology can fully consider the needs of different disciplines at the beginning of the design to generate 3D printing products with higher quality and better performance. Modern products are mostly complex products involving multiple disciplines, such as electricity, light, and magnetism. 3D printing has brought new technical means to producing and manufacturing these products. MDO is an effective method to ensure the performance of complex products and reduce their cost. Combining the two will bring new prospects for the design of complex products. From the product's concept to the detailed design process, the 3D printing design technology based on UBMDO has the following paramount development directions.

10.2.3.1 Multidisciplinary topology optimization design for 3D printing

This technology belongs to the conceptual design stage of the product. Topology optimization is a load-driven structural design method designed to help designers find the best material layout that meets specific load conditions and constraints in the initial stage of structural design. The results of topology design make a crucial difference in the final performance of the structure. As the most sensitive preliminary design, it has great potential for improving the design quality of 3D printing.

Traditional topology optimization methods are mostly related to the mechanical properties of structures. With the development of acoustic, optical, electrical, magnetic, thermal, and other computing technologies and the development needs of multi-functional complex products, topology design optimization has begun to consider the influence of multiple disciplines. With the support of 3D printing technology, the multidisciplinary topology optimization design method related to the functional characteristics of the structure will become the cutting-edge technical problem for 3D printing design. The main technical difficulties in its development are as follows. The first is the multidisciplinary decoupling method from the macro and micro perspectives. The second is how to achieve efficient global sensitivity equation solving. The third is how to effectively integrate the existing MDO algorithms into the multidisciplinary topology optimization process.

10.2.3.2 UBMDO technology integrated with topology optimization

This technology belongs to the detailed design stage of the product concept. The traditional MDO mainly optimizes the size of the product, which belongs to the optimization of the detailed design stage. Topology optimization mainly optimizes the internal topology and layout of the product, which belongs to the optimization technology in the conceptual design stage. The integration of topology optimization to MDO can effectively improve the automation of design. In the past, due to the limitation of manufacturing technology, the topology design optimization results could be directly used for manufacturing in most cases. This means that the design optimization results usually generated by topology optimization still need to be processed further to meet the needs of traditional manufacturing processes. The development of 3D printing technology has dramatically reduced this limitation, and the development of multi-UBMDO technology with integrated topology optimization is bound to become a hot issue for 3D printing design.

The MDO technology integrating topology optimization does not simply perform the two stages of topology optimization and MDO in sequence. It is a process of iterative optimization in each stage and repeated iteration in two stages. It is necessary to consider whether the conceptual design based on topology optimization can be printed and how to realize the optimization connection between the topology design results and the multidisciplinary design in the detailed stage. It is also necessary to consider how to reduce the amount of calculation.

10.2.3.3 The integrated design optimization technology of 3D printing based on multidisciplinary

3D printing is a technology opposed to traditional material processing methods. It manufactures products by adding layers of materials based on three-dimensional CAD model data. 3D printing technology involves CAD modeling, interface software, numerical control, laser, materials, and other disciplines, which belong to the high-tech of multidisciplinary integration. The integration of multiple disciplines makes 3D printing technology to have a variety of uncertainties in the product manufacturing process, such as material type, manufacturing temperature, and so on. Therefore, the design for 3D printing still has certain constraints on the manufacturing process.

The multidisciplinary 3D printing integrated design optimization technology is required to incorporate the product performance and the requirements of the 3D printing process into the design model. Then, MDO is used to achieve the common goal of design and manufacturing. Simply considering the performance constraints and uncertainties of multiple disciplines has put forward a great test for this technology. The addition of uncertainties and constraints in the 3D printing process makes this technology more complicated. In particular, the coupling analysis of various performance disciplines and 3D process disciplines is the main. The global sensitivity equation with process constraints and the integrated design optimization based on distributed MDO are the difficulties that must be solved.

10.2.4 MDO modeling of Model-Based System Engineering (MBSE)

Traditional complex product design adopts a document-based system engineering method. As the complexity of the R&D design process continues to increase, the number and version of documents have also increased significantly. This makes document management and information search and change extremely difficult, and ensuring the consistency of design information in different documents is difficult. Therefore, the concept of MBSE came into being, which is an essential development direction in the field of system engineering. The model has the advantages of intuitiveness, non-ambiguity, and module reusability. It is the development of traditional document-based system engineering theory and practice [5]. With the continuous development of computer technology, it becomes easier to describe the system through graphical modeling language. Therefore, the computer processable model plays an increasingly important role in system development, which also brings significant advantages to the MBSE method.

MDO is optimized from the system-level perspective, which coincides with the idea of system engineering. The MBSE method is used for MDO modeling, so designers only need to focus on establishing a system functional behavior model. The system model can be automatically converted to a multidisciplinary optimization model, which will help improve the efficiency of optimization verification and reduce the programming burden of designers. MBSE supports the integration of various models, and the detailed design process of each field can be associated with the abstract system layer. Therefore, the optimization information can be integrated into the design information in the system design process of complex products to build a formal optimization model to achieve automatic optimization of the system. In addition, MDO can be applied in all stages of product design. The system model can define the subject interface, data flow, and policy process information and establish abstract design parameter information. Establishing an MDO model based on MBSE is conducive to extracting the optimization model but can reuse the established MDO model in the iterative design process to carry out the hierarchical design.

With the advancement of MBSE technology, MDO based on MBSE will be further developed. The MDO under the MBSE system is not only the application of the existing MDO technology but also promotes the innovation of the basic theory of MDO. MBSE-based MDO is mainly to optimize the multidisciplinary design of complex systems more efficiently and objectively. It draws on the technical characteristics and connotations of MBSE. For example, the modeling characteristics are the whole process of system design, the whole object, and the whole parameter. The automation of complex system design is characterized by model-driven and mapping in different modeling parts. Starting from the demand analysis, an MDO model for complex systems is established and solved. This helps to achieve the original innovation and automation of system optimization.

MDO based on MBSE has the following advantages.

a. MDO modeling is more objective and aligns with design habits. For example, from the beginning of demand analysis, the mapping between demand-function-structure is established, which is extended to more detailed design processes such as function decomposition and structure decomposition

b. Parameter model changes are automatically transmitted and updated. For example, DMBSE uses unified modeling languages such as SysML to eliminate model heterogeneity between disciplines. Thus, the problem of parameter transfer between disciplines in the process of MDO modeling is simplified. MBSE emphasizes the close relationship between disciplines, which also aligns with the characteristics of multidisciplinary coupling relationship expression in MDO. These characteristics of MBSE can not only simplify the modeling process of MDO but also promote the reuse and transfer update of MDO models
c. The advancement of MBSE technology can facilitate the automation of MDO. The automatic mapping and decomposition of requirement-function-structure is essential in MBSE research. It is foreseeable that with the development of these technologies, MDO based on MBSE will also develop toward the semi-automated and automated direction

10.2.5 MDO modeling method of Modelica language

Modelica language is an object-oriented physical modeling language. It uses concepts such as class, instance, and inheritance to describe physical objects, which aligns with people's thinking habits of understanding things. At the same time, Modelica language is an equation-based physical modeling language. It uses mathematical equations rather than assignment statements to define the behavior of classes, so it has the characteristics of non-causal declarative modeling. Modelica language has a powerful component connection mechanism used to establish the coupling relationship between different domain models, so it has the characteristics of multi-domain unified modeling [6].

10.2.5.1 Non-causal declarative modeling

The equation has non-causal characteristics. When the equation is declared, the direction of solving the equation is not limited. Therefore, the equation has greater flexibility and a more robust function than the assignment statement. This feature of the equation dramatically improves the reusability of the Modelica model. At the same time, the solution direction of the equation is finally determined automatically by the numerical solver according to the data flow environment of the equation system, which is transformed into a causal assignment form without modeling. This can significantly reduce the modeling workload and make the model more robust. The declarative design idea supports the construction of simulation models based on the physical topology of the actual system so that the constructed physical model has a hierarchical structure similar to the actual system.

10.2.5.2 Multi-domain unified modeling

The Modelica language describes the behavior of components in any field uniformly using mathematical equations and defines the communication interface between components and the outside world as connectors. The connector is usually composed of matched potential variables and flow variables. The component connection mechanism based on the connector lays a theoretical foundation for multi-domain unified modeling using Modelica language. Components in the same domain communicate using connectors between the same domains.

10.2.5.3 Top-down modeling

Top-down modeling refers to the modeling method of starting from the whole design, determining the composition of each part, and then completing the system model using the model library components. Because the Modelica language uses an object-oriented idea to describe the model, it is particularly suitable for hierarchical top-down modeling. Top-down modeling mainly includes the following three steps. First, the system is decomposed into subsystems. Second, the relationship between the subsystems must be determined. Finally, the model design of each subsystem is completed by connecting the model library component model.

According to the characteristics of the Modelica language discussed above and many problems existing in MDO, the MDO modeling method based on Modelica can make up for the shortcomings of the current MDO method. This is because it has the following characteristics.

a. It is beneficial to the extraction of the optimization model. Unified physical modeling languages, such as Modelica support equation-based modeling mechanisms conducive to extracting MDO mathematical models with physical meaning. Thus, the problem of information loss in the model established by human experience can be avoided
b. Support simulation optimization iteration. The sensitivity analysis of the MDO model can be carried out through the unified physical modeling and simulation environment. The design parameters with low influence on the performance function are deleted to effectively determine reasonable design variables and performance function
c. It can more reasonably define the coupling relationship between multiple disciplines. The system discipline divided from the physical meaning is more conducive to the designer's understanding than the artificially divided system. The coupling relationship between MDO models based

on physical models is obtained using connectors, which is simpler and more natural

d. It supports the reuse and inheritance of model knowledge. Developing a unified physical modeling-language-support domain library can encapsulate the experience and knowledge of related fields to support the reuse and inheritance of MDO model knowledge

REFERENCES

[1] Meng D., Xie T., Wu P., He C., Hu Z., Lv Z. (2021). An uncertainty-based design optimization strategy with random and interval variables for multidisciplinary engineering systems. Structures 32: 997–1004.

[2] Meng D., Xie T., Wu P., Zhu S. P., Hu Z., Li Y. (2020). Uncertainty-based design and optimization using first order saddle point approximation method for multidisciplinary engineering systems. ASCE-ASME Journal of Risk and Uncertainty in Engineering Systems, Part A: Civil Engineering, 6(3): 04020028.

[3] Zhang J., Du X. (2015). Time-dependent reliability analysis for function generation mechanisms with random joint clearances. Mechanism and Machine Theory, 92: 184–199.

[4] Wang Z., Wang P. (2012). A nested extreme response surface approach for time-dependent reliability-based design optimization. Journal of Mechanical Design, 134(12): 121007.

[5] Ramos A. L., Ferreira J. V., Barceló J. (2011). Model-based systems engineering: An emerging approach for modern systems. IEEE Transactions on Systems, Man, and Cybernetics, Part C (Applications and Reviews), 42(1): 101–111.

[6] Wu Y., Chen L. (2011). Simulation Optimization Method for Multi-domain Physical Systems. Beijing, China: Science Press.

Index

www.ingramcontent.com/pod-product-compliance
Lightning Source LLC
LaVergne TN
LVHW020612110826
845149LV00002B/455

* 9 7 8 1 0 3 2 7 3 5 6 3 4 *